Ergebnisse der Mathematik und ihrer Grenzgebiete

Band 20

Herausgegeben von
P. R. Halmos · P. J. Hilton · R. Remmert · B. Szőkefalvi-Nagy

Unter Mitwirkung von
L.V. Ahlfors · R. Baer · F. L. Bauer · R. Courant
A. Dold · J. L. Doob · S. Eilenberg · M. Kneser · G. H. Müller
M. M. Postnikov · B. Segre · E. Sperner

Geschäftsführender Herausgeber: P. J. Hilton

Richard Hubert Bruck

A Survey of Binary Systems

Third Printing, Corrected

Springer-Verlag Berlin Heidelberg GmbH 1971

Professor RICHARD HUBERT BRUCK
University of Wisconsin, Van Vleck Hall, Madison, Wi/USA

AMS Subject Classifications (1970): 20-02, 20 N 05, 20 M 10, 05 B 15, 05 B 30

ISBN 978-3-662-42837-5 ISBN 978-3-662-43119-1 (eBook)
DOI 10.1007/978-3-662-43119-1

Originally published by Springer-Verlag Berlin Heidelberg New York in 1971.

Library of Congress Catalog Card Number 79-143906.

Foreword

The area of mathematics surveyed by the present report might be described as a little too broad to be treated in an encyclopedic manner in a reasonable space and yet not sufficiently developed in all its parts to warrant several distinct reports. Group theory and lattice theory are both regarded as outside the scope of the report. Subject to this restriction, the area surveyed is the algebraic theory of systems with a single binary operation. The operation is understood to be single-valued except in the case of multigroups and allied systems. Since quasigroups, for example, can be studied either in terms of a single binary operation or in terms of three, a little of the theory of general algebras enters quite naturally. A diagram indicating the interrelations of the various systems will be found at the end of Chapter II.

The more highly developed parts of the subject have been treated in detail. Here the emphasis is on theorems and proofs; the work of several mathematicians has been fused into a single whole with only a mild attempt to indicate their several contributions. This is the case, for example, with the main body of the theory of loops and with the theory of ideals in semigroups. At other points, significant theorems have been stated without proof but with precise references to the literature. The author is well aware that many special topics, for example in the theory of semigroups and quasigroups, have not been specifically mentioned in the text. The bibliography should partly make up for such omissions.

I wish to acknowledge with thanks research grants from the John Simon Guggenheim Foundation for the year 1946—47 and from the University of Wisconsin Research Committee for the fall of 1946 and the summers of 1952, 1954 and 1955 — grants which were used partly to expand the subject matter and partly to delineate it.

I am indebted to many persons for aid in the present task: to my wife, HELEN; to the many authors (above all, to R. BAER and B. H. NEUMANN) whose papers and letters have strongly influenced my work; to my student, DONALD W. MILLER, now of the University of Nebraska, for a painstaking survey of the literature of semigroups which I used freely and according to my own judgement in connection with Chapter II; to MILLER, to A. H. CLIFFORD and to others for contributions to Part D of the Bibliography; to BAER for a detailed and helpful criticism of the manuscript; to R. ARTZY for preparation of the Index; and, less specifically but no less sincerely, to the various students in my seminars who have helped to shape or correct much of the present text.

March 13, 1958

R. H. BRUCK
Madison, Wisconsin

Contents

Note on the Bibliography

For the convenience of the reader, the bibliography has been divided and subdivided. Books are specified in the text by author and title or, occasionally, merely by author; precise references are given in Part A. Part B contains a few necessary papers on subjects outside the scope of the text; references to these take the form: Author [Bn], where n is an integer. Part C, the main body of the bibliography, is divided according to subjects, authors being listed alphabetically under each heading; Part D is a supplement to Part C. Papers are numbered consecutively throughout Parts C and D; references take the form: Author [n]. No paper appears under two headings; the narrower classification has generally been preferred (e. g., semigroups as against groupoids) but there are exceptions to the rule in doubtful cases. The papers listed are those reviewed in the first seventeen volumes of Mathematical Reviews together with a few earlier or more recent papers. Wherever possible a reference has been appended in form [$MR\ v, p$], v being the volume number and p, the page, of the review in Mathematical Reviews. The criterion for selection was a vague one; completeness seemed desirable but was clearly beyond reach or even definition.

Note on Symbols

We define here certain symbols used consistently throughout the text. The symbols $=$ and $\neq$ denote equality and inequality respectively. $\in$ denotes membership in a set or system: "$a \in B$" means that a is an element (or member) of (the set or system) B. $\notin$ denotes lack of membership: "$a \notin B$" means that a is not an element of B. $\subset$ and $\supset$ describe the subset relation: each of "$A \subset B$", "$B \supset A$" means that every member of A is a member of B. The relations "$A \subset B$, $B \supset A$", taken together, and the relation "$A = B$", each mean that A and B have the same elements. $\cap$ denotes intersection or meet or common part; $\cup$ denotes union or join or logical sum: if A, B are sets, $A \cap B$ is the set of elements common to A and B, whereas $A \cup B$ is the set of elements belonging to A or B or both. $\rightarrow$ is used for two distinct purposes. The most frequent use is for informal specification of a (single-valued) mapping: "the mapping $x \rightarrow x^2$ of the group G" specifies the mapping of G which maps each element of G upon its square. $\rightarrow$ and $\leftrightarrow$ are used (very sparingly) as symbols of logical implication: "$a = b \rightarrow a^2 = b^2$" means that if $a = b$, then $a^2 = b^2$; and "$x \in A \cap B \leftrightarrow x \in A$ and $x \in B$" means that $x \in A \cap B$ if and only if $x \in A$ and also $x \in B$. $\sum$ sometimes is used as the sign of summation and sometimes to denote a set; both uses are rare and easily distinguished. Π occurs only in connection with free products.

I. Systems and their Generation

1. Groupoids

In this and the next two sections we adopt the viewpoint of BATES [63] with modifications suggested independently by PEREMANS [98] and EVANS [81].

A single-valued mapping α of a set G into a set H is a correspondence which assigns to each g in G a unique element $g\alpha$ in H. If α, β are single-valued mappings of G into H then $\alpha = \beta$ on G if and only if $g\alpha = g\beta$ for every g in G. We allow the possibility that G may be the empty set. If G, H are non-empty sets, the logical product $G \times H$ is the set of all ordered pairs (a, b), a in G, b in H, where $(a, b) = (c, d)$ if and only if $a = c$, $b = d$. By a (single-valued) *binary operation* α on the (non-empty) set G we mean a single-valued mapping α from some subset $R(\alpha)$ of $G \times G$ into G. Here $R(\alpha)$ is the *range* of α; we allow the possibility that $R(\alpha)$ may be empty. Two binary operations α, β on G are equal ($\alpha = \beta$) if and only if $R(\alpha) = R(\beta)$ and $\alpha = \beta$ on $R(\alpha)$. We make the following conventions in connection with a binary operation α on G: (1) If (a, b) is in $R(\alpha)$, we usually write the "product" ab instead of $(a, b)\,\alpha$. (2) The statement "ab is defined in G" means that (a, b) is in $R(\alpha)$. (3) The statement "$ab = c$ in G" means (a, b) is in $R(\alpha)$, c is in G and $(a, b)\,\alpha = c$. It will be obvious from the context what operations are in question.

A *halfgroupoid* G is a system consisting of a non-empty set G and a binary operation on G. A (proper or improper) subset H of the half-groupoid G is a subhalfgroupoid of G provided H is a halfgroupoid such that $ab = c$ in G whenever $ab = c$ in H. A *groupoid* G is a halfgroupoid G such that ab is defined in G for all a, b in G. A *subgroupoid* H of a half-groupoid G is a groupoid H which is a subhalfgroupoid of G.

The *order* of a halfgroupoid G is the cardinal number of the set G.

Let A be any (non-empty) index set and let $\{H_\alpha;\ \alpha \in A\}$ be a collection of halfgroupoids H_α. The collection will be termed *compatible* if $ab = c$ in H_α, $ab = d$ in H_β implies $c = d$. The collection will be termed *disjoint* if $a \in H_\alpha$, $a \in H_\beta$ implies $\alpha = \beta$. Every disjoint collection of halfgroupoids is of course compatible. A compatible collection of halfgroupoids $\{H_\alpha;\ \alpha \in A\}$ determines uniquely a halfgroupoid $S = \cup H_\alpha$, its *union*, defined as follows: (i) $a \in S \leftrightarrow a \in H_\alpha$ for at least one $\alpha \in A$. (ii) $ab = c$ in $S \leftrightarrow ab = c$ in H_α for at least one $\alpha \in A$. In particular, each H_α is a subhalfgroupoid of S. For special cases of A we use notations such as $H \cup K$ or $\bigcup_{i=0}^{\infty} H_i$. A compatible collection of halfgroupoids $\{H_\alpha;\ \alpha \in A\}$

will be termed *intersecting* if there exists an element a such that $a \in H_\alpha$ for every α in A. Such a collection determines a unique halfgroupoid $M = \cap H_\alpha$, its *intersection*, defined as follows: (i) $a \in M \leftrightarrow a \in H_\alpha$ for every $\alpha \in A$. (ii) $ab = c$ in $M \leftrightarrow ab = c$ in H_α for every $\alpha \in A$. In particular, M is a subhalfgroupoid of each H_α. The meanings of $H \cap K$, $\bigcap_{i=0}^{\infty} H_i$ should be clear.

Let G be a subhalfgroupoid of the halfgroupoid E. We say that G is *closed in* E if $a, b \in G$, $ab = c$ in E implies $ab = c$ in G. We say that E is an *extension* of G if (i) $ab = c$ in E implies $a, b \in G$; (ii) $c \in E$, $c \notin G$ implies $c = ab$ in E for at least one pair a, b in G. An extension E of G is *complete* if ab is defined in E for all a, b in G; in particular, E could be a complete extension of G and consist of the same elements as G. An extension E of G is *open* if (iii) $ab = c$ in E, $c \in G$ implies $ab = c$ in G; (iv) $c \in E$, $c \notin G$, $ab = a'b' = c$ in E implies $a = a'$, $b = b'$. Every halfgroupoid G is an open extension of itself, but is a complete extension of itself only if it is a groupoid. If G is a halfgroupoid but not a groupoid then, as is obvious from the definitions, G has extensions, distinct from itself, which are complete or open or both. To construct a complete open extension, let E be the set consisting of G and those elements (a, b) of $G \times G$ for which ab is not defined in G and, for all a, b in G, define $ab = c$ or $ab = (a, b)$ in E according as $ab = c$ in G or ab is not defined in G.

A collection $\{L_i; i = 0, 1, 2, \ldots\}$ of halfgroupoids L_i is called an *extension chain* if L_{i+1} is an extension of L_i for each i; and is open (or complete) if L_{i+1} is an open (or complete) extension of L_i for each i. If G is a subhalfgroupoid of the halfgroupoid H, the ***maximal extension chain of G in H*** is defined as follows: $G_0 = G$ and, for each nonnegative integer i, (a) G_i is a subhalfgroupoid of H; (b) G_{i+1} is an extension of G_i; (c) if $x, y \in G_i$ and if $xy = z$ in H, then $xy = z$ in G_{i+1}. The halfgroupoid $K = \bigcup_{i=0}^{\infty} G_i$ may be characterized as follows: (i) G is a subhalfgroupoid of K; (ii) K is a closed subhalfgroupoid of H; (iii) if G is a subhalfgroupoid of the closed subhalfgroupoid L of H, then K is a subhalfgroupoid of L. We call K the subhalfgroupoid of H *generated* by G. In particular, G generates H if $K = H$. The following criterion is convenient: *G generates H if (and only if) the only closed subhalfgroupoid of H containing G as a subhalfgroupoid is H itself.* We shall use this criterion as a definition.

Lemma 1.1. *If G is a generating subhalfgroupoid of H and H is a generating subhalfgroupoid of a halfgroupoid K, then G generates K.*

Proof. Let G be a subhalfgroupoid of a closed subhalfgroupoid C of K. Let $D = H \cap C$. Then G, D are subhalfgroupoids of D, H respectively. If $a, b \in D$ and $ab = c$ in H then (since $a, b \in C$ and $ab = c$ in K) $ab = c$

in C. Consequently, $ab = c$ in D. That is, D is closed in H. Then, since G generates H, $D = H$; in other words, H is a subhalfgroupoid of C. Thus, finally, $C = K$, and the proof of Lemma 1.1 is complete.

A *homomorphism* θ of a halfgroupoid G into (upon) a halfgroupoid K is a single-valued mapping of G into (upon) K such that $ab = c$ in G implies $(a\theta)(b\theta) = c\theta$ in K. An isomorphism θ of G upon K is a one-to-one homomorphism θ of G upon K such that θ^{-1} is a homomorphism of K upon G. An endomorphism (automorphism) of G is a homomorphism of G into (isomorphism of G upon) G. If G is a subhalfgroupoid of a halfgroupoid H and if θ, φ are homomorphisms of G, H respectively into the same halfgroupoid K, we say that *φ extends θ to H* (or that *φ induces θ on G*) provided $g\theta = g\varphi$ for each g in G.

Lemma 1.2. *If G is a generating subhalfgroupoid of a halfgroupoid H and if θ is a homomorphism of G into a halfgroupoid K, then θ can be extended in at most one way (and, possibly, in none) to a homomorphism of H into K.*

Proof. Let φ, ψ be homomorphisms of H into K which induce θ on G. Let S be the set of all s in H such that $s\varphi = s\psi$. Define $ab = c$ in S if and only if $a, b, c \in S$ and $ab = c$ in H. Then S is a subhalfgroupoid of H and G is a subhalfgroupoid of S. If $a, b \in S$ and $ab = d$ in H, then $d\varphi = (a\varphi)(b\varphi) = (a\psi)(b\psi) = d\psi$ in K, so $d \in S$. Hence S is closed in H, $S = H$, $\varphi = \psi$.

We say that a halfgroupoid H is *free over* its subhalfgroupoid G if, for every groupoid K and homomorphism θ of G into K, θ can be extended to a homomorphism of H into K. And H is *freely generated* by G if H is both free over and generated by G.

Lemma 1.3. *Let the halfgroupoid H be an extension of the halfgroupoid G. Then H is free over G if and only if H is an open extension of G.*

Proof. (I) Let H be an open extension of G and let θ be a homomorphism of G into a groupoid K. Define $a\varphi = a\theta$ for each a in G. If $a \in H$ and $a \notin G$ then $a = bc$ in H for a unique ordered pair b, c in H; and $b, c \in G$. In this case, define $a\varphi = (b\theta)(c\theta)$. Then φ is a homomorphism of H into K which induces θ on G.

(II) Let H be an extension of G which is not open. At least one of the following must hold: (i) There exist a, b, c in G such that $ab = c$ in H but ab is not defined in G. (ii) There exist a, b, c, d in G, e in H such that $ab = cd = e$ in H, ab is not defined in G and either $a \neq c$ or $b \neq d$. If (i) is true, let K be the set obtained by adjoining one new element, say x, to G. Define multiplication in K as follows: for all u, v in G, $uv = w$ or $uv = x$ in K according as $uv = w$ in G or uv is not defined in G; for all u in G, $ux = xu = xx = x$ in K. Then K is a groupoid. The identity mapping I_G of G is a homomorphism of G into K. If φ extends I_G to H then, since $ab = c$ in H, $ab = c$ in K; however, ab is not defined in G so $ab = x$ in K, a contradiction. Hence H is not free over G. If (i) is false then (ii) is true (with e not in G). We suppose for definiteness that $a \neq c$. This time let K be the same set as G with multiplication as follows: for all u, v in G, $uv = w$ or $uv = u$ in K according as $uv = w$ in G or uv is not defined in G. Then K is a groupoid and I_G is a homomorphism of G upon K. If φ extends I_G to H then $e\varphi = (ab)\varphi = a$ in K and (since cd cannot be defined in G) $e\varphi = (cd)\varphi = c$ in K, whereas $a \neq c$, a contradiction. Hence H is not free over G. This completes the proof of Lemma 1.3.

Theorem 1.1. (Existence and Uniqueness.) *Every halfgroupoid G freely generates at least one groupoid H. If G freely generates two groupoids H, H' there exists an isomorphism θ of H upon H' which induces the identity mapping on G.*

Proof. (I) There exists a complete open extension chain $\{L_i\}$ with $G = L_0$. Let $H = \bigcup_{i=0}^{\infty} L_i$. Completeness ensures that H is a groupoid. G is a subhalfgroupoid of H and $\{L_i\}$ is the maximal extension chain of G in H; hence G generates H. Let $\varphi(0)$ be a homomorphism of G into a groupoid K. By Lemma 1.3, openness ensures the existence of a set $\{\varphi(i)\}$ such that, for each nonnegative integer i, $\varphi(i)$ is a homomorphism of L_i into K and $\varphi(i+1)$ extends $\varphi(i)$. Let φ be the mapping of H which induces $\varphi(i)$ on L_i for each i. Then φ is an extension of $\varphi(0)$ to a homomorphism of H into K. Hence H is free over G.

(II) Let H, H' be groupoids freely generated by G. The identity mapping I_G of G, regarded as a homomorphism of G into H', can be extended (uniquely, by Lemma 1.2) to a homomorphism θ of H into H'. Likewise, I_G can be extended to a homomorphism θ' of H' into H. Since $\theta\theta'$ and I_H both extend I_G to an endomorphism of H, $\theta\theta' = I_H$. Similarly, $\theta'\theta = I_{H'}$. Hence θ is an isomorphism of H upon H' and $\theta' = \theta^{-1}$. This completes the proof of Theorem 1.1.

If $a = bc$ in a halfgroupoid H we say that b and c *divide a in H*. An element of H is *prime in H* if it has no divisors in H. A (finite or infinite) sequence $\{a_i\}$ of elements $a_1, a_2, \ldots$ of H is a *divisor chain* of H if each member a_{i+1} divides the preceding member a_i in H. A divisor chain $\{a_i\}$ of H is *finite over* a subhalfgroupoid G of H if a_k and all subsequent terms are in G for some integer k; the chain has *length n over G* if n is the least such integer k.

Lemma 1.4. *The following conditions are necessary and sufficient in order that a halfgroupoid H be freely generated by a given subhalfgroupoid G:* (i) *If a is prime in G, then a is prime in H.* (ii) *If $a \in H$, $a \notin G$, then $a = bc$ in H for one and only one ordered pair $b, c \in H$.* (iii) *Every divisor chain of H is either finite or finite over G.*

Proof. (I) Let H be freely generated by G. By Theorem 1.1 and its proof, if $\{L_i\}$ is the maximal extension chain of G in H then $\{L_i\}$ is a complete open extension chain and $H = \bigcup_{i=0}^{\infty} L_i$. Openness ensures that if $a \in L_i$ and $a = bc$ in L_{i+1}, then $a = bc$ in L_i. Since $a = bc$ in H if and only if $a = bc$ in some L_i, we deduce (i) by finite descent to $G = L_0$. Again, openness together with finite descent ensures that if $a = bc = xy$ in H and if either $b \neq x$ or $c \neq y$, then $a = bc = xy$ in G. Moreover if $a \in H$ and $a \notin G$, then $a \in L_{n+1}$, $a \notin L_n$ for some $n \geqq 0$ and so $a = bc$ in L_{n+1} for some (unique) ordered pair $b, c \in L_n$. This proves (ii). Finally, let $\{a_i\}$ be a divisor chain of H and let $n(i)$ be the least integer k such that $a_i \in L_k$. If a_{i+1} is a member of the chain and if $n(i) > 0$, then $a_i = bc$ in H for a unique ordered pair $b, c \in H$ (by (ii)), where in fact $b, c \in L_m$, $m = n(i) - 1$. Thus a_{i+1} is one of b, c, so $n(i+1)$ is at most $n(i) - 1$. Hence we see readily that the chain has length at most $n(1) + 1$ over G. This proves (iii) and also shows that the elements of L_n are those which are first terms of no divisor chains of length exceeding $n + 1$ over G.

(II) Let G, H satisfy (i)—(iii) and let L_i be the set of all elements of H which are first terms of no divisors chains of H of length exceeding $i + 1$ over G. In particular, $L_0 = G$. Define $ab = c$ in L_{i+1} if and only if $a, b \in L_i$ and $ab = c$ in H. It is then easy to see that $\{L_i\}$ is the maximal extension chain of G in H, that $H = \bigcup_{i=0}^{\infty} L_i$, and that L_{i+1} is an open extension of L_i for each i. Thence, by Lemma 1.1, 1.2, 1.3, H is freely generated by G. This proves Lemma 1.4.

A non-empty set B may be regarded as a halfgroupoid with no products defined. By Theorem 1.1, B freely generates a groupoid G, unique to within an isomorphism. A non-empty subset B of a halfgroupoid H is called a *free basis* of H if B freely generates H. And a halfgroupoid is *free* if it has a free basis. Existence of free groupoids is ensured. Two of the most significant properties of free systems are indicated in the next two theorems.

Theorem 1.2. (Free Representation Theorem.) *Every groupoid G is the homomorphic image of at least one free groupoid F.*

Proof. Let B be a generating subset of G; for example, the set G itself. Let F be the free groupoid with free basis B. The identity mapping of B may be extended to a homomorphism θ of F into G. Let H be the image of F under θ and define $ab = c$ in H if and only if $a, b, c \in H$ and $ab = c$ in G. Since F is a groupoid, H is a subgroupoid of G and hence H is closed in G. But B is a subhalfgroupoid of H and B generates G. Hence $H = G$, θ is upon G, and the proof is complete.

Theorem 1.3. *If θ is a homomorphism of the groupoid G upon the free groupoid F, then there exists an isomorphism φ of F upon a subgroupoid H of G such that $\varphi\theta = I_F$.*

Proof. Let B be a free basis of F. By the Axiom of Choice, there exists at least one single-valued mapping $\varphi(0)$ of B into G such that $\varphi(0)\theta = I_B$. Then $\varphi(0)$ can be extended to a homomorphism φ of F into G. And φ is a homomorphism upon some subgroupoid H of G. Since $\varphi\theta$ is an endomorphism of F extending $\varphi(0)\theta = I_B$, necessarily $\varphi\theta = I_F$. Let θ induce θ' on H, so that $\varphi\theta' = I_F$. If $h \in H$ then $h = f\varphi$ for at least one f in F and $h\theta' = f$. Hence φ is one-to-one and $\theta' = \varphi^{-1}$. This completes the proof.

The following lemma is obvious from Lemma 1.4:

Lemma 1.5. *The following conditions are necessary and sufficient in order that a halfgroupoid H be free:* (i) *If $a \in H$ and if a is not prime in H, then $a = bc$ in H for one and only one ordered pair $b, c \in H$.* (ii) *Every divisor chain in H is finite. Moreover, if H is free then H has one and only one free basis, namely the set of all primes in H.*

The *rank* of a free halfgroupoid is the cardinal number of its (unique) free basis. We have as corollaries of Lemma 1.5:

Theorem 1.4. *Every subhalfgroupoid of a free halfgroupoid is free.*

Lemma 1.6. *If a non-empty subset B_1 of a free halfgroupoid H generates a closed subhalfgroupoid H_1 of H and if no proper subset of B_1 generates H_1, then B_1 is a free basis of H_1.*

Theorem 1.5. *Every free groupoid H contains a free subgroupoid of countable rank.*

Proof of Theorem 1.5. Choose any $a \in H$, let B_1 be the countable sequence $\{a_i;\ i = 0, 1, 2, \ldots\}$, where $a_0 = aa$, $a_{i+1} = aa_i$; let H_1 be the subgroupoid of H generated by B_1; and apply Lemma 1.6.

Let $\{H_\alpha;\ \alpha \in A\}$ be any compatible collection of halfgroupoids and let $S = \cup H_\alpha$ be its union. By Theorem 1.1, S freely generates a groupoid G, unique to within an isomorphism. If the collection is disjoint, G is called the *free product* of the H_α and is denoted by $\Pi^* H_\alpha$. If the collection is not assumed to be disjoint, G is called the *generalized free product* of the H_α and is denoted by $\Pi \underline{*} H_\alpha$. For two compatible halfgroupoids H, K we write $H * K$ or $H \underline{*} K$ for their free or generalized free product, and similarly for any finite number. The proof of the next theorem is straightforward if we consider the "trees" formed by the divisor chains beginning with the various elements.

Theorem 1.6. *A groupoid G is free if and only if it is a free product of free groupoids of rank* 1.

Theorem 1.7. *Let the groupoid G be freely generated by a subhalfgroupoid S and let H be a subgroupoid of G. If P is the set of all primes of H which are not in $S \cap H$ (the latter being either the empty set or a subhalfgroupoid of G) then one of the following possibilities must occur:* (i) *P is empty, $S \cap H$ is not empty and H is freely generated by $S \cap H$.* (ii) *$S \cap H$ is empty, P is not empty and P is a free basis of H.* (iii) *Neither P nor $S \cap H$ is empty and $H = F * K$ where F is a free subgroupoid of H*

with free basis P and K is a subgroupoid of H which is freely generated by $S \cap H$.

Proof. Let $\{a_i\}$ be a divisor chain of H. Then $\{a_i\}$, as a divisor chain of G, is either finite or finite over S. In the latter case, $a_n \in S \cap H$ for some integer n, so $S \cap H$ is not empty. On the other hand, if $S \cap H$ is empty every divisor chain of H must be finite, so P cannot be empty. Hence the set $W = P \cup (S \cap H)$ is never empty. We introduce multiplication in W by the requirement that $ab = c$ in W if and only if a, b, $c \in W$ and $ab = c$ in H. Every divisor chain of H is finite or finite over W. If $a \in H$ and $a \notin W$ then a is not a prime in H and $a \notin S$. Then $a = bc$ in H for at least one pair b, $c \in H$ and also $a = bc$ in G for exactly one ordered pair $b, c \in G$. Therefore H is freely generated by W. This disposes of cases (i), (ii). In case (iii) we have proved that $H = P * (S \cap H)$. Now let F, K be the subgroupoids of H generated by P, $S \cap H$ respectively. Since P, $S \cap H$ are disjoint and $P \cup (S \cap H)$ freely generates $F \cup K$, we see that F, K must be disjoint. Clearly $F \cup K$ freely generates H; hence $H = F * K$. This completes the proof of Theorem 1.7. We note that (i), (ii), (iii) can be incorporated into the single statement that $H = F * K$, provided that we make the convention that an empty factor is to be deleted.

Theorem 1.7 allows a refinement theorem for free decomposition. Suppose that the groupoid G has subgroupoids P, Q, R, S such that

$$G = P * Q = R * S .$$

Since P, Q are disjoint, $(P \cup Q) \cap R = (P \cup R) \cap (Q \cap R)$ and $P \cap R$, $Q \cap R$ are disjoint (possibly empty). Thus the groupoid freely generated by $(P \cup Q) \cap R$ is $(P \cap R) * (Q \cap R)$, with the previously mentioned convention as to empty factors. Application of Theorem 1.7 therefore gives

$$P = F_1 * (P \cap R) * (P \cap S) , \quad Q = F_2 * (Q \cap R) * (Q \cap S) ,$$
$$R = F_3 * (P \cap R) * (Q \cap R) , \quad S = F_4 * (P \cap S) * (Q \cap S) ,$$

where each F is either empty or a free groupoid. Thus, if P, R are distinct but not disjoint, they have a common free-decomposition factor $P \cap R$. These remarks have obvious generalizations.

A halfgroupoid G is said to be *imbeddable* in a halfgroupoid H if there exists an isomorphism θ of G upon a subhalfgroupoid of H; then θ will be called an *imbedding* of G in H. Much of what has gone before can be rephrased in terms of imbedding.

Theorem 1.8. *A finite or countable halfgroupoid G can be imbedded in at least one groupoid H with a single generator.* (Cf. Evans [6], [7].)

Proof. Let $\{g_i; i = 0, 1, 2, \ldots\}$ be the set of all elements of G in some arbitrary ordering. Let K be obtained by adjoining one new element a to G. Define multiplication in K as follows: (i) if $uv = w$ in G, then

$uv = w$ in K; (ii) $aa = g_0$ in K; (iii) if g_i, g_{i+1} are in G, then $ag_i = g_{i+1}$ in K; (iv) no other products are defined in K. Then K is a halfgroupoid generated by a and I_G is an imbedding of G in K. It suffices, therefore, to take for H a groupoid freely generated by K. This completes the proof of Theorem 1.8. If we assume merely that G has a well-ordering of ordinal type τ we may show in similar fashion that G is imbeddable in a groupoid H with a generator corresponding to each non-successor ordinal $\alpha (\alpha \leqq \tau)$.

We close the section with a discussion of *generalized free products with amalgamated subgroupoids*. Let $\{H_\alpha; \alpha \in A\}$ be a disjoint collection of groupoids. Suppose that, for each ordered pair $\alpha, \beta \in A$, with $\alpha \neq \beta$, $K(\alpha, \beta)$ is a subgroupoid of H_α and $K(\alpha, \beta)$, $K(\beta, \alpha)$ are isomorphic. Suppose further that it is possible to define a compatible collection $\{K_\alpha; \alpha \in A\}$ of groupoids and a set $\{\varphi(\alpha); \alpha \in A\}$ such that $\varphi(\alpha)$ is an isomorphism of H_α upon K_α which maps $K(\alpha, \beta)$ upon $K_\alpha \cap K_\beta$ for each $\beta \neq \alpha$. Then $\Pi * K_\alpha$ is called the generalized free product of the H_α with amalgamated subgroupoids $K(\alpha, \beta)$. Clearly, if $\theta(\alpha, \beta)$ is the mapping induced by $\varphi(\alpha)\varphi(\beta)^{-1}$ on $K(\alpha, \beta)$:

(i) $\theta(\alpha, \beta)$ is an isomorphism of $K(\alpha, \beta)$ upon $K(\beta, \alpha)$;

(ii) $\theta(\alpha, \beta)^{-1} = \theta(\beta, \alpha)$;

(iii) if $x \in K(\alpha, \beta)$ and $x\theta(\alpha, \beta) \in K(\beta, \gamma)$, then $x \in K(\alpha, \gamma)$ and $x\theta(\alpha, \gamma) = x\theta(\alpha, \beta)\theta(\beta, \gamma)$.

Conversely, if we are given a set of mappings $\theta(\alpha, \beta)$ which satisfy (i)—(iii), we may replace equality $(=)$ in $S = \cup H_\alpha$ by $(\equiv)$, defined as follows: $a \equiv b$ in S if and only if either $a = b$ in S or, for some pair $\alpha, \beta \in A, \alpha \neq \beta$, we have $a \in K(\alpha, \beta)$, $b \in K(\beta, \alpha)$ and $a\theta(\alpha, \beta) = b$. Then we may redefine multiplication in S as follows: $xy \equiv z$ in S if and only if $x'y' = z'$ in S for some x', y', z' with $x' \equiv x$, $y' \equiv y$, $z' \equiv z$. The groupoid freely generated by the modified S is (isomorphic to) the generalized free product of the H_α with amalgamated subgroupoids $K(\alpha, \beta)$.

2. Quasigroups

The following laws (L), (R), which may or may not hold in a given halfgroupoid G, will be known as the *left-* and *right-cancellation* laws, respectively:

(L) If $ab = ac$ in G, then $b = c$. (R) If $ba = ca$ in G, then $b = c$.
Clearly (L) could be rephrased as follows:

(L') If $a, d \in G$, there is at most one x such that $ax = d$ in G.
It should be observed that if a halfgroupoid G satisfies (L) (or (R)) then so does every subhalfgroupoid of G and also every open extension of G. Given a left-cancellation halfgroupoid G (one satisfying (L)) we define the operation of *left-division* ($\backslash$) in G as follows: $a \backslash b = c$ if and only if

$ac = b$. Denoting the operation of multiplication by $\cdot$, we note that if $(G, \cdot)$ is a left-cancellation halfgroupoid, so is $(G, \backslash)$; moreover, $\cdot$ is the left-division operation of $(G, \backslash)$. We now must distinguish between the halfgroupoids $(G, \cdot)$, $(G, \backslash)$ and the (special) halfalgebra $(G, \cdot, \backslash)$ with two binary operations. Although most of what we have to say is valid (with obvious simplications) for left-cancellation halfgroupoids, we shall use the pair of laws (L), (R) henceforth.

A *halfquasigroup* (or *cancellation halfgroupoid*) G is a halfgroupoid G which satisfies both cancellation laws. A *quasigroup* G is a groupoid G such that, for each ordered pair $a, b \in G$, there is one and only one x such that $ax = b$ in G and one and only one y such that $ya = b$ in G. In a halfquasigroup G, left-division $(\backslash)$ and right-division $(/)$ are defined by the requirement that the equations $ab = c$, $a\backslash c = b$, $c/b = a$ are equivalent; all hold or none hold. More symmetrically, a halfquasigroup G is a halfalgebra $(G, \cdot, \backslash, /)$, forming a halfgroupoid with respect to each of the three binary operations, such that, in the halfalgebra:

(i) If $a \cdot b$ is defined, then $a\backslash(a \cdot b) = b$ and $(a \cdot b)/b = a$.

(ii) If $a\backslash b$ is defined, then $a \cdot (a\backslash b) = b$ and $b/(a\backslash b) = a$.

(iii) If a/b is defined, then $(a/b) \cdot b = a$ and $(a/b)\backslash a = b$.

Here we follow EVANS [81]. In these terms, a quasigroup G is a halfquasigroup such that each of $(G, \cdot)$, $(G, \backslash)$, $(G, /)$ is a groupoid.

A non-empty subset H of a halfquasigroup G is a subhalfquasigroup of G if H is a halfquasigroup which is a subhalfgroupoid of G with respect to each operation. And H is closed in G if H is closed as a subhalfgroupoid with respect to each operation. The latter statement means that if $ab = c$ in G and if two of a, b, c are in H, the third is in H and $ab = c$ in H. The term "generate" takes on the corresponding strong meaning. Again, a halfquasigroup G is free over a subhalfquasigroup H if every homomorphism of H into a quasigroup K can be extended to a homomorphism of G into K. We proceed to parallel the theorems of § 1, indicating necessary changes in the proofs.

Theorem 2.1. *A halfquasigroup G freely generates at least one quasigroup H. If G freely generates two quasigroups H, H′, there exists an isomorphism θ of H upon H′ which induces the identity mapping on G.*

Proof. We think in terms of the three operations $(\cdot, \backslash, /)$. The halfgroupoid $(G, \cdot)$ freely generates a groupoid $(H_1, \cdot)$. Since $(G, \cdot)$ is a cancellation halfgroupoid and since the cancellation properties are preserved under open extensions and under unions, $(H_1, \cdot)$ is a cancellation groupoid. Introducing the operations $\backslash, /$, we make H_1 into a (three-operation) halfquasigroup with G as a generating subhalfquasigroup. Similarly, H_1 generates a halfquasigroup H_2 such that $(H_2, \backslash)$ is a groupoid freely generated by $(H_1, \backslash)$. Singling out in cyclic order the three operations, we may define a chain $\{H_i;\ i = 0, 1, \ldots\}$ of halfquasigroups such that, for each i: (a) $H_0 = G$; (b) H_i is a subhalfquasigroup of H_{i+1}; (c) $(H_{3i+1}, \cdot)$ is a groupoid freely generated by $(H_{3i}, \cdot)$; (d) $(H_{3i+2}, \backslash)$ is a groupoid freely generated by $(H_{3i+1}, \backslash)$; (e) $(H_{3i+3}, /)$ is a groupoid freely generated by $(H_{3i+2}, /)$. Finally we set $H = \bigcup_{i=0}^{\infty} H_i$, with the obvious definition of the three operations, and verify that H is a quasigroup and that G generates H.

Let θ be a homomorphism of $(G, \cdot)$ into $(K, \cdot)$, where K is a quasigroup. We readily verify that θ is also a homomorphism of $(G, \backslash)$ into $(K, \backslash)$ and of $(G, /)$ into $(K, /)$. Indeed, owing to the fact that K is a quasigroup, each of the three "homomorphism" conditions on θ implies the other two. Moreover, θ can be extended uniquely to a homomorphism $\varphi(1)$ of $(H_1, \cdot)$ into $(K, \cdot)$; and $\varphi(1)$ is a three-operation homomorphism of H_1 into K. Proceeding in this way, we see finally that θ can be extended uniquely to a homomorphism of H into K. This proves the first half of Theorem 2.1. The second half depends on the appropriate analogue of Lemma 1.2, which holds with the present strong meaning of "generate".

In the above construction, some simplification results if the chain $\{H_i\}$ is replaced by a chain $\{L_i\}$ satisfying conditions obtained from (a)—(e) by replacing "groupoid freely generated by" in (c), (d), (e) by "complete open extension of". The chain $\{L_i\}$ was used by BATES [63].

A non-empty subset B of a halfquasigroup G may be regarded a subhalfquasigroup with no operations defined. We say that B is a *free set of generators* of G if B freely generates G. We say that G is *free* if G has a free set of generators. Observe that a free quasigroup G is not a free groupoid with respect to any of its operations; for example, every element of $(G, \cdot)$ is a product.

Theorem 2.2. *Let* $(G, \cdot)$ *be a groupoid. If (and only if)* $a \cdot G = G \cdot a = G$ *for every* a *in* G *then there exists a free quasigroup* H *such that the groupoid* $(H, \cdot)$ *possesses a homomorphism upon* $(G, \cdot)$ (BATES and KIOKEMEISTER [65], EVANS [81]).

Corollary. *Every quasigroup* G *is a homomorphic image of at least one free quasigroup.*

Remark. If G is the set of all nonnegative integers under the product operation $a \cdot b = |a - b|$, then $a \cdot G = G \cdot a = G$ for every a but G is not a quasigroup.

Proof. If H is a quasigroup then $x \cdot H = H \cdot x = H$ for every x in H and hence a similar condition must hold for every homomorphic image of $(H, \cdot)$. Conversely, let $(G, \cdot)$ satisfy $a \cdot G = G \cdot a = G$ for every a in G. We may avoid the explicit use of extension chains as follows. There exists a free quasigroup H with a free basis B and a singlevalued mapping ψ of B upon G. (For example, let B consist of the elements of G and let $\psi = I_G$.) We consider the set of all pairs $\{K, \theta\}$ where K is a subhalfquasigroup of H containing B such that the complete extension of $(K, \cdot)$ in $(H, \cdot)$ is open and θ is a homomorphism of $(K, \cdot)$ upon $(G, \cdot)$ which extends ψ. We partially order the pairs by defining $\{K,\theta\} \subset \{L, \varphi\}$ if and only if K is a subhalfquasigroup of L and φ extends θ. By ZORN's Lemma, there exists a maximal pair, say $\{K, \theta\}$. We prove that $K = H$ by showing that K is closed in H. Let $ab = c$ in H. If a, b are in K but c is not, let L be obtained from K by adjoining the element c and the equation $ab = c$; let φ be the mapping which induces θ on K and satisfies $c\varphi = (a\theta)(b\theta)$. Then $\{L, \varphi\}$ is a pair, contradicting the maximality of $\{K, \theta\}$. If a, c are in K but b is not, let L be similarly obtained by adjoining b and $ab = c$; let x be an arbitrarily chosen element of G such that $(a\theta)x = b\theta$ and let φ induce θ and satisfy $b\varphi = x$. Again we get a contradiction; similarly, we get a contradiction if b, c are in K but not a. Therefore K is closed in H and the proof of Theorem 2.2 is complete.

It is customary, and usually convenient, to consider a quasigroup in terms of a single binary operation. But then, as is clear from the remark following Theorem 2.2, it is usually essential to specify of a homomorphism whether it is into a quasigroup or merely a groupoid. For cases where this is unnecessary see [65] and also Chapter VII.

Let G be a free quasigroup with a free basis A. We define the ***rank*** of G to be $|A|$, where $|A|$ denotes the cardinal number of A. If $|A| > 1$, let a, b be two distinct elements of A and let A' be the subset of G obtained from A by replacing the selected element a by a' where a' is some one of ab, ba, $a\backslash b$, $b\backslash a$, a/b, b/a. It is easy to see that A' is a free basis of G; consequently, G has infinitely many free bases. The question as to whether one can get from any free basis A to any other free basis B by iteration of transformations of the type just described remains open. (For a positive answer to a like question for monogenic loops see EVANS [83].) However, the notion of rank is justified by the following theorem, for which we give a proof which seems instructive:

Theorem 2.3. *If A, B are two free sets of generators of the free quasigroup G, there exists an automorphism θ of G such that $A\theta = B$.*

Proof. We shall show that $|A| = |B|$, and this will ensure the existence of θ. Let R be the additive group of rational integers and let $\mathfrak{H}$ be the set of all homomorphisms of G into R. For each $a \in A$, let $D(a)$ be the unique member of $\mathfrak{H}$ which maps a upon 1 and every other element of A upon 0. We define $x \equiv y$, for x, y in G, if and only if $x\varphi = y\varphi$ for every φ in $\mathfrak{H}$. Clearly, $\equiv$ is an equivalence relation. Moreover, $a, a' \in A$, $a \neq a'$ implies $a \not\equiv a'$, since $aD(a) = 1$, $a'D(a) = 0$. We denote by x^* the equivalence class of the element x and verify readily that the mapping $x \to x^*$ is a homomorphism of G upon an additive abelian group G^* (a subdirect sum of copies of R) where $x^* + y^* = (xy)^*$ in G^*. We prove as follows that G^* is a free abelian group with A^* as free basis: First we note that each φ in $\mathfrak{H}$ determines a homomorphism φ^* of G^* into R by the requirement that $x^*\varphi^* = x\varphi$ for each x. Next, if an element x of G has a "finite representation" $x^* = n_1a_1^* + n_2a_2^* + \cdots + n_ka_k^*$, where the n_i are integers and the a_i are distinct elements of A, then the represention is unique since, for a in A, $xD(a) = n_i$ if $a = a_i$ and $xD(a) = 0$ if a is distinct from a_1, $a_2, \ldots, a_k$. The set of all elements with "finite representations" is a closed subhalfquasigroup of G containing A and hence coincides with G. Therefore A^* is a free basis of G^*. For like reasons, B^* is a free basis of G^*. By a well known result, $|A^*| = |B^*|$. Therefore $|A| = |A^*| = |B^*| = |B|$ completing the proof of Theorem 2.3.

A collection of halfquasigroups is compatible if it is compatible as a collection of halfgroupoids for each of the three operations. With this understood the definitions of a union, intersection, free product or generalized free product of a collection of halfquasigroups should be clear.

Theorem 2.4. *Let G be a quasigroup freely generated by a subhalfquasigroup S and let H be a subquasigroup of G. Then there is a subset B of H, depending only on S and H, such that one of the following possibilities must occur:* (i) *B is empty, $S \cap H$ is not empty and $S \cap H$ freely generates H.* (ii) *$S \cap H$ is empty, B is not empty and B is a free basis of H.* (iii) *Neither B nor $S \cap H$ is empty and $H = F * K$ where F, K are subquasigroups of H, B is a free basis of F and $S \cap H$ freely generates K.*

Corollary. *Every subquasigroup of a free quasigroup is free.*

Proof. For each nonnegative integer i, let 0_i denote the operation $\cdot$, $\backslash$, or $/$ according as i is congruent modulo 3 to 0, 1 or 2. Then (compare the proof of Theorem 2.1) we may assume that $G = \bigcup_{i=0}^{\infty} G_i$ for subhalfquasigroups G_i such that $G_0 = S$ and $(G_{i+1}, 0_i)$ is a groupoid freely generated by $(G_i, 0_i)$. We define $L_i = H \cap G_i$. Then, trivially, $L_0 = H \cap S$ is freely generated by $S \cap H$. We define B_0 to be the empty set. Considering some nonnegative integer n, we suppose inductively that, for $0 \leqq i \leqq n$, an increasing sequence of subsets B_i has been defined such that B_i, $S \cap H$ are disjoint and L_i is either empty or freely generated (as a halfquasigroup) by $B_i \cup (S \cap H)$. If L_{n+1} is empty, let B_{n+1} be the empty set. Otherwise $(L_{n+1}, 0_n)$ is a subgroupoid of $(G_{n+1}, 0_n)$. In this case, let P_{n+1} be the set of all primes of $(L_{n+1}, 0_n)$ which are not in $L_{n+1} \cap G_n = L_n$. By Theorem 1.7, since $(G_{n+1}, 0_n)$ is freely generated by $(G_n, 0_n)$, $(L_{n+1}, 0_n)$ is freely generated by $(P_{n+1} \cup L_n, 0_n)$. Consequently, $P_{n+1} \cup B_n \cup (S \cap H)$ freely generates L_{n+1}. Clearly P_{n+1}, B_n, $S \cap H$ are disjoint. We define $B_{n+1} = B_n \cup P_{n+1}$, completing the inductive definition, and then define

$$B = \bigcup_{i=0}^{\infty} B_i = \bigcup_{i=0}^{\infty} P_i\,.$$

Clearly B and $S \cap H$ are disjoint and, since H is the union of the L_i, $B \cup (S \cap H)$ freely generates H. This suffices for the proof of Theorem 2.4; we merely proceed as in the proof of Theorem 1.7.

Theorem 2.4 allows a refinement theorem for free decomposition of quasigroups; see the remarks following the proof of Theorem 1.7.

Instead of paralleling the proof of Theorem 1.5 we shall first consider the *commutator-associator subquasigroup*, G', of a quasigroup G. For arbitrary elements x, y, z of G the commutator, (x, y), and associator, (x, y, z), are defined by

$$xy = (yx)(x, y), \quad (xy)z = [x(yz)](x, y, z). \tag{2.1}$$

And G' is the subquasigroup of G generated by all commutators and associators. In view of this definition,

$$xy \in G' \leftrightarrow yx \in G'; \tag{2.2}$$

$$(xy)z \in G' \leftrightarrow x(yz) \in G'. \tag{2.3}$$

Thus $(xy)z \in G' \leftrightarrow x(yz) \in G' \leftrightarrow (yz)x \in G'$, so

$$(xy)z \in G' \leftrightarrow (yz)x \in G'. \tag{2.4}$$

Now write $k = (x, y)$, so that $k \in G'$ and $xy = (yx)k$. Then $(xy)z \in G' \leftrightarrow [(yx)k]z \in G' \leftrightarrow [z(yx)]k \in G' \leftrightarrow z(yx) \in G' \leftrightarrow (yx)z \in G'$, or

$$(xy)z \in G' \leftrightarrow (yx)z \in G'. \tag{2.5}$$

We note that (2.4), (2.5) imply that $(xy)z \in G'$ if and only if $(uv)w \in G'$ where u, v, w are x, y, z in some order. If x is a given element of G the equation $xa = bx$ sets up a one-to-one mapping $a \to b$ of G upon G. Choose any x' such that $x'x \in G'$. Then $x'(xa) \in G' \leftrightarrow (x'x)a \in G' \leftrightarrow a \in G'$ and, similarly, $x'(bx) \in G' \leftrightarrow b \in G'$. Since $xa = bx$, we see that a, b are both or neither in G'. Thus,

$$xG' = G'x \tag{2.6}$$

for every x in G. Next, for any fixed x, y in G, write $p = a(xy) = (bx)y = x(cy)$ and choose any w such that $w(xy) \in G'$. Then $x(yw) \in G'$, $y(xw) \in G'$. Moreover, $pw \in G' \leftrightarrow [a(xy)]w \in G' \leftrightarrow [w(xy)]a \in G' \leftrightarrow a \in G'$, and, similarly, $pw \in G' \leftrightarrow b \in G'$ and $pw \in G' \leftrightarrow c \in G'$. From this we deduce that

$$G'(xy) = (G'x)y = x(G'y) \tag{2.7}$$

for all x, y in G. It now follows rapidly from (2.6), (2.7) and (2.1) that *the mapping $x \to G'x$ is a homomorphism of G upon an abelian group G/G' in which multiplication is defined by $(G'x)(G'y) = G'(xy)$.* Thus, in particular,

$$x \backslash x \in G' \quad \text{and} \quad x/x \in G' \tag{2.8}$$

for every x in G. From the equations (2.1) we may verify readily that every homomorphism of G upon an additive abelian group maps G' into 0. In particular, in view of the proof of Theorem 2.3: *if G is a free quasigroup, then G/G' is a free abelian group of the same rank.*

Theorem 2.5. *The commutator-associator subquasigroup, G', of a free quasigroup G has infinite rank. Moreover, rank $G' \geqq$ rank G.*

Corollary. *Every free quasigroup contains a subquasigroup of infinite rank.*

Proof. Denote G' by H and let S be a free basis of G; then we may use the proof of Theorem 2.4 without change of notation. Hence $B \cup (S \cap H)$ is a free basis of H. For each s in S, let $D(s)$ be the endomorphism of G into the additive group of integers which maps s upon 1 and the remaining elements of S upon 0. Note that an element x of G is in $H = G'$ if and only if x is mapped into 0 by each $D(s)$. We first prove that $L_i = H \cap G_i$ is empty for $i = 0, 1$. Indeed, since $(G_1, \cdot)$ is a free groupoid with free basis S, a straightforward mathematical induction shows that for each x in G_1 there is at least one s in S such that $xD(s)$ is positive. In particular, $L_0 = S \cap H$ is empty,

$$B = \bigcup_{i=2}^{\infty} B_i$$

is a free basis of H and $B_2 = P_2$ is the set of all primes of $(L_2, \backslash)$. We complete the proof by showing that P_2 is infinite and that $|P_2| \geqq |S|$. We choose some s in S and define positive right and left powers s^n, ${}^n s$ inductively by $s^1 = {}^1 s = s$ and $s^{n+1} = s^n \cdot s$, ${}^{n+1}s = s \cdot {}^n s$. Then we define $s(n) = {}^n s \backslash s^n$ for each positive integer n; in particular, $s(3) = (s, s, s)$. For each positive integer n, s^n and ${}^n s$ are in G_1 and hence not in H; but $s(n)$ is in H and hence in L_2. If $s(n) = y \backslash z$ in $(G_2, \backslash)$, then, since $s(n)$ is not in G_1, the ordered pair is unique and hence $z = s^n$, $y = {}^n s$. Consequently, $s(1), s(2), \ldots$ are infinitely many distinct elements of P_2 for each s in S. Moreover, as s ranges over S, the element $s(n)$ ranges over distinct elements of P_2 for each positive integer n. This completes the proof of Theorem 2.5 and Corollary. Note that the proof readily furnishes information about the ranks of important subquasigroups of G'; for example, any subquasigroup containing the elements $s(1) = s \backslash s$ or the elements $s(3) = (s, s, s)$. Moreover, if S contains two distinct elements s, t, the elements (s^m, t^n), where m, n range over the positive integers, are distinct elements of P_2.

The proofs of the following theorems will be omitted:

Theorem 2.6. *A quasigroup G is free if and only if G is a free product of free quasigroups of rank 1.*

Theorem 2.7. *If θ is a homomorphism of the quasigroup G upon the free quasigroup F, there exists an isomorphism φ of F upon a subquasigroup of G such that $\varphi\theta = I_F$.*

Theorem 2.8. *Every finite or countable quasigroup G can be imbedded in a quasigroup H generated by a single element.*

The proofs should be sufficiently clear from the foregoing; but see also Bates [63], [64] and Evans [82], [83], [5], [6], [7].

The notion of a generalized free product of quasigroups with amalgamated subquasigroups carries over from groupoids with obvious modifications.

3. Loops

The element e of the groupoid $(G, \cdot)$ is a *left identity* of $(G, \cdot)$ if $e \cdot x = x$ for every x in G. Similarly for right identities and (two-sided) identities. A groupoid may have many left identities; let G be any non-empty set and define $x \cdot y = y$ for all x, y in G. But if e is a left and f is a right identity of a groupoid then $f = e \cdot f = e$. Hence a groupoid can have at most one identity. If the identity exists it will usually be denoted by 1. A *halfloop* is a halfquasigroup with an identity; a *loop* is a quasigroup with an identity. Thus a halfloop G is a halfquasigroup such that

$$x \cdot 1 = 1 \cdot x = x \quad \text{and} \quad x\backslash x = x/x = 1, \quad 1\backslash x = x/1 = x \tag{3.1}$$

for every x in G; here 1 is an identity for $(G, \cdot)$, a left identity for $(G, \backslash)$, a right identity for $(G, /)$.

Because of the cancellation laws, a subhalfloop G of a halfloop H automatically has the same identity element as H. The identity element plays an exceptional rôle throughout. For example, a collection of halfloops is said to be compatible if it is compatible as a collection of halfquasigroups and if the halfloops of the collection have the same identity element. And such a collection is called disjoint if it is disjoint aside from the identity element. In dealing with extensions we automatically assert (3.1) for each new element x. For example, if G is a halfloop, we say that $(H, \cdot)$ is an open extension of $(G, \cdot)$ if $(H, \cdot)$ has the same identity 1 as G and moreover: (i) if $a \cdot b = c$ in $(H, \cdot)$ and if $a \cdot b$ is not defined in $(G, \cdot)$, then c is not in G; (ii) if c is in H but not in G then $c = a \cdot b$ in $(H, \cdot)$ for exactly one ordered pair a, b of elements of H satisfying the inequalities $a \neq 1$, $b \neq 1$; and a, b are in G. Similar, but slightly different, modifications are made for each of the division operations.

For convenience in carrying over the material of § 2 to loops we restrict attention to generating halfloops (and hence, in particular, to generating subsets) containing the identity element. The loop consisting only of the identity element is said to be free of rank 0. More generally, the rank of a free loop with free basis A is defined to be the cardinal number of the set obtained from A by deleting the identity element. With minor exceptions, Theorem 2.i and its proof give rise to a valid

Theorem 3.i and proof if "quasigroup" is everywhere replaced by "loop". We will note a few necessary changes or improvements:

Theorem 3.2. *Let $(G, \cdot)$ be a groupoid. If (and only if) $(G, \cdot)$ has an identity element and $a \cdot G = G \cdot a = G$ for every a in G then there exists a free loop H such that the groupoid $(H, \cdot)$ possesses a homomorphism upon $(G, \cdot)$.*

Corollary. *Every loop is a homomorphic image of at least one free loop.*

Theorem 3.4. *Let G be a loop freely generated by a subhalfloop S and let H be a subloop of G. Then $H = F{*}K$ where F, K are subloops of H, F is free and K is freely generated by $S \cap H$.*

Corollary 1. *Every subloop of a free loop is free.*

Corollary 2. *If $G = P * Q = R * S$ where P, Q, R, S are subloops of the loop G, then*

$$P = F_1{*}(P \cap R) * (P \cap S)\,, \quad Q = F_2{*}(Q \cap R) * (Q \cap S)\,,$$
$$R = F_3{*}(P \cap R) * (Q \cap R)\,, \quad S = F_4{*}(P \cap S) * (Q \cap S)\,,$$

where the F_i are free loops.

Theorem 3.5. *If G is a free loop of positive rank, the commutator-associator subloop, G', has infinite rank. Moreover, rank $G' \geqq$ rank G.*

Corollary. *Every free loop of positive rank contains a subloop of infinite rank.*

In most of the proofs we have to deal with cancellation groupoids with a one-sided or two-sided identity element; the definition of a prime must be modified according to the case. Again, in the proof of Theorem 3.5, the elements $s(1), s(2), \ldots$ will not be distinct if $s = 1$.

Now let us turn to other concepts of freeness. We first introduce the concept of identical relations on a loop. Let F_n be the free loop of rank n freely generated by $X_1, X_2, \ldots, X_n$ (together with the identity element.) Let L be any loop and $a_1, a_2, \ldots, a_n$ be any elements of L. The mapping $X_1 \rightarrow a_1, X_2 \rightarrow a_2, \ldots, X_n \rightarrow a_n$ can be extended uniquely to a homomorphism θ of F_n into L. For each element W_n of F_n we denote the image $W_n\theta$ by $W_n(a_1, a_2, \ldots, a_n)$. This definition has the effect of turning each element W_n of F_n into a function defined on every loop. In particular, $W_n = W_n(X_1, X_2, \ldots, X_n)$; consequently, the definition is essentially a substitution principle. From the present point of view we shall call each element of F_n a *loop word*. It is important to note that if φ is a homomorphism of the above-mentioned loop L into a loop then $W_n(a_1, a_2, \ldots, a_n)\,\varphi = W_n(a_1\varphi, a_2\varphi, \ldots, a_n\varphi)$, since $\theta\varphi$ maps X_i on $a_i\varphi$. We shall say that the loop L *satisfies the identical relation* W_n provided $W_n(a_1, a_2, \ldots, a_n) = 1$ for all a_i in L.

Now let $\mathfrak{C}$ be any set of loop words. For example, $\mathfrak{C}$ might be empty or might contain the commutator word (X_1, X_2) and the associator word (X_1, X_2, X_3). Then a loop may be called a $\mathfrak{C}$-*loop* provided it satisfies all the identical relations of $\mathfrak{C}$. Consider an arbitrary free loop F and let θ be a mapping of F subject to the following conditions: (i) θ is a homomorphism of F upon a loop $F\varphi$. (ii) If W_n is in $\mathfrak{C}$ and if $a_1, a_2, \ldots, a_n$ are in F, then $W_n(a_1, a_2, \ldots, a_n)\,\theta = 1$. (iii) If x is in F, then $x\theta = 1$ only as required by (i) and (ii). (The existence of θ may be established most concretely in terms of normal subloops; see Chapter IV.) The loop $F\theta$ is a $\mathfrak{C}$-loop and is called a *free $\mathfrak{C}$-loop*. A subset of $F\theta$ is called a free basis of $F\theta$ if it has the form $B\theta$ where B is a free basis of F. It follows immediately that free $\mathfrak{C}$-loops have the characteristic property: *If φ is a single-valued mapping of a free basis of a free $\mathfrak{C}$-loop $F\theta$ into a $\mathfrak{C}$-loop L such that $1\varphi = 1$, then φ may be extended uniquely to a homomorphism of $F\theta$ into L.* In particular, $F\theta$ is uniquely defined, aside from an isomorphism, by F and $\mathfrak{C}$.

The theory of free $\mathfrak{C}$-loops cannot be expected to parallel the theory of free loops or free groups too closely except for judicious choices of $\mathfrak{C}$. As one example (details of which will be found in Chapter VIII) every free commutative MOUFANG loop of positive rank is infinite but, if its rank exceeds 2, contains subloops with exactly 3 elements. Hence subloops of free $\mathfrak{C}$-loops need not be free. We may note that if $\mathfrak{C}$ contains only the associator word, $\mathfrak{C}$-loops are simply groups, whereas if $\mathfrak{C}$ contains additional words which restrict the class, $\mathfrak{C}$-loops are "reduced" groups. (See, e. g., HALL [B15] and the references therein.) Thus free $\mathfrak{C}$-loops are somewhat vague objects.

The concept of a free $\mathfrak{C}$-product of $\mathfrak{C}$-loops is readily defined and always exists but amalgamation of subloops can give trouble. The approach by which one defines "half" $\mathfrak{C}$-loops and extends them to $\mathfrak{C}$-loops can also give trouble. In both cases, consider the theory of groups.

4. Loops, nets and projective planes

We shall touch briefly on certain aspects of geometry which are relevant to the present survey. The theory of free planes herein discussed was originally due to MARSHALL HALL in 1943 and both preceded and strongly influenced the theory of free loops (BATES [63]). For a detailed alternative treatment as well as an extensive bibliography, see G. PICKERT, Projektive Ebenen.

A *partial plane* G is a system consisting of a non-empty set G partitioned into two disjoint subsets (one of which may be empty), namely the *point-set* and the *line-set* (the elements of which are called *points* and *lines*, respectively), together with a binary relation, called *incidence*, such that: (i) (Disjuncture) if x is incident with y in G then one of x, y is a line of G and the other is a point; (ii) (Symmetry) if x is incident with y in G then y is incident with x in G; (iii) if x, y are distinct elements of G there is at most one z in G such that x and y are both incident with z in G.

A *projective plane* G is a partial plane G such that: (iv) if x and y are distinct points or distinct lines of G, there exists a z in G such that x and y are both incident with z in G; (v) there exists at least one set of four distinct points of G no three of which are incident in G with the same element. — It is easy to show that, in the presence of (i)—(iv), postulate (v) is equivalent to: (vi) there exists at least one set of four distinct lines of G no three of which are incident in G with the same element.

A *degenerate projective plane* G is a partial plane G which satisfies (iv) but not (v). For purposes of contrast, a projective plane is often termed nondegenerate.

A partial plane G may be turned into a halfgroupoid by defining $xy = z$ in G if and only if x, y are distinct elements of G and are incident in G with the element z of G. This definition does not turn projective planes into groupoids but it can be used as a guide for a theory of "freeness". There is, moreover, no particular difficulty in writing down a set of axioms (with or without mention of incidence) which characterizes partial planes and projective planes among halfgroupoids. Nevertheless, a few cautionary remarks are appropriate in connection with free projective planes, free products of projective lanes and generalized free products. First of all, we want elements to retain their "sex". For example, a sub-partial plane G of a partial plane H is a subhalfgroupoid (relative to the operation defined above) such that an element of G is a point or line of G according as it is a point or line of H. A similar condition must be imposed in connection with homomorphisms and with the concept of compatibility of a collection of partial planes. Secondly, freeness is best defined in terms of openness of extension chains rather than in terms of the possibility of extending homomorphisms. For if the element z of the partial plane H is incident in H with exactly two distinct elements x, y of H, both of which are in a sub-partial plane G which does not contain z, and if θ is a homomorphism of G into a projective plane K such that $x\theta = y\theta$, an extension φ of θ to H may perhaps exist but there is not enough information to determine $z\varphi$ uniquely. (Nevertheless, every projective plane is a homomorphic image of at least one free projective plane.) And, finally, although a partial plane G is always contained in at least one projective plane, G will only freely generate a projective plane if (v) holds at some finite stage of the construction. Equivalently, G must be *nondegenerate* in the sense that G contains one of the following partial planes as a sub-partial plane whose complete extension in G is open: (1) four distinct points, no lines; (2) three distinct points and a line incident with none of these; (3) three distinct points, two distinct lines, both lines incident with just one of the points but not with the same point; and the dual partial planes obtained from (1)—(3) by interchanging "point" and "line". Here "openness" ensures that the incidences in the sub-partial plane will not change when viewed from G.

A projective plane G is called *free* if it is freely generated by a sub-partial plane H consisting of four or more points and a single line incident in H with all but two of the points. A partial plane K is called *confined* provided K is finite and every element of K is incident in K with three or more distinct elements of K. It turns out that the free projective planes are precisely those without any confined sub-partial planes; this was proved by M. HALL for finitely generated planes and by L. KOPEIKINA for infinitely generated planes. In particular, if a projective plane G is freely generated by a (nondegenerate) sub-partial plane B and if B has no confined sub-partial planes, then G is free. If, in addition, B is finite, then G may be assigned a *rank*, namely $2b - i$ where b is the number of elements (points and lines) of B and where i is the number of incidences of B, defined as the number of ordered pairs (x, y) such that the point x is incident with the line y in B. The rank is an invariant: if G has rank r then $r \geqq 8$ and G can be freely generated by a sub-partial plane H consisting of $r - 4$ distinct points and a single line incident in H with exactly $r - 6$ of the points. Moreover, finitely generated free planes are isomorphic if and only if they have the same rank; every subplane of a free plane is free; and every free plane contains subplanes of every finite rank.

Free projective planes, since they contain no confined sub-partial planes, violate in the strongest possible sense all the classical postulates of projective geometry: Pappus' Theorem, Desargues' Theorem and so on. We may also note (EVANS [6]) that every countable projective plane is imbeddable in a projective plane generated by four points.

Let k be a cardinal number not less than 3. A *k-net* N is a partial plane N whose line-set has been partitioned into k disjoint classes such that: (a) N has at least one point; (b) each point of N is incident in N with exactly one line of each class; (c) every two lines of distinct classes in N are both incident in N with exactly one point. The definition of a half-k-net will be omitted. Free 3-nets are studied in BATES [63] and the results may be extended to k-nets. If some line of a k-net N is incident with exactly n distinct points in N, so is every line of N; moreover, every line class of N contains exactly n lines and either $n = 1$ or $n + 1 \geqq k$. The cardinal number n is called the *order* of N.

Since a net is a partial plane, every net may be imbedded in at least one projective plane. Every projective plane contains nets and, of these, two types have special significance. First consider some definite line L of a projective plane P and let S be a set of k distinct points of P all incident with L, $k \geqq 3$. Let the points of N be the points of P not incident with L in P; let each line-class of N consist of the lines of P incident in P with exactly one point of S, with a class for each point of S; and let elements a, b of N be incident in N if and only if they are incident in P. Then N is a k-net. If S contains every point incident with L in P, N is called an *affine plane*. If $k = 3$, N is called an *additive net* of P. Not every k-net can be obtained from a projective plane in this manner but, as is easily seen, every k-net is a sub-k-net of one so obtained; we first adjoin S and L, throw in additional points if necessary in order to make the resultant partial plane nondegenerate, and then freely generate a plane. Our second type is a 3-net N, a *multiplicative net* of a projective plane P. We select three distinct points A, B, C of P which form a triangle; that is, the lines AB, BC, CA are distinct. The points of N are the points of P not incident in P with AB, BC or CA; the lines of N are the lines incident in P with exactly one of A, B, C; and elements a, b of N are incident in N if and only if they are incident in P. Again, not every 3-net is a multiplicative net of some projective plane but every 3-net is a sub-3-net of such a multiplicative net.

Every 3-net N of order n gives rise to a class of quasigroups of order n as follows: First, in any fixed one of six possible ways, we designate the line-classes of N as the classes of 1-lines, 2-lines and 3-lines. Next we choose a set Q of n distinct elements and, for $i = 1, 2, 3$, a one-to-one mapping $\theta(i)$ of Q upon the class of i-lines of N. For x in Q, the i-line $x\theta(i)$ is assigned the coordinates (i, x); the unique point of N which is incident with both $(1, x)$ and $(2, y)$ is assigned the coordinates $[x, y]$. The operation o is defined on Q by the requirement that the line $(3, x o y)$ be incident in N with the point $[x, y]$. Then (Q, o) is a quasigroup. Conversely, any quasigroup (Q, o) defines a net whose lines and points are given by coordinates as stated, such that the point $[x, y]$ is incident with and only with the lines $(1, x)$, $(2, y)$, $(3, x o y)$.

Two quasigroups obtainable from the same 3-net N by different choices of the set Q or of the mappings $\theta(i)$ are said to be *isotopic*. For any Q, the $\theta(i)$ can always be chosen so that (Q, o) is a loop with a prescribed element e of Q as identity element or, more generally, with a prescribed point E of N as the point $[e, e]$. To do this, let L_1, L_2 be the 1-line and 2-line, respectively, incident with E and let φ be a one-to-one mapping of Q upon the points incident with L_1, such that $e\varphi = E$. Define $\theta(2)$, $\theta(3)$ so that $x\varphi$ is incident with $x\theta(2)$ and $x\theta(3)$; in particular, $L_2 = e\theta(2)$. Then define $\theta(1)$ so that $x\theta(1)$, L_2 and $x\theta(3)$ are incident with the same point; in particular, $L_1 = e\theta(1)$. A pictorial representation is illuminating; see PICKERT p. 43 and the various references there given.

If the 3-net N is an additive (or multiplicative) net of a projective plane P, each loop defined by N is called an *additive* (or *multiplicative*) *loop* of P. An important open question is this: *What loops are additive (or multiplicative) loops of projective planes?* At present we have information mainly about the finite case and even the possible finite orders of such loops are only partly known. See PICKERT, Chapter 12. HUGHES [352] has answered the question for countably infinite loops.

By a construction similar to the above, a k-net N may be used to define in many ways a system Q which is a quasigroup with respect to $k-2$ distinct but interrelated operations. When N and hence k is finite, the multiplication tables of these quasigroups form a set of $k-2$ mutually orthogonal latin squares.

5. General algebras

Let G, A be non-empty sets, let G^A be the set of all single-valued mappings of A into G and let $|A|$ be the cardinal number of A. By an *operation of cardinal* $|A|$ *on* G we mean a single-valued mapping of a subset of G^A into G. The operation is *completely defined on* G if it is a mapping of the whole of G^A into G. Properly speaking, our definition should depend not on A but only on $|A|$. The case that A has been well-ordered is sometimes of interest. If $|A|$ is finite the operation is called *finitary*; otherwise, *infinitary*. If $|A| = n$, where n is a positive integer, the operation is called n-ary; in this case it is customary to represent G^A as the set of all ordered n-tuples $(x_1, x_2, \ldots, x_n)$ where the x_i range over G. A unary operation ($n = 1$) is then a mapping of a subset of G into G. Binary and ternary operations correspond to $n = 2$, $n = 3$ respectively.

A *halfalgebra* (or *partial algebra*) G is a system consisting of a non-empty set G together with a specified non-empty set of operations on G and a (possibly empty) set of specified relations connecting the operations. An *algebra* G is a halfalgebra G each of whose operations is completely defined on G. These definitions obviously take us very far afield from binary systems and little space can be devoted to algebras. We shall limit discussion (here and later) to a few topics closely connected with binary systems. Two general references are BIRKHOFF, Lattice Theory and (in Japanese) SHÔDA, General Algebra. A few recent papers on the subject are listed in Part C, I of the bibliography; linear algebras, rings, lattices have largely been excluded.

In §§ 2, 3, it was convenient to consider quasigroups as algebras with three binary operations. Similarly, groups can be considered as algebras with a binary operation (multiplication) and a unary operation (the inverse operation). Again, groups have been defined by various authors as algebras with a single ternary operation satisfying a specified identity. An allied subject is the theory of *polyadic groups*; see DÖRNTE [333], POST [26], TVERMOES [31] and also TCHOUNKINE [28].

Aside from the various theories of convergence of sequences in analysis or topology, the literature of infinitary operations is very sparse; see, for example, LYAPIN [20], [21] and a few remarks in JÔNSSON and TARSKI [15]. We shall assume henceforth that all operations in question are finitary and speak in terms of an n-ary operation f written as a function $f(x_1, x_2, \ldots, x_n)$.

Let α be a binary relation on a non-empty set G and write $x \sim y(\alpha)$, for x, y in G, if and only if x is in the relation α to y. Then α is an *equivalence* (or *equivalence relation*) on G provided that, for all x, y in G: (i) if $x = y$ then $x \sim y(\alpha)$; (ii) if $x \sim y(\alpha)$ then $y \sim x(\alpha)$; (iii) if $x \sim y(\alpha)$ and $y \sim z(\alpha)$ then $x \sim z(\alpha)$. The *equivalence class* of x mod α is the set $[x]$ of all y such that $y \sim x(\alpha)$; if $y \in [x]$ then $[y] = [x]$. And G/α is the set of all equivalence classes mod α. Two equivalences α, β on G are said to *commute* provided that, if $x \sim y(\alpha)$ and $y \sim z(\beta)$ for some x, y, z in G, then there exists a w in G such that $x \sim w(\beta)$ and $w \sim y(\alpha)$. If G is an algebra, an equivalence α on G is said to be a *congruence* provided that, for each operation f of the algebra, if x_i, $y_i \in G$ and $x_i \sim y_i(\alpha)$ for $i = 1, 2, \ldots, n$, then $f(x_1, x_2, \ldots, x_n) \sim f(y_1, y_2, \ldots, y_n)\ (\alpha)$. In this case G/α may be turned into an algebra with the "same" operations as G by defining $f([x_1], \ldots, [x_n]) = [f(x_1, \ldots, x_n)]$ for each operation f of G and all x_i in G. The mapping θ defined by $x\theta = [x]$ is a homomorphism of G upon G/α; conversely, each homomorphism of G upon an algebra of the same type as G uniquely determines a congruence. A satisfactory theory of homomorphisms or congruences has been developed for those algebras which have the property that every two congruences commute. This is along lattice-theoretic lines (see Birkhoff, loc. cit.). If we define a *primitive class* of algebras to be the set of all algebras with a prescribed set of (finitary) operations and identical relations, Malcev [24] has given a necessary and sufficient condition that all congruences should commute for every algebra of a primitive class: *There must exist a polynomial* $P(x, y, z)$ (a function defined by iteration of the operations) *such that* $P(x, x, y) = y$, $P(x, y, y) = x$ *are identities for each of the algebras*. Note that if P exists and if $x \sim y\ (\alpha)$, $y \sim z\ (\beta)$, then, since $w = P(x, y, z)$ is built up from x, y, z by iteration of the operations, we have $x \sim P(x, y, y) \sim w\ (\beta)$ and $w \sim P(x, x, z) \sim z\ (\alpha)$. As an illustration, consider the set of all three-operation quasigroups $(G, \cdot, /, \backslash)$ and define $P(x, y, z) = \{(x \cdot y)/x\}\backslash(x \cdot z)$; then $P(x, x, y) = \{(x \cdot x)/x\}\backslash(x \cdot y) = x\backslash(x \cdot y) = y$ and, if $k = x \cdot y$, $P(x, y, y) = (k/x)\backslash k = x$. Thus *all congruences on a three-operation quasigroup commute.* Malcev's result implies that *the free groupoid of rank* 3 (and hence any free groupoid of higher rank) *has non-commuting congruences*. For if the contrary were true then $P(x, y, z)$ would exist with the stated properties and hence all congruences on an arbitrary groupoid would commute. It is known however that this is false. Specifically, *the free quasigroup of rank* 4 (and hence any free quasigroup of higher rank) *has non-commuting "multiplicative" congruences* (Thurston [112], Trevisan [115]). Malcev's result does not apply to multiplicative quasigroups and the facts for free quasigroups of rank 1, 2 or 3 seem to be unknown; similarly for free loops of arbitrary positive rank. (But see COWELL [349].) Another question is whether a congruence on an algebra is uniquely determined by a single congruence class; the answer is no in general (MALCEV [23], JABUKIK [14]) but MALCEV [24] gives conditions for a positive answer. — It is easily seen from first principles that the answer is yes for groups and three-operation quasigroups.

LYNDON [22] exhibits a finite algebra G (a groupoid) such that no finite set of identical relations of G implies every identical relation of G. In a different direction, EVANS and NEUMANN [43] show that no finite set of identical relations in a single variable will ensure that a groupoid is power-associative; using this fact they prove that the number of varieties of groupoids (or loops) of countable order has the power of the continuum.

EVANS [5] studies a class of algebras (with finitary operations) subject to the requirement that every halfalgebra of the type considered is imbeddable in an algebra. Thus the class can be the class of groupoids or (three-operation) quasigroups or loops, but not the class of groups. The methods are general. He takes as given a finite set of generators and a finite set of relations between the generators, expressed in terms of the operations of the algebras. He constructs "words" inductively by asserting that each generator is a word and that, if f is an operation and $w_1, \ldots, w_n$ are words, then $f(w_1, \ldots, w_n)$ is a word. The elements of the corresponding algebra are the classes of equivalent words, two words being equivalent only on the basis of specified rules involving the relations between the generators. He shows the existence of an algorithm by which the equivalence or nonequivalence of two words can be decided in finitely many steps. Thus *the word problem for the class of algebras is recursively solvable.* In the case of groupoids, quasigroups and loops, these methods form an alternative to those of §§ 1—3; they are carried further in EVANS [82], [83].

EVANS [7] complements the preceding result by proving that the word problem is recursively solvable for *every* finitely generated and finitely related algebra of a primitive class $\mathfrak{A}$ if and only if the following imbedding problem is recursively solvable: To decide whether a finite halfalgebra can be imbedded in an algebra belonging to $\mathfrak{A}$.

II. The Associative Law

1. Semigroups

A *semigroup* S is an associative groupoid; that is, a groupoid such that the *associative law*

$$(xy)z = x(yz) \tag{1.1}$$

holds for all x, y, z in S. [Many authors, including most of those writing in French, use the term "demigroup" for an associative groupoid;

these authors reserve "semigroup" for what we shall call a cancellation semigroup. Other terms are "monoid" (BOURBAKI) and "associative system" (Russian authors). The present terminology is standard in English and German.]

For the rôle of semigroups in analysis see HILLE, Functional Analysis and Semigroups. Aside from this reference, the literature on the application of semigroups to analysis or topology has been omitted from the bibliography. Likewise, papers applying semigroup theory to arithmetic (unique factorization) and lattice theory have largely been omitted.

The homomorphism theory of semigroups has been studied rather thoroughly without many interesting results. If θ is a homomorphism of a semigroup S into a groupoid, the image $S\theta = S'$ is also a semigroup. If s' is an element of S', let $K = s'\theta^{-1}$ be the *inverse image* of s'; that is, K is the set of all s in S such that $s\theta = s'$. Clearly K has the following properties: (i) if $k \in K$, $x \in S$ and $xk \in K$, then $xK \subset K$; (ii) if $k \in K$, $x \in S$ and $kx \in K$, then $Kx \subset K$; (iii) if $k \in K$, $x, y \in S$ and $xky \in K$, then $xKy \subset K$. [Here, for example, xKy is the set of all elements $xk'y, k' \in K$.] A non-empty subset K of S with properties (i), (ii), (iii) is called a *normal* subset. To a given normal subset K of S there may correspond two (or more) distinct homomorphisms of S with K as an inverse image; herein resides, perhaps, the relative poverty of the theory. LYAPIN [217], [218], [219] associates with each normal subset K of S a "weakest" homomorphism θ of S as follows: He defines a binary relation R on S by the requirement that aRb, for a, b in S, if and only if either $a = b$ or $a, b \in K$. He shows that R may be extended to a congruence relation ($\equiv$) on S with the property that, for $k \in K$, $x \equiv k$ if and only if $x \in K$. The mapping $x \to [x]$ of S into the corresponding set of congruence classes $[x]$ is a homomorphism θ of S into a semigroup $S\theta$ such that K is the inverse image of some element of $S\theta$. Defining a semigroup S to be simple if every homomorphism of S is either an isomorphism or a homomorphism upon a semigroup with one element, LYAPIN shows that the only *commutative* simple semigroups are the group of order one and the cyclic groups of prime order.

Various authors, both before and after LYAPIN, have investigated special types of normal subsets. Suppose, for example, that θ is a homomorphism of S upon a semigroup $S\theta$ with identity element 1 and let K be the inverse image of 1. Then K is a subsemigroup of S. Such a *normal subsemigroup* may be characterized by the following properties: (a) if $k \in K$, $x \in S$ and either $kx \in K$ or $xk \in K$, then $x \in K$; (b) if $k \in K$ and $x, y \in S$ and if one of xky, xy is in K, then so is the other. If $S\theta$ is to have further properties; for example, if $S\theta$ is to be a group, then further restrictions must be placed on K. See, for example, LEVI [213], [214], DUBREIL [185], [187], [188], LYAPIN [217], EILENBERG and MACLANE [79], STOLL [266]. Again, suppose that $S\theta = S'$ has a *zero* element; that is, an element 0 such that $0S' = S'0 = 0$; and let K be the inverse image of 0. Then K is an *ideal* of S, namely a non-empty subset of S such that $SK \subset K$, $KS \subset K$. Ideals have been studied, from more than one point of view, by SUŠKEVIČ [267], [269], BAER and LEVI [160], KAWADA and KONDO [205], REES [247], [248], [249], CLIFFORD [168]—[172], CLIFFORD and MILLER [174], SCHWARZ [253] to [256], RICH [251], COTLAR and ZARONTONELLO [176], GREEN [197], TEISSIER [289], VOROBEV [320], ANDERSEN [154], ISEKI [202], AUBERT [157], [158], HASHIMOTO [200] and others. Some of the more interesting theorems on ideals are developed in § 8 below. Since normal subsets of a semigroup S are equivalence classes of S with respect to congruences, the theory of normal subsets and the theory of congruences may reasonably be lumped together; further references are: DUBREIL and DUBREIL-JACOTIN [189], VOROBEV [319], SIVERCEVA [261], THURSTON [312], CHAMBERLIN and WOLFE [162], CROISOT [179], THIERRIN [302]—[306] and [308], [309], PIERCE [239], IVAN [203].

Free semigroups are studied by LEVI [213], BAER [159].

Let S be a semigroup given in terms of a finite set of generators and a finite set of relations between the generators. Then the elements of S are classes of equivalent words in the generators, the rules for equivalence being explicitly set up in terms of the relations between the generators. The *word problem* for S consists in describing an algorithm which will show in a finite number of steps whether two (arbitrarily chosen) words are equivalent. POST [240] constructed a semigroup for which the word problem is recursively unsolvable; another example with two generators was later given by HALL [199]. MARKOV [227], [228] developed a method of showing the recursive insolvability of a number of allied problems; for example, the problem of deciding whether two semigroups (given by generators and relations) are isomorphic. The word problem for free semigroups and free commutative semigroups on a finite number of generators is always solvable.

Many authors have studied semigroups subject to one or more conditions generalizing well known properties of groups. The most interesting of these topics will be deferred to subsequent sections; many others will be omitted. Here we shall outline the main concepts in the theory of *inverse* semigroups; relevant or closely allied papers are: CLIFFORD [164], [166], MANN [226], PRACHAR [390], FUCHS [194], THIERRIN [295]—[298] and [310], VAGNER [314], CROISOT [181], [183], PRESTON [241]—[243], LIBER [216], MUNN and PENROSE [230]. An element e of a semigroup S is called *idempotent* if $e^2 = e$. An element x of S is called *regular* if $x \in xSx$; idempotent elements are regular. If elements x, x' of S satisfy $x = xx'x$, $x' = x'xx'$, then each is called a *relative inverse* of the other. Now let x be a regular element of S and let A be the set of all a in S such that $x = xax$; clearly A is non-empty.

We observe that $(ax)(bx) = ax$ for all a, b in A. Hence: (i) Ax is a subsemigroup; (ii) every element of Ax is idempotent; (iii) Ax is commutative if and only if Ax consists of a single element. Similar remarks hold for xA. Since $x(AxA)x = xAx = x$, then $AxA \subset A$. If $a, b \in A$, then $(axb)x(axb) = axb$; hence every element of AxA is a relative inverse of x. If x' is a relative inverse of x then $x' \in A$ and $x' = x'xx' \in AxA$. Hence AxA is the set of all relative inverses of x. If Ax, xA are both commutative, then, for each a in A, $Ax = ax$, $xA = xa$ and $AxA = axA = axa$; hence x has a unique relative inverse. If x has a unique relative inverse x', then, for all a, b in A, $axb = x'$, $x'x = axbx = ax$, $xx' = xaxb = xb$; thus $Ax = x'x$, $xA = xx'$, whence Ax, xA are both commutative. That is, x has a unique relative inverse if and only if Ax, xA are both commutative. Next let e, f be idempotent elements of S such that ef has a unique relative inverse p. Since $(ef)(pe)(ef) = ef$ and $(pe)(ef)(pe) = pe$, then $p = pe$. Similarly, $p = fp$. Hence $p^2 = (pe)(fp) = p$. Then $ppp = p$, so p and ef are relative inverses of p. If p also has a unique relative inverse, then $p = ef$ and $p = fpe = (fe)^2$. If, in addition, fe has a unique relative inverse q and q has a unique relative inverse, then $q = fe = q^2$ and $p = (fe)^2 = q^2 = q$. That is, $ef = fe$. We may draw the following conclusion: *If the regular elements of S form a subsemigroup R and if each element of R has a unique relative inverse in S, then every two idempotent elements of S commute.* Finally, let us assume that the set R of regular elements of S is non-empty and that every two idempotent elements of S commute. If $x, y \in R$ then (by the earlier discussion) x, y have unique relative inverses x', y' (in S) respectively. Since the idempotents yy', $x'x$ commute, then $(xy)(y'x')(xy) = (xx'x)(yy'y) = xy$ and $(y'x')(xy)(y'x') = (y'yy')(x'xx') = y'x'$. Hence $xy \in R$ and $(xy)' = y'x'$. Thus: *If the set R of regular elements of S is non-empty and if every two idempotent elements of S commute, then R is a subsemigroup of S, each element x of R has a unique relative inverse x' in S, and the mapping $x \to x'$ is an (involutary) anti-automorphism of R.* A semigroup R is called an *inverse* semigroup provided R is regular (that is, each element of R is regular in R) and R satisfies one of the equivalent conditions: (I) each x in R has a unique relative inverse in R; (II) every two idempotent elements of R commute. For further details see the papers cited.

2. Cancellation semigroups

A *cancellation* semigroup is a semigroup satisfying the two cancellation laws (I § 2). [In most French writing, "semigroup" means "cancellation semigroup".] We may note that every free semigroup is a cancellation semigroup (Levi [213]).

Every subsemigroup of a group is a cancellation semigroup. Every *commutative* cancellation semigroup can be imbedded in a group just as the multiplicative semigroup of positive integers is imbedded in the multiplicative group of positive integers; for a slight generalization see VANDIVER [316]. An example due to MALCEV shows that not every cancellation semigroup can be imbedded in a group; CHEHATA [163] proves (as does VINOGRADOV [420]) that MALCEV's example can be simply ordered. MALCEV [224] gives an infinite set of conditions which are necessary and sufficient that a cancellation semigroup be imbeddable in a group and shows that no finite set of conditions can be necessary and sufficient. LAMBEK [211] gives an equivalent set of necessary and sufficient conditions in geometric language. DOSS [184] gives a clear discussion of the necessary and sufficient conditions of MALCEV and uses these conditions to establish the sufficiency of a condition which we now describe: Let S be a cancellation semigroup. Call an element a of S *regular on the left* if, for every $b \in S$, a and b have a common left multiple in S; that is, $Sa \cap Sb$ is non-empty. Call S *quasi-regular on the left* if, whenever two elements a, b of S have a common left multiple, there exist elements c, d in S such that $ca = db$ and at least one of c, d is regular on the left. DOSS shows that a cancellation semigroup which is quasi-regular on the left can be imbedded in a group. This result contains as a special case an earlier one of DUBREIL [186] which uses the Öre condition: every element of S is regular on the right. Additional literature: PTAK [244], [245], DUBREIL-JACOTIN [190], KONTOROVIC [209].

LEVI [214] shows that every groupoid has a unique "maximal" homomorphic image which is a semigroup; MALCEV [224] shows that a semigroup which can be imbedded in a group can be imbedded in one which is "absolutely minimal"; for precise statements and proofs of comparable theorems see § 4 below. BAER [159] studies a generalization of a semigroup called an *add*; the work is intimately connected with the imbedding problem. The paper of TAMARI [280] is related both to MALCEV [224] and to BAER [159].

TURING [239] gives an example of a cancellation semigroup with an unsolvable word problem (compare the discussion in § 1). In his (unpublished) University of Wisconsin thesis (1954) ADDISON uses TURING's example together with an imbedding theorem of EVANS [193] to carry over the methods of MARKOV [228] to cancellation semigroups. In particular, ADDISON shows that the isomorphism problem and the imbedding problem are recursively unsolvable. The work of ADDISON is almost identical with independent work of K. A. BOONE; a joint paper is planned but has not yet been written.

One further topic seems worthy of note. Let n, k be positive integers. Let $S(n, k)$ denote the semigroup with n generators which satisfies the identical relation $x^{k+1} = x$ but is otherwise free; and let $B(n, k)$ be the (BURNSIDE) group with n generators which satisfies the identical relation $x^k = 1$ but is otherwise free. GREEN and REES [198] prove that, for each k, a necessary and sufficient condition that $S(n, k)$ be finite for every n is that $B(n, k)$ be finite for every n. Thus, by known results on BURNSIDE groups, $S(n, 1)$, $S(n, 2)$, $S(n, 3)$ are finite for every n. The authors determine the order of $S(n, 1)$; see also MCLEAN [223].

3. Groups

Although the theory of groups is outside the scope of this report, we wish to relate groups to other types of system. It seems desirable to begin with two familiar results.

We shall define a *group* as an associative quasigroup. If G is a group and if $a, b \in G$, elements e, f of G are uniquely defined by $ae = a$, $fb = b$. Then $(ae)b = ab = a(fb) = (af)b$, so $ae = af$, $e = f$. Hence a group G has a unique identity element. That is: *a group is an associative loop.* The identity of G will be denoted by 1. If $a \in G$, elements a', a'' of G are uniquely defined by $aa' = 1$, $a''a = 1$. Then $a'' = a''1 = a''(aa') = (a''a)a' = 1a' = a'$. The element $a'' = a'$ is called the *inverse* of a and denoted by a^{-1}. As is easily verified, $(a^{-1})^{-1} = a$ and $(ab)^{-1} = b^{-1}a^{-1}$.

Theorem 3.1. *Let G be a semigroup with a right identity element e such that to each $a \in G$ there corresponds at least one element $a' \in G$ satisfying $aa' = e$. Then (and only then) G is a group with identity e.*

Proof. Consider any $a \in G$, any $a' \in G$ such that $aa' = e$ and any $a'' \in G$ such that $a'a'' = e$. Then $a = ae = a(a'a'') = (aa')a'' = ea''$. Hence $a = ea''$ and therefore $ea = e(ea'') = (ee)a'' = ea'' = a$. That is, e is an identity element for G. In particular, $a = ea'' = a''$. Hence $aa' = e$ implies $a'a = e$. If also $aa_1 = e$ then $a' = a'e = a'(aa_1) = (a'a)a_1 = ea_1 = a_1$. Thus a' is uniquely defined by a and we write $a' = a^{-1}$. If $a, b \in G$, $a(a^{-1}b) = (aa^{-1})b = eb = b$. Hence the equation $ax = b$ has at least one solution x in G. Conversely, if $ax = b$, then $a^{-1}b = a^{-1}(ax) = (a^{-1}a)x = ex = x$, so the solution is unique. Similarly, the equation $ya = b$ has one and only one solution in G, namely $y = ba^{-1}$. Therefore G is an associative loop; that is, a group.

Theorem 3.2. *Let G be a semigroup such that $aG = Ga = G$ for every a in G. Then (and only then) G is a group.* (Cf. HUNTINGTON [201].)

Proof. If $a, b \in G$ there exist e, f, x, y in G such that $ae = a$, $fb = b$, $xa = f$, $by = e$. Then $fe = (xa)e = x(ae) = xa = f$ and $fe = f(by) = (fb)y = by = e$. Hence $e = f$. Therefore G has an identity $e = f$, which we denote by 1. Since, for each b in G, $by = 1$ has a solution y in G, G is a group by Theorem 3.1.

A subsemigroup G of a groupoid $(S, \cdot)$ is called a subgroup of S if $(G, \cdot)$ is a group.

Theorem 3.3. *Let A be a non-empty subset of the semigroup S such that (i) $a \in A \Rightarrow a \in a^2S \cap Sa^2$; (ii) $a \in A \Rightarrow A \subset aS \cap Sa$. Then there exists a unique subgroup G of S, maximal in the property that $A \subset G$. Specifically, if $a \in A$, the set aS contains exactly one element e such that $ae = a$;*

and G consists of all elements of eSe which have inverses in eSe with respect to e.

Proof. By (ii) clearly we have: (iii) $a \in A$, $u \in S$, $au = a \Rightarrow bu = b$, all $b \in A$ and: (iv) $a \in A$, $v \in S$, $va = a \Rightarrow vb = b$, all $b \in A$.

Let $b, c \in A$. Since $b \in b^2S$, $c \in Sc^2$, we have $b(bs) = b$ and $(s'c)c = c$ for at least one choice of s, s' in S. Then, by (iii), $(s'c)(bs) = s'[c(bs)] = s'c$ and, by (iv), $(s'c)(bs) = [(s'c)b]s = bs$. Hence $s'c = bs = e$, say, where (v) $e = e^2$, (vi) $ea = ae = a$ for all $a \in A$ and (vii) $e \in aS \cap Sa$ for all $a \in A$.

Now set $T = eSe$. Then, by (v), T is a subsemigroup with identity e and, by (vi), T contains $A = eAe$. Since, by (v), (vii), $e = ee \in aSe = aeSe = aT$, $e = at$ for some $t \in T$. Since $e = ee \in eSa = Ta$, $e = t'a$ for some $t' \in T$. Then $t' = t'e = t'(at) = (t'a)t = et = t$ so $t' = t$ is the unique inverse a^{-1} of a in T, relative to e. Finally, let G be the set of all $g \in T$ such that g has an inverse g^{-1} in T relative to e. If $g, h \in G$ we verify that $(gh)^{-1} = h^{-1}g^{-1}$, and if $g \in G$ we verify that $(g^{-1})^{-1} = g$. Moreover, $e^{-1} = e$. Hence, by Theorem 3.1, G is a subgroup of S containing A.

If H is a subgroup of S containing A, let f be the identity of H. If $a \in A$, then $af = a$ and f is in aS, so $f = e$. Hence e is the identity element of H, so $H \subset eSe = T$ and consequently $H \subset G$.

Corollary 1. *If the subset A of Theorem 3.3 is maximal in the properties* (i), (ii), *then A is a subgroup of S.*

Corollary 2. (J. A. Green [197]). *A necessary and sufficient condition that the element a of the semigroup S be contained in a subgroup of S is that $a \in Sa^2 \cap a^2S$.*

Various authors have shown that the class of all groups can be defined as the class of all groupoids (in terms of one of the group division operations) satisfying a small number of identical relations. Higman and Neumann [45] have reduced the number of identical relations to one. And they have gone further, by showing that the subclass of all groups which satisfy any specified finite set of identical relations can also be characterized as the class of all groupoids satisfying a single identity. Some preparation is needed. Suppose $W(X_1, \ldots, X_m)$ is an (associative) word, expressed in terms of the group operation, the symbols $X_1, \ldots, X_m$, and the inverses X^{-1} of these. In particular, then, the group of order one satisfies $W = 1$. Let $W'(Y_1, \ldots, Y_n)$ be another such word. The class of all groups satisfying both $W = 1$ and $W' = 1$ is identical with the class of all groups satisfying

$$W(X_1, \ldots, X_m)\, W'(Y_1, \ldots, Y_n) = 1\,, \tag{3.1}$$

where the X_i and Y_j are allowed to range independently over the elements of such a group. For, by putting $Y_1 = Y_2 = \cdots = Y_n = 1$ in (3.1), we get $W(X_1, \ldots, X_m) = 1$. Similarly, the class of all groups satisfying any finite set of identical relations is identical with the class of all groups satisfying a single suitably chosen one.

Next, let $G = (G, \cdot)$ be a group and define (G, o) by $x o y = x y^{-1}$ where y^{-1} is the inverse of y in G. Then $x o x = 1$ for all x and $1 o y = y^{-1}$ or $y^{-1} = (z o z) o y$ for all z, y. Hence also $x y = x o y^{-1} = x o [(z o z) o y]$. Therefore any associative word $W(X_1, \ldots, X_n)$ can be transformed into a "groupoid" word involving perhaps more variables but no inverses and only the single groupoid operation (here the operation (o)). In particular, if G satisfies $W(X_1, \ldots, X_n) = 1$, the transformed word W appears as a fixed element of (G, o) for all values of the variables. To avoid the "empty" word we can use $X X^{-1}$ in connection with G and $X o X$ in connection with (G, o).

For the group G, define mappings $\varrho(x)$, $\lambda(x)$ of G into G by $y\varrho(x) = yx$, $y\lambda(x) = xy$ for all x, y of G. Here $\varrho(x)$ is the right multiplication by x and $\lambda(x)$ the left multiplication by x. They are permutations of G; that is, one-to-one mappings of G upon G. In particular, since $z\varrho(xy) = z(xy) = (zx)y = z\varrho(x)\varrho(y)$, then $\varrho(xy) = \varrho(x)\varrho(y)$ for all x, y. Thus the mapping $x \to \varrho(x)$ is an isomorphism of G upon a group of permutations. Also define the permutation J of G by $xJ = x^{-1}$. For the groupoid (G, o) define right and left mappings $R(x)$, $L(x)$ by $yR(x) = yox$, $yL(x) = xoy$. Then $R(x) = \varrho(x^{-1}) = \varrho(x)^{-1}$ and $L(x) = J\lambda(x)$. If G satisfies $w = 1$ where $w = W(X_1, \ldots, X_n)$, then

$$L((xox)ow)R(z)R([(xox)ox]oz)L(x) = L(1)R(z)R(x^{-1}oz)L(x)$$
$$= J\varrho(z^{-1})\varrho(zx)J\lambda(x) = J\varrho(x)J\lambda(x) = I\,,$$

where I is the identity mapping, since $yJ\varrho(x)J\lambda(x) = x(y^{-1}x)^{-1} = xx^{-1}y = y$. We interchange the roles of $(\cdot)$ and (o) and embody the converse of this result in the following theorem.

Theorem 3.4. *Let $w = W(X_1, \ldots, X_n)$ be a "groupoid" word and let $G = (G, \cdot)$ be a groupoid satisfying the identical relation*

$$x([\{((xx)w)y\}z]\,[((xx)x)z]) = y \tag{3.2}$$

for all $x, y, z, X_1, \ldots, X_n$ in G. Then G is a quasigroup and $(G, /)$ is a group satisfying the identical relation corresponding to $w = 1$, where 1 is the identity of $(G, /)$. Moreover, as $(G, \cdot)$ ranges over the class of all groupoids satisfying (3.2), $(G, /)$ ranges over the class of all groups satisfying $w = 1$.

Proof. In terms of the right and left multiplications $R(x)$, $L(x)$ of G, regarded as operating on y, (3.2) becomes

$$L(xx \cdot w)\,R(z)\,R((xx \cdot x)z)\,L(x) = I\,, \tag{3.3}$$

where, to save parentheses, we have written, for example, $xx \cdot w$ instead of $(xx)w$. If S, T are two single-valued mappings of G such that $ST = P$, where P is a permutation of G, then T is upon G and S is one-to-one. Applying this principle to (3.3), we see that $L(x)$ is upon G for all x

and that $L(xx \cdot w)$ is one-to-one. Then $L(xx \cdot w)$ is a one-to-one mapping of G upon G; that is, a permutation. Hence it has an inverse and (3.3) becomes

$$R(z)\, R((xx \cdot x)z)\, L(x) = L(xx \cdot w)^{-1}. \tag{3.4}$$

Therefore $R(z)$ is one-to-one for all z. If, for some y, we take $x = yy \cdot w$ in (3.4), then $L(x) = L(yy \cdot w)$ is also a permutation and thus $R((xx \cdot x)z)$ is a permutation. Hence, finally, $R(z)$ is a permutation for all z. But then, from (3.4), with x again arbitrary, $L(x)$ is a permutation for all x. Therefore, in an equation $ab = c$, $a, b, c \in G$, each two of a, b, c uniquely determine the third. That is, *G is a quasigroup.*

Comparing (3.4) with the corresponding equation in which z is replaced by y, we deduce that

$$R(z)\, R((xx \cdot x)z) = R(y)\, R((xx \cdot x)y) \tag{3.5}$$

for all x, y, z. For fixed x, we set $a = xx \cdot x$ in (3.5) and operate with both sides on a. Then $(az)(az) = (ay)(ay)$ for all y, z. Since G is a quasigroup, this means that pp has a fixed value, say e, for all p in G. Thus

$$xx = ee = e \tag{3.6}$$

for all x. From (3.6) in (3.5), $R(z)\, R(ex \cdot z) = R(y)\, R(ex \cdot y)$, whence, if x is chosen so that $ex = y$,

$$R(z)\, R(yz) = R(y)\, R(e) \tag{3.7}$$

for all y, z.

In (3.3) we take $x = z = e$, apply both sides to ew, and use (3.6), getting $e = ew$. Thus $ew = ee$, so

$$w = e. \tag{3.8}$$

Via (3.8), (3.6), now (3.3) becomes $L(e)\, R(z)\, R(ex \cdot z)\, L(x) = I$; we operate on x and get $x[(ex \cdot z)(ex \cdot z)] = x$ or, by (3.6), $xe = x$. That is,

$$R(e) = I. \tag{3.9}$$

From (3.7) with $y = e$, $R(z)\, R(ez) = R(e)\, R(e) = I$. We replace z by ez in (3.7) and get $R(y \cdot ez)^{-1} = R(y)^{-1} R(ez) = R(y)^{-1} R(z)^{-1}$. Thus, if the operation (o) is defined by

$$x o y = x(ey), \tag{3.10}$$

we have $R(yoz)^{-1} = R(y)^{-1} R(z)^{-1}$. That is: *the mapping $x \to R(x)^{-1}$ is an isomorphism of the groupoid (G, o) upon a group of permutations of G.*

Consequently, (G, o) is a group with, by (3.9), identity element e. Moreover, (G, o) satisfies the identity $w = e$, by (3.8). Let x^{-1} denote the inverse of x in (G, o). Then $e = x^{-1} o x = x^{-1}(ex)$, whence, by (3.6), $x^{-1} = ex$. In particular, $e(ex) = (x^{-1})^{-1} = x$. Hence $xy = x[e(ey)] = xo(ey) = xoy^{-1}$. Also, if $x/y = z$, where $(/)$ is defined, as in I.2, relative to $G = (G, \cdot)$, then $x = zy = zoy^{-1}$, so $xoy = (zoy^{-1})\ oy = zo(y^{-1}oy) = zoe = z$. That is:

$$xy = xoy^{-1}, \quad xoy = x/y\,. \tag{3.11}$$

And (3.11) completes the proof of Theorem 3.4.

An identical relation singling out the abelian groups is given by (3.2) with $w = [p(pp \cdot q)]\,[q(qq \cdot p)]$ for arbitrary $p, q \in G$: a relation involving 5 distinct variables and 17 variables including repetitions. However, as HIGMAN and NEUMANN show by a proof similar to that of Theorem 3.4, the simpler identical relation

$$x[(yz)\,(yx)] = z\,, \quad \text{or} \quad L(y)R(yx)L(x) = I\,,$$

does equally well.

4. Homomorphic imbedding

By *I*, § 2, a necessary and sufficient condition that a groupoid G be imbeddable in a quasigroup is that G satisfy the two cancellation laws. As shown by MALCEV [224] there exist cancellation semigroups which cannot be imbedded in any group. Thus, if G is a groupoid which does not satisfy the cancellation laws or a cancellation semigroup which is not imbeddable in a group, the best we can expect is the existence of a homomorphic image of G which is in some sense maximal in the properties with which we are concerned. This is provided by the following theorems:

Theorem 4.1. *Let G be a groupoid. There exists one and (to within an isomorphism) only one cancellation groupoid (semigroup, cancellation semigroup) H with the following properties:* (i) *G possesses a homomorphism θ upon H;* (ii) *if φ is a homomorphism of G upon a cancellation groupoid (semigroup, cancellation semigroup) K, there exists a homomorphism α of H upon K such that $\varphi = \theta\alpha$.*

Theorem 4.2. *Let G be a groupoid. There exists one and (to within an isomorphism) only one quasigroup (loop, group) Q with the following properties:* (i) *G possesses a homomorphism θ into Q such that $G\theta$ generates Q;* (ii) *if φ is a homomorphism of G into a quasigroup (loop, group) K, there exists a homomorphism α of $G\theta$ into K such that* (a) $\varphi = \theta\alpha$ and (b) *α can be extended to a homomorphism β of Q into K.*

Proof of Theorem 4.1. (I) *Uniqueness.* Let H, H' have the properties (i), (ii) of Theorem 4.1, θ' being the homomorphism of G upon H'. Then $\theta' = \theta\alpha$, $\theta = \theta'\beta$ where α is a homomorphism of H upon H', β is a homomorphism of H' upon H. Hence $\theta = \theta\alpha\beta$, $\theta' = \theta'\beta\alpha$. If x is in H, $x = a\theta$ for at least one a in G. Then $x\alpha\beta = a\theta\alpha\beta = a\theta = x$, so $\alpha\beta = I_H$. Similarly, $\beta\alpha = I_{H'}$. Consequently α and β are one-to-one upon. Therefore α is an isomorphism of H upon H'.

(II) *Existence.* We give two conceptually different proofs for the case of homomorphisms upon cancellation groupoids. Similar proofs hold for homomorphisms upon semigroups or upon cancellation semigroups.

The first proof requires us to accept the existence of the class Φ of all homomorphisms of G upon cancellation groupoids. Since G possesses a homomorphism upon a group of order one, Φ is non-empty. For each x in G we define a "function" f_x on Φ by $f_x(\varphi) = x\varphi$, all $\varphi \in \Phi$. Let H be the set of all functions f_x, $x \in G$, and define multiplication in H by $(f_x f_y)(\varphi) = f_x(\varphi) f_y(\varphi)$. Then $f_x f_y = f_{xy}$ for all $x, y \in G$. Clearly H is a cancellation groupoid. The mapping $\theta: x \to f_x$, is a homomorphism of G upon H. If φ is a homomorphism of G upon a cancellation groupoid K then $\varphi \in \Phi$ and hence $\varphi = \theta\alpha$ where $\alpha: f_x \to x\varphi$, is a homomorphism of H upon K.

The second proof uses internal properties of G. We define a congruence relation $(\equiv)$ on G by means of the following rules:

(E_1) If $a = b$ in G then $a \equiv b$.

(E_2) If $a \equiv b$ then $b \equiv a$.

(E_3) If $a \equiv b$ and $b \equiv c$ then $a \equiv c$.

(C_1) If $a \equiv b$ and if $c \in G$ then $ac \equiv bc$.

(C_2) If $a \equiv b$ and if $c \in G$ then $ca \equiv cb$.

(S_1) If $ac \equiv bc$ then $a \equiv b$.

(S_2) If $ca \equiv cb$ then $a \equiv b$.

(L) If $a, b \in G$ then $a \equiv b$ only as required by (E), (C), (S).

Here the rules (E) are the properties of an equivalence relation; the rules (E), (C) are the properties of a congruence relation. The special rules (S) and the limiting rule (L) specify a particular congruence relation. (In discussing homomorphisms of G upon semigroups we delete (S_1), (S_2) and use only (S_3): if $a, b, c \in G$ then $(ab)c \equiv a(bc)$; in discussing homomorphisms of G upon cancellation semigroups, we use (S_i) for $i = 1, 2, 3$.)

For each $a \in G$ let $[a]$ denote the congruence class of a; that is, the set of all $b \in G$ such that $b \equiv a$. The rules (E) ensure that b is in $[a]$ if and only if $[b] = [a]$. The rules (C) ensure that the set H of all classes $[a]$ is a groupoid under the product operation $[a]\,[b] = [ab]$, and the rules (S) ensure that H is a cancellation groupoid. Moreover, the mapping $\theta: a\theta = [a]$, is a homomorphism of G upon H. Now let φ be any homomorphism of G upon a cancellation groupoid $G\varphi$. Suppose we have shown (*) if $a \equiv b$ then $a\varphi = b\varphi$. Then the mapping α defined by $[a]\alpha = a\varphi$ is a homomorphism of H upon $G\varphi$ such that $\varphi = \theta\alpha$. To prove (*) we use a course-of-values induction over the height of the "proof scheme", (involving finitely many applications of (E), (C), (S),) which demonstrates that $a \equiv b$. (See KLEENE, Metamathematics). This completes the proof of Theorem 4.1.

Proof of Theorem 4.2. (I) *Uniqueness.* Let Q, Q' have the properties (i), (ii) of Theorem 4.2, θ' being the homomorphism of G into Q'. Then, by Theorem 4.1, $G\theta$ is isomorphic to $G\theta'$. Therefore, without loss of generality, we may assume that $\theta' = \theta, G\theta = H$. Then the identity mapping I_H of H can be extended to a homomorphism β of Q into Q' and to a homomorphism γ of Q' into Q. Thus $\beta\gamma$ is an extension of I_H to an endomorphism of Q. However H generates Q, so $\beta\gamma = I_Q$. Similarly, $\gamma\beta = I_{Q'}$. Consequently, β is an isomorphism of Q upon Q'.

(II) *Existence.* Consider first the case of homomorphisms of G into quasigroups. There exists a cancellation groupoid H_0 with the properties of Theorem 4.1, and H_0 freely generates a quasigroup Q_0. Let θ_0 be the homomorphism of G upon H_0. If φ is a homomorphism of G into a quasigroup K then $G\varphi$ is a cancellation groupoid. Therefore $\varphi = \theta_0\alpha$ where α is a homomorphism of H_0 upon $G\varphi$. Since α is a homomorphism of H_0 into the quasigroup K, then α can be extended (uniquely) to a homomorphism β of Q_0 into K. Therefore we may take $Q = Q_0$, $\theta = \theta_0$.

For the case of homomorphisms into loops we first establish the existence of a "maximal" homomorphic image $Q_0\theta_1$ of Q_0 which is a loop. This is done along the lines of the proof of Theorem 4.1. Then we set $Q = Q_0\theta_1$, $\theta = \theta_0\theta_1$. Similarly for the case of groups. This completes the proof of Theorem 4.2.

There is no analogue of Theorem 4.1 for homomorphisms of groupoids *upon* quasigroups or of semigroups *upon* groups. For example, if G is the additive semigroup of positive integers, the quasigroup images of G are the groups of prime order.

5. Brandt groupoids. Mixed groups

A BRANDT *groupoid* G is a halfgroupoid (See I.1) subject to the following postulates:

(i) To each a in G there corresponds one and only one ordered pair e, f of elements of G such that $ea = fa = a$.

(ii) If $ea = a$ or $ae = a$ for some a, e in G, then $ee = e$.

(iii) If a, b are in G, then ab is defined in G if and only if $ae = a$, $eb = b$ for some e in G.

(iv) If a, b, c are in G and if ab, bc are defined in G, then $(ab)c$ and $a(bc)$ are defined and $(ab)c = a(bc)$.

(v) If $ea = a$, $af = a$ for a, e, f in G, there exists b in G such that $ab = e$ and $ba = f$.

(vi) If $ee = e$ and $ff = f$ in G, there exists a in G such that $ea = a$, $af = f$.

Let P be any non-empty set, H be any group, and define products in $P \times H \times P$ as follows:

$$(p, h, q)\,(q, h', r) = (p, hh', r) \tag{5.1}$$

for all p, q, r in P, h, h' in H, and $(p, h, q)\,(r, h', s)$ is *not* defined if $q \neq r$. The resulting halfgroupoid is a BRANDT groupoid.

An element satisfying $ee = e$ is called an idempotent. Let P be the set of all idempotents of the BRANDT groupoid G and, for a fixed element e of P, let H be the subset of G consisting of all a such that $ea = ae = a$. By (vi), H is non-empty. By (iii), (iv), (v), H is a group. For arbitrary p, q in P, let $G(p, q)$ be the set of all a in G such that $pa = aq = a$. By (vi), $G(p, q)$ is not empty. For each p, select an element $a(p)$ in $G(p, e)$. By (v), there exists an element $b(p)$ in $G(e, p)$ such that $a(p)\,b(p) = p$, $b(p)\,a(p) = e$. By (iii), (iv), (v), the mapping $h \to a(p)\,hb(q)$ can be shown to be a one-to-one mapping of H upon $G(p, q)$. Furthermore, $a(p)\,hb(q) \cdot a(q)\,h'b(r) = a(p)\,hh'b(r)$, and no other products are defined. Hence *every* BRANDT *groupoid is isomorphic to one of form* (5.1).

The literature of BRANDT groupoids may be found by beginning with BRANDT [36]. See also JACOBSON, Theory of Rings; in particular, Chapter 6, §§ 11—14.

CLIFFORD [167] notes that a BRANDT groupoid G can be imbedded in a semigroup S with a single additional zero element, 0, such that $a0 = 0a = 00 = 0$ for all a in G and $ab = 0$ in S if a, b are in G and ab is not defined in G. The semigroup S belongs to the class of completely simple semigroups to be considered in § 8 below.

A *mixed group* M is a halfgroupoid M containing a non-empty subset K (the *kernel* of M) such that:

(a) If $x, y \in M$, then xy is defined in M if and only if $x \in K$.

(b) If $k, k' \in K$ and $x \in M$, then $(kk')\,x = k(k'x)$ in M.

(c) If $k \in K$, then $K \subset kK \cap Kk$ in M.

(d) If $k \in K$, $x \in M$ and $kx = x$ in M, then $ky = y$ in M for every y in M.

The original postulates of LOEWY [329], although slightly different, are equivalent to (a)—(d).

We shall give three examples of mixed groups. (I) Let K be any group, E be any non-empty set and let 1 be a distinguished element of E. Define multiplication in $M = K \times E$ as follows: $(k, 1)\,(k', e') = (kk', e')$ for all $k, k' \in K$, $e' \in E$; and $(k, e)\,(k', e')$ is not defined if $e \neq 1$. (II) Let G be a group and K be a subgroup of G. Let M consist of the elements of G with multiplication defined as follows: $xy = z$ in M if and only if $x \in K$ and $xy = z$ in G. (III) Let G be a group, let H be any subgroup of G and let K be the normalizer of H in G. That is, K consists of all

$k \in G$ such that $kH = Hk$. Let M consist of the elements of G with muliplication defined as follows: $xy = z$ in M if and only if $(Hx)(Hy) = Hz$ in G and $xy = z$ in G.

In each of the examples, M is a mixed group with kernel K. Example (I) is equivalent to the special case of (II) in which G is the direct product of K and a group E. And (III) is also a special case of (II).

Now let M be a mixed group with kernel K. Consider K as a closed subhalfgroupoid of M (see I.1). By (a), (b), K is a semigroup; by (c), K is a group. Let 1 be the identity element of K. By (d), since $1.1 = 1$, then $1x = x$ for every x in M. If $k, k' \in K$ and $x \in M$ and $kx = k'x$, then we see readily that $(k^{-1}k')x = x$. Hence, by (d) with $y = 1$, $k^{-1}k' = 1$, so that $k = k'$. Again, if $k \in K$ and $x, y \in M$ and $kx = ky$, then $x = k^{-1}(kx) = k^{-1}(ky) = y$. At this stage we see that there must exist at least one non-empty subset E of M (which we may and do assume to contain the identity 1 of K) such that each x in M has a unique representation $x = ke$ in M, where $k \in K$, $e \in E$. It is now clear that M has a representation of type (I) and therefore a representation of type (II).

We shall leave aside the more difficult problem of determining all representations of type (II) (and hence all of type (III).) For this see Baer [325], [326].

6. Polyadic groups. Flocks

Let $n \geqq 2$ be a positive integer, G be a non-empty set and $f = f_n$ be an n-ary operation completely defined on G, so that (G, f) is an algebra in the sense of I.5. The algebra (G, f) is called a *polyadic group*, or, more specifically, an *n-group*, provided the following axioms are satisfied:

(I) If x_{n+1} and any $n-1$ of the symbols $x_1, x_2, \ldots, x_n$ are specified as elements of G, the equation

$$f(x_1, x_2, \ldots, x_n) = x_{n+1}$$

has at least one solution in G for the remaining symbol.

(II) f is associative. That is,

$$f(f(x_1, \ldots, x_n), x_{n+1}, \ldots, x_{2n-1}) \\ = f(x_1, \ldots, x_i, f(x_{i+1}, \ldots, x_{i+n}), x_{i+n+1}, \ldots, x_{2n-1}) \tag{6.1}$$

for $i = 1, 2, \ldots, n-1$ and for all $x_1, \ldots, x_{2n-1}$ in G.

It will be clear later that an n-group satisfies (I) in the stronger form with "at least one solution" replaced by "exactly one solution".

An important concept in the theory of polyadic groups is that if a covering group. An (ordinary) group $K = (K, \cdot)$ is said to be a *covering group* of an n-group (G, f) provided K has the following properties:

(i) The set G is a generating subset of K.

(ii) $f(x_1, \ldots, x_n) = x_1 x_2 \ldots x_n$ in K for all x_i in G.

First let us assume that the n-group (G, f) has a covering group K. Let $G_0 = G^{n-1}$ be the subset of K consisting of all products of $n-1$

elements of G. We shall show that G_0 *is a normal subgroup of* K, G *is a coset of* K *modulo* G_0, *and* K/G_0 *is a finite cyclic group whose order divides* $n-1$. By (ii) and the fact that the values of f are in G, we see that $G_0G_0 \subset G_0$. If x, y are in G, then, by (I), there exist elements $x_2, \ldots, x_n$ in G such that $f(x, x_2, \ldots, x_n) = y$. Moreover, if $x, x_2, \ldots, x_n$ are arbitrarily given elements of G, the same equation determines a y in G. Considered in K, the equation is equivalent to $x_2 \ldots x_n = x^{-1}y$. Therefore G_0 is also the set of all elements $x^{-1}y$ where x, y range over G. Since $(x^{-1}y)^{-1} = y^{-1}x$, we conclude that G_0 is a subgroup of K. If x is in G and p is in G_0, then $p = y_2 \ldots y_n$ is a product of $n-1$ elements of G; thus $x^{-1}px = (x^{-1}y_2)(y_3 \ldots y_n x)$ is the product of two elements of G_0 and hence is in G_0. That is, $x^{-1}G_0x \subset G_0$. Similarly, we may show first that G_0 contains all elements yx^{-1} with x, y in G and thence that $xG_0x^{-1} \subset G_0$ for every x in G. Now it clear from (i) that $k^{-1}G_0k = G_0$ for every k in K. This means (See IV.1 or any book on group theory) that G_0 is a normal subgroup of K. Consequently we can form the quotient group K/G_0. In addition, $G = xG_0$ for every x in G; showing that G is a coset and hence a single element of K/G_0. Since the elements of G generate K, the group K/G_0 is cyclic. And since x^{n-1} is in G_0 for each x in G, the order of K/G_0 is a divisor of $n-1$.

The above discussion suggests how to define a class of n-groups. We select any group K which possesses a normal subgroup G_0 such that K/G_0 is a finite cyclic group of order dividing $n-1$. We choose some element x of K such that the coset xG_0 generates K/G_0. Then we use (ii) to define the n-ary operation f on the coset $G = xG_0$. This makes (G, f) into an n-group with K as a covering group.

Dörnte [333] was the first to study polyadic groups as abstract algebras. The theory was later developed at great length by Post [26]. Futher references and a broad range of topics will be found in Post [26]. We shall be content to prove Post's Coset Theorem (loc. cit., p. 218). This may be stated as follows:

Theorem 6.1. (Post's Coset Theorem.) *Every polyadic group has a covering group.*

Outline of proof. We suppose given an n-group (G, f). If $n = 2$, (G, f) is an ordinary group and hence is its own covering group. Therefore we shall assume $n > 2$.

We work in terms of the free semigroup S on the elements of G as free generators. The elements of S are the ordered sequences

$$(a_1, a_2, \ldots, a_i)$$

of elements $a_1, a_2, \ldots, a_i$ of G, every finite length i being permitted. Multiplication is performed by juxtaposition:

$$(a_1, a_2, \ldots, a\,)\,(b_1, b_2, \ldots, b_j) = (a_1, \ldots, a_i, b_1, \ldots, b_j)\,. \qquad (6.2)$$

In particular, S is a cancellation semigroup. Whenever convenient, we adjoin the empty sequence, ( ), of length 0, as an identity element of S.

We must define a congruence relation θ on S (see I.5 for notation) such that:

(i) S/θ is a group, which we shall call G^*.

(ii) If x, y are in G and if $(x) \sim (y) \bmod \theta$, then $x = y$ in G.

(iii) $(f(x_1, x_2, \ldots, x_n)) \sim (x_1)(x_2) \ldots (x_n) \bmod \theta$ for all x_i in G.

Once θ has been constructed with properties (i)—(iii), we identify each element x of G with the equivalence class of $(x) \bmod \theta$, and we replace equivalence mod θ by equality. Then the group G^* has the properties of a covering group for (G, f).

Proof. In what follows, any Greek letter (other than θ) denotes a sequence or, to be more precise, an element of the free semigroup S defined above.

We call such a sequence α *compatible* (with f) provided α has length $k(n-1)+1$ for some positive integer k. If α is in S, there exists β in S such that $\alpha\beta$ is compatible; and then $\beta\alpha$ is compatible also. We need to define $f(\alpha)$ for every compatible α. We do this inductively as follows: If

$$\alpha = (x_1, x_2, \ldots, x_n)$$

has length n, we define

$$f(\alpha) = f(x_1, x_2, \ldots, x_n)\,.$$

If α is compatible, if $f(\alpha)$ has been defined, and if β has length $n-1$, we define

$$f(\alpha\beta) = f((f(\alpha))\,\beta)\,. \tag{6.3}$$

This completes the definition.

In addition, we need the following generalized associative law:

$$f(\lambda\alpha\mu) = f(\lambda(f(\alpha))\,\mu) \tag{6.4}$$

provided α and $\lambda\alpha\mu$ are both compatible. (Here we allow one of λ, μ to be empty.) For the case that α and $\lambda\alpha\mu$ have lengths n and $2n-1$ respectively, (6.4) is equivalent to (6.1). The general case comes from (6.1) by an inductive argument.

We may observe that if $m = k(n-1)+1$ for some positive integer k and if we define f_m so that $f_m(\alpha) = f(\alpha)$ for every α of length m, then (G, f_m) is an m-group (a so-called *extension* of (G, f)). We need the following instances of this fact:

(A) *If x is in G, α is in S and $(x)\alpha$ is compatible, there exist y, z in G such that $x = f((y)\alpha) = f(\alpha(z))$.*

(B) *If x, y are in G, there exist α, β in S such that $(y)\alpha$, $\beta(y)$ are compatible and $x = f((y)\alpha) = f(\beta(y))$.*

Next we must prove the following:

(C) *If the sequence* α *is such that either* $f(\alpha(y)) = y$ *for some* y *in* G *or* $f((z)\alpha) = z$ *for some* z *in* G, *then*

$$f(\alpha(x)) = x = f((x)\alpha) \tag{6.5}$$

for every x *in* G; *and*

$$f(\lambda\alpha\mu) = f(\lambda\mu) \tag{6.6}$$

for every two sequences λ, μ *such that* $\lambda\mu$ *is compatible.*

We shall prove (C) on the hypothesis that $f(\alpha(y)) = y$ for some y in G. (Note, by (B), that at least one α satisfying this hypothesis exists for each y in G.) By (B), there exists for each x in G a φ such that $x = f((y)\varphi)$. Let λ, μ be any sequences such that $\lambda\alpha(x)\mu$ is compatible. Then

$$\begin{aligned} f(\lambda\alpha(x)\mu) &= f(\lambda\alpha(f((y)\varphi))\mu) = f(\lambda(f(\alpha(y))\varphi\mu) \\ &= f(\lambda(y)\varphi\mu)\,. \end{aligned}$$

Since $f((y)\varphi) = x$, we first assume that λ, μ are empty and deduce that $f(\alpha(x)) = x$. Next we assume that $\lambda(x)\mu$ is compatible and deduce that $f(\lambda\alpha(x)\mu) = f(\lambda(x)\mu)$, thus proving (6.6) for the case that μ is not empty. In particular, for any z in G, $f((z)\alpha\alpha) = f((z)\alpha)$. Hence, since $x = f((z)\alpha)$ for some z, $f((x)\alpha) = x$; which completes the proof of (6.5). And, finally, if $\lambda(x)$ is compatible, $f(\lambda(x)\alpha) = f(\lambda(f((x)\alpha)) = f(\lambda(x))$; whence (6.6) holds when μ is empty. This proves (C) on the hypothesis that $f(\alpha(y)) = y$ for some y; the proof on the other hypothesis is quite similar.

A sequence α which satisfies the hypothesis of (C) is called an *identity*. Note that the length of an identity is $k(n-1)$ for some positive integer k.

Our final proposition is the following:

(D) *If* α, α' *are sequences such that*

$$f(\lambda\alpha\mu) = f(\lambda\alpha'\mu) \tag{6.7}$$

holds for some pair of sequences λ, μ, *then* (6.7) *holds for every pair of sequences* λ, μ *such that* $\lambda\alpha\mu$ *and* $\lambda\alpha'\mu$ *are both compatible.*

To see this, suppose (6.7) holds for some pair λ, μ. By (B), (C), there exist λ', μ' such that $\lambda'\lambda$, $\mu\mu'$ are identities. Hence, if $\beta\alpha\gamma$, $\beta\alpha'\gamma$ are compatible, (6.6), (6.7) give

$$\begin{aligned} f(\beta\alpha\gamma) &= f(\beta\lambda'\lambda\alpha\mu\mu'\gamma) = f(\beta\lambda'(f(\lambda\alpha\mu))\mu'\gamma) \\ &= f(\beta\lambda'(f(\lambda\alpha'\mu))\mu'\gamma) = f(\beta\lambda'\lambda\alpha'\mu\mu'\gamma) \\ &= f(\beta\alpha'\gamma)\,. \end{aligned}$$

Now we define the relation θ as follows:

$$\alpha \sim \alpha' \mod \theta \tag{6.8}$$

if and only if (6.7) holds for some λ, μ. In view of (D), θ is certainly an equivalence relation. Moreover, if $\alpha \sim \alpha'$ and $\beta \sim \beta'$, then $f(\lambda\alpha\beta\mu)$

$= f(\lambda\alpha'\beta\mu) = f(\lambda\alpha'\beta'\mu)$; whence $\alpha\beta \sim \alpha'\beta'$. Again, if $\alpha \sim \alpha'$ and $\alpha\beta \sim \alpha'\beta'$, then $f(\lambda\alpha\beta\mu) = f(\lambda\alpha'\beta'\mu) = f(\lambda\alpha\beta'\mu)$; so $\beta \sim \beta'$. Similarly, if $\beta \sim \beta'$ and $\alpha\beta \sim \alpha'\beta'$, then $\alpha \sim \alpha'$. This is enough to show that θ is a congruence relation on S and that S/θ is a cancellation semigroup. In addition, by (C), the identities form an equivalence class which contains the empty sequence. Moreover, for every α, there is a β such that $\alpha\beta$ is an identity. Consequently, S/θ is a group. This proves (i) of the outline.

Next suppose that $(x) \sim (y)$ for some x, y in G. Then, if α is an identity, $x = f(\alpha(x)) = f(\alpha(y)) = y$. This proves (ii).

Finally, let α have length n and set $\alpha' = (f(\alpha))$. Then, if $\alpha\beta$ is compatible and β non-empty, $f(\alpha\beta) = f((f(\alpha))\beta) = f(\alpha'\beta)$. Hence $\alpha \sim \alpha'$. This proves (iii) and completes the proof of Theorem 6.1.

It is easy to prove that G^* is the *free* covering group. That is: if K is a covering group, there exists a homomorphism of G^* upon K which induces the identity mapping on G.

Although the Coset Theorem by no means exhausts the subject of polyadic groups, it seems a proper stopping point in a survey of binary systems. For many interesting topics, see the references cited above.

Baer [328] has investigated a connection linking Brandt groupoids and mixed groups with the flocks of Prüfer [332]. We shall briefly describe the latter. A *flock* (German: Schar; also called an *imperfect brigade* or an *abstract coset*) may be defined as a ternary algebra (G, f) subject to the identities

$$f(f(u, v, w), x, y) = f(u, v, f(w, x, y)), \tag{6.9}$$

$$f(x, y, y) = x = f(y, y, x). \tag{6.10}$$

For various other systems of axioms and a study of their independence, see Certaine [329]. Certaine (loc. cit.) also applies flocks to affine geometry, as does Baer in his book Linear Algebra and Projective Geometry.

The theory of flocks is essentially a theory of groups in which the role of the identity element has been minimized. Indeed (Baer [328], Certaine [329]), to each element e of a flock (G, f) there corresponds a group $G(e)$ consisting of the elements of G under the multiplication $xy = f(x, e, y)$. The inverse of x in $G(e)$ is given by $x^{-1} = f(e, x, e)$. Moreover,

$$f(x, y, z) = xy^{-1}z \tag{6.11}$$

for all x, y, z in G. In addition, the groups $G(e)$, as e ranges over G, are all isomorphic. Conversely, if f is defined on a multiplicative group by (6.11), then (G, f) is a flock. In this case, (H, f) is a subflock of (G, f) if and only if H is a coset of (some subgroup of) G — whence the alternative name "abstract coset". Baer [328] shows that the automorphism group of (G, f) is the so-called holomorph of the group G; he also investigates allied groups of permutations of G.

When is a flock (G, f) a (ternary) polyadic group? Since f can be defined by (6.11) in terms of a group operation, axiom (I) for 3-groups is automatically satisfied. However, (6.11) implies the identity

$$f(f(u, v, w), x, y) = f(u, f(x, w, v), y) = f(u, v, f(w, x, y)), \tag{6.12}$$

which differs from the associative law (II) for 3-groups in that $f(x, w, v)$ should be $f(v, w, x)$. Hence: *A flock (G, f) is a 3-group if and only if it satisfies the commutative law*

$$f(x, y, z) = f(z, y, x). \tag{6.13}$$

On the other hand, the commutative 3-groups (in the sense of (6.13)) are precisely those whose covering groups are commutative. A commutative 3-group is a flock if and only if it satisfies the neutral law

$$f(x, x, x) = x. \tag{6.14}$$

Finally we must observe that no collection of identities which are "homogeneous" in the sense of (6.9), (6.12) or (6.13) can characterize a class of ternary algebras all of whose members are flocks or 3-groups. For let G be a set consisting of a distinguished element e and at least one other element. If we define $f(x, y, z) = e$ for all x, y, z in G, then f will satisfy every "homogeneous" identity but (G, f) will be neither a 3-group nor a flock.

7. Multigroups, hypergroups, etc.

Let us understand by a *multigroupoid* M a system consisting of a non-empty set M together with a binary operation (multiplication) which is a mapping of $M \times M$ not into M but into the set of non-empty subsets of M. That is, for each ordered pair a, b of elements of M, $a \cdot b$ is a non-empty subset of M. By a homomorphism θ of a multigroupoid M upon a multigroupoid M' we mean a single-valued mapping of M upon M', extended in the natural way to a mapping of the non-empty subsets of M upon the non-empty subsets of M', such that $(ab)\theta = (a\theta)(b\theta)$ for all a, b in M. An element e of M is a *left identity* element of M if $a \in ea$ for every a in M; similarly for right identity elements. The associative law has two forms: (A_1) $(ab)c \subset a(bc)$ for all a, b, c in M; (A_2) $a(bc) \subset (ab)c$ for all a, b, c in M. Similarly, the commutative law has two forms. If e is a left or right or two-sided identity element of M and if a, b are elements of M such that $e \in ab$, then b is a *right inverse* of a with respect to e and a is a *left inverse* of b with respect to e. A *multigroup* or *hypergroup* (the terms seem to be synonomous) is a multigroupoid endowed with one or both associative laws, an identity element and inverses for each element. A *cogroup* is a multigroup obeying a stricter set of axioms which are satisfied by (but do not characterize) the set of all right cosets of a group G with respect to a fixed subgroup H, the operation being that induced by the operation of G.

The rather extensive literature on multigroups and hypergroups (see the bibliography) has not progressed much beyond an analysis of the axioms required to give an elementary theory of homomorphisms or of congruences. It appears to the author that those multigroupoids which arise naturally can best be studied from other points of view. This seems to be born out by the following facts, which we feel impelled to discuss in detail:

(I) *Every multigroupoid determines and is determined by a partially ordered groupoid of special type and can be studied entirely in terms of the groupoid.*

(II) *Many multigroupoids are naturally endowed with a multiplicity function which states how often the element c occurs in the product ab. If the multiplicity function is of interest, the multigroupoid can best be studied in terms of a ring.*

In connection with (I), if M is a multigroupoid, let G be the set of all non-empty subsets of M, let ($\leqq$) be the relation of inclusion (as subsets of M) among the elements of G and, for each ordered pair (x, y) of elements of G, let xy be the element of G consisting of all elements c in M for which there exist elements a, b in M with a in x, b in y and c in ab. We note that multiplication in G is single-valued, so that G is a groupoid. Moreover, for x, x', y, y' in G, if $x \leqq x'$, $y \leqq y'$ then $xy \leqq x'y'$. Hence G is a partially ordered groupoid. As a partially ordered set, G may be characterized abstractly as a complete atomic lattice; the postulates merely ensure the existence of a set M such that G possesses a one-to-one order-preserving mapping upon the set of all subsets of M (for present purposes, we exclude the null subset of M). From this point of view, we may call G a lattice-ordered complete atomic groupoid. If such a groupoid G is given, let us suppose for convenience that G is the set of all non-empty subsets of a set M. Then we turn M into a multigroupoid by defining the ordered product ab of two elements a, b of M as the element (a) (b) of G considered as a subset of M. Hence there is a one-to-one correspondence between multigroupoids M and lattice-ordered complete atomic groupoids G. It is easy to see that the associative law (A_1) for M becomes the law $(xy)z \leqq x(yz)$ for G; that homomorphisms of M correspond to order-homomorphisms of G; and so on. This completes the discussion of (I). For a similar discussion see CROISOT [121]. The fact that single-valued multiplication could be restored to a multigroupoid was earlier noted by GRIFFITHS [127], KUNTZMAN [140].

In connection with (II) we wish merely to underline a well-known fact. As we shall show, there exist many *multigroupoids M* with *multiplicity*, in the following sense: (i) The subset ab is finite for all a, b in M. (ii) To each ordered triple a, b, c of elements of M there corresponds a non-negative integer $n(a, b, c)$ (the multiplicity function) such that $n(a, b, c)$ is positive if and only if c is in ab. When (i), (ii) hold we may imbed M in a ring F as follows: Let F be the additive free abelian group with the elements of M as free basis. Define multiplication in M by the distributive laws (of multiplication with respect to addition) together with the definition

$$ab = \sum_c n(a, b, c)c \tag{7.1}$$

for all a, b, c in M. (In general, F will not be associative.) Conversely, if F is a ring such that $(F, +)$ is a free abelian group and if M is a free basis of $(F, +)$ such that (7.1) holds for non-negative integers $n(a, b, c)$, not all zero, we make M into a multigroupoid by defining ab to be the subset of M consisting of all c in M for which $n(a, b, c) > 0$.

For example, let R be the ring defined as above from a quasigroup G. Certain subrings of R give rise quite naturally to multigroupoids with multiplicity functions. The left nucleus, N_λ, of R is the associative subring consisting of all k in R such that $(kx)y = k(xy)$ for x, y in R. It is easily seen that if N_λ is nonzero (as for example, when G is finite or a loop) one basis of N_λ consists of what might be called *right-conjugate* class sums. We define a non-empty subset S of G to be *right-invariant* if $(Sa)b = S(ab)$ for all a, b in G and to be a *right-conjugate class* if no proper subset of S is right-invariant. Then, for each finite right-conjugate class of S, the sum of the elements of S is one of the basis elements for N_λ. If G is not a group, the finite right-conjugate classes clearly yield a multigroup with multiplicity. The condition that $(kx)y = k(xy)$ for all x, y in R may be replaced by any one of a variety of such conditions which yield a subring of R with similar properties. In particular we may use the condition that $kx = xk$ and $k(xy) = (kx)y = x(ky)$ for all x, y in R; this defines the centre of R and brings in the *conjugate classes* of G. As another example, if A is a finite subgroup of G contained in the left nucleus of R, the cosets Ab, b in G, give rise in a similar fashion to a basis of a subring of R and hence to a multigroupoid with multiplicity. When G is a group, the latter is a so-called *cogroup*.

We shall not discuss in detail the papers on multigroupoids which are listed in the bibliography. It is interesting to observe, however, that a long series of notes by KRASNER on class-field theory at first makes systematic use of multigroups but ends by removing all trace of multigroups from the theory. (KRASNER [128]—[133].)

For *cogroups* see EATON [125], UTUMI [150].

For *ultragroups* see VIKHROV [151].

For geometry in terms of multigroups see PRENOWITZ [145]—[148].

8. Ideal structure of semigroups

A non-empty subset A of the semigroup S is called a *left ideal* if $SA \subset A$, a *right ideal* if $AS \subset A$ and an *ideal* of S if $SA \subset A$, $AS \subset A$. If S contains a one-element ideal (z) then $xz = zx = z$ for every x in S and hence (z) is contained in every left, right or two-sided ideal of S. Such an element z is called a zero of S and is usually denoted by 0. A semigroup can have at most one zero. If a semigroup S has no zero we may adjoin a zero, 0, and make $S' = S \cup 0$ into a semigroup by defining $x0 = 0x = 00 = 0$ for every x in S. If the semigroup S has an ideal A we may define, following Rees [247], a quotient semigroup S/A as follows: We replace equality $(=)$ in S by $(\equiv)$ where $a \equiv b$ if and only if either $a, b \in A$ or $a, b \notin A$ and $a = b$. Then S/A is S with $(=)$ replaced by $(\equiv)$. Equivalently, we collapse the elements of A into a single element, A. Then S/A is a semigroup with A as zero.

The following definitions are stated for semigroups with zero. To obtain the analogue for a semigroup S without zero, first adjoin a zero and then interpret the definition in terms of S.

An element e of the semigroup S is called *idempotent* if $e^2 = e$ and *nilpotent* if $e^n = 0$ for some positive integer n. An idempotent e of S is called *primitive* if $e \neq 0$ and if the only idempotents f of S such that $ef = fe = f$ are $f = e$ and $f = 0$. A non-empty subset T of S is called a *nil* subset if every element of T is nilpotent and is called *nilpotent* if $T^n = 0$ for some positive integer n; that is, if $t_1 t_2 \ldots t_n = 0$ for all t_i in T.

A semigroup S is called *simple* (left simple, right simple) if (i) the only ideals (left ideals, right ideals) of S are S itself and the zero ideal and (ii) $S^2 \neq 0$. A semigroup S is called *completely simple* if (a) S is simple; (b) every nonzero idempotent of S is primitive; (c) to each x in S there corresponds at least one pair of nonzero idempotents e, f of S such that $ex = xf = x$.

A left (right, two-sided) ideal A of the semigroup S is called *minimal* if $A \neq 0$ and if the only left (right, two-sided) ideals of S contained in A are A itself and the zero ideal.

A semigroup S with zero is called a *group with zero* if the nonzero elements of S form a subgroup.

Lemma 8.1. *Let S be a semigroup with zero. Then every nil (nilpotent) left or right ideal of S is contained in a nil (nilpotent) ideal of S.*

Proof. If A is a left ideal of S then $B = A \cup AS$ is an ideal of S. If A is nil and $a \in A$, $s \in S$ then $sa \in A$; so $(sa)^n = 0$ for some positive integer n and hence $(as)^{n+1} = a(sa)^n s = a0s = 0$. Therefore B is nil. For any left ideal A, $B^n \subset A^n \cap A^n S$ for every positive integer n. If A is nilpotent, $A^n = 0$ and hence $B^n = 0$ for some positive integer n. Similarly for right ideals. This completes the proof.

The *nil radical* $N = N(S)$ is the set of all properly nilpotent elements of S. An element a of S is properly nilpotent if aS is a nil subset. Since $a^2 \in aS$, a itself is nilpotent. Since every nilpotent subset of S is a nil subset, we see from Lemma 8.1 that N contains every nil or nilpotent left or right or two-sided ideal. Moreover N is itself a nil ideal and hence is the (unique) maximal nil ideal of S.

The semigroup S is said to satisfy the *minimal condition* for left ideals if every non-empty set Σ of left ideals of S contains a minimal element, say M, in the sense that M is in Σ and, if L is in Σ and $L \subset M$, then $M = L$. Similarly for right ideals, ideals, etc. The *descending chain condition* may be regarded as identical with the minimal condition.

Theorem 8.1. (Hopkins-Brauer.) *Let S be a semigroup with zero which satisfies the minimal condition for nil ideals. Then the nil radical, N, of S is nilpotent.*

Proof. The set $\{N^n; n = 1, 2, \ldots\}$ consists of nil ideals and hence has a minimal element, say A. Since $N^{n+1} \subset N^n$ for $n = 1, 2, \ldots$, then $A = N^m = N^{m+1} = \cdots$, for some integer m. In particular, $A^2 = A$. Assume that $A \neq 0$, and consider the set Σ of all ideals B of S such that $B \subset A$ and $ABA \neq 0$. Since $AAA = A$ and $A \neq 0$ by assumption, $A \in \Sigma$. Hence Σ has a minimal element, say C. Since $ACA \neq 0$ then $AcA \neq 0$ for some c in C. Then $A(AcA)A = AcA \neq 0$, so $AcA \in \Sigma$. And $AcA \subset C$, whence, by the minimality of C, $C = AcA$. Therefore $c = acb$ for some a, b in A, and hence $c = a^n c b^n$ for $n = 1, 2, \ldots$. Since A is a nil ideal, we conclude that $c = 0$, contradicting $AcA \neq 0$. Therefore $N^m = A = 0$ and the proof of Theorem 8.1 is complete.

Morse and Hedlund [229] have indicated the existence of a semigroup S with three generators such that $x^2 = 0$ for every x in S but S is not nilpotent.

If A is an ideal of a semigroup S, an element x of S is called A-nilpotent if $x^n \in A$ for some positive integer n. Equivalently either x is in A or x is a nilpotent element of S/A. Thus Lemma 8.1 and Theorem 8.1 have obvious generalizations to A-nil and A-nilpotent ideals. Lemma 8.1 and Theorem 8.1 are a little better than their analogues for (associative) rings. On the other hand, in a ring satisfying the minimal condition for left ideals, every non-nilpotent left ideal contains a nonzero idempotent, whereas a like result is false for semigroups. Indeed, Baer and Levi [160] have constructed a left-simple semigroup which contains no idempotents whatever.

Theorem 8.2. (Suschkewitz-Rees-Clifford.) *Let M be a nonzero ideal of the semigroup S with zero. Then the following statements are equivalent:*

(i) *M is a completely simple semigroup.*

(ii) *M is simple and contains a primitive idempotent of S.*

(iii) *M contains a minimal left ideal L of S and a minimal right ideal R of S such that $LR = M$ and $RL \neq 0$.*

(iv) *$M^2 \neq 0$, M is a minimal ideal of S and M is both a union of minimal left ideals of S and a union of minimal right ideals of S.*

Proof. First assume (i). By definition, M contains an idempotent e, primitive relative to M. If f is an idempotent of S such that $fe = ef = f$, then f is in M. Hence (i) implies (ii).

Now assume (ii) and let $e = e^2 \neq 0$ be a primitive idempotent of S contained in M. Set $L = Me$, $R = eM$. Then L (R) is a left (right) ideal of S contained in M and containing $e^2 = e \neq 0$. Also M^2 and $LR = MeM$ are ideals of M contained in M and containing $e = e^2 = e^3 \neq 0$, so, by the simplicity of M, $M = M^2 = LR$. Also RL contains e, so $RL \neq 0$. Let A be the set of all a in M such that $MaM = 0$. Then A is an ideal of M and $MAM = 0 \neq MMM$, so $A = 0$. Therefore if a is a nonzero element of M, MaM is a nonzero ideal of M, so $M = MaM$. Suppose a is a nonzero element of $R = eM$. Since e is in $eMe = e(MaM)e$, $e = bac$ for some (nonzero) b in eMe, c in Me. Now set $f = acb$. Since $bf = bacb = eb = b$, then $f \neq 0$. Since $f^2 = acbf = acb = f$ and $ef = fe = f$ and e is primitive, $f = e$. Thus $R = eM = acbM \subset aS \subset R$, so $R = aS$. Therefore R is a minimal right ideal of S and, similarly, L is a minimal left ideal of S. Hence (ii) implies (iii).

Next assume (iii). Let A be the set of all elements a of L such that $Sa = 0$. Then A is a left ideal of S contained in L. Since $RL \neq 0$, then $A \neq L$, and since L is minimal, $A = 0$. Thus, and similarly,

$$a \in L, a \neq 0 \Rightarrow Sa = L; \quad b \in R, b \neq 0 \Rightarrow R = bS. \tag{8.1}$$

Now let c be any nonzero element of M. Since $M = LR$, $c = ab$ where $a \in L$, $b \in R$ and $a \neq 0$, $b \neq 0$. Thus $ScS = (Sa)(bS) = LR = M$. Hence M is a minimal ideal of S. Moreover, $c = ab \neq 0$ is contained in Lb. Any left ideal of S contained in Lb must be of form Nb where N is a left ideal of S contained in L. If Nb contains $c \neq 0$, then $N \neq 0$ and hence $N = L$, $Nb = Lb$. Therefore Lb is a minimal left ideal of S contained in M and containing c. Similarly, aR is a minimal left ideal of S contained in M and containing c. Hence (iii) implies (iv).

Finally, assume (iv) and let x be any nonzero element of M. Since M is a union of minimal left ideals of S, $x \in L$ where L is a minimal left ideal of S contained in M. Since M^2 and $L \cup LS$ are nonzero ideals of S contained in M, and since M is minimal, $M = M^2 = L \cup LS$. Thus $M = M^2 \subset L^2 \cup L^2S$, so $M = L^2 \cup L^2S$. This implies $L^2 \neq 0$. Therefore, by the minimality of L, $L^2 = L$. Hence LM is a nonzero ideal of S contained in M, so $LM = M$. Since M is a union of minimal right ideals and since $LM \neq 0$, there exists a minimal right ideal R of S contained in M such that $LR \neq 0$. Then LR is a nonzero ideal of S contained in M, so $LR = M$. Since ML contains $L^2 = L$ and is contained in L, then $L(RL) = ML = L$. Hence $RL \neq 0$ and also $(RL)^2 = RL$. Now (iii) holds, so we may apply (8.1). Let A be the set of nonzero elements of RL. Since $A \subset R \cap L$, (8.1) implies $L = Sa$, $R = aS$ for every a in A. Then $M = LR = Sa^2S$, so $a^2 \neq 0$. Hence $a^2 \in A$ and $A \subset aS \cap Sa$ for every a in A. Therefore, by Theorem 3.3, there exists a unique nonzero idempotent e and a subgroup G of S, consisting of all elements of eSe which have inverses relative to e, such that $A \subset G$. If the element a of A has inverse a^{-1} in G, then $L = Sa$ contains $a^{-1}a = e$. Hence, and similarly,

$$L = Se\,, \quad R = eS\,, \quad M = LR = SeS\,. \tag{8.2}$$

Clearly $L \cap R = eSe$. Since SL contains $e \neq 0$, $SL = L$ and therefore $RL = eSL = eL = eSe$. Thus $RL = 0 \cup A \subset 0 \cup G \subset eSe = RL$, so

$$RL = L \cap R = eSe \quad \text{is a group with zero.} \tag{8.3}$$

If f is an idempotent of S such that $ef = fe = f$, then f is in eSe. Hence either $f = e$ or $f = 0$. Thus

$$e \text{ is primitive.} \tag{8.4}$$

We now recall that L was a minimal left ideal of S containing the preassigned nonzero element x of M. Since $L = Se$, we have $xe = x$ for a nonzero idempotent e of M. Similarly, by starting with a minimal right ideal containing x, we deduce the existence of a nonzero idempotent e' in M such that $ex' = x$. Then, also, MxM contains $e'xe = x \neq 0$, so $M = MxM$. Consequently, M is simple. Finally, let g be a nonzero idempotent contained in M. Then g is contained in L, R and LR where L, R are suitable minimal left and right ideals, respectively, of S contained in M. Since $LR \neq 0$, we have (8.2), (8.3), (8.4) for a suitable idempotent e. Since g is in $L \cap R = eSe$ and since $g \neq 0$, then $g = e$. Thus (iv) implies (i), and the proof of Theorem 8.2 is complete.

Rich [251] contributed to the proof of Theorem 8.2 and also showed the independence of the postulates $LR \neq 0$, $RL \neq 0$ by exhibiting (a) a semigroup S of order four in which $LR \neq 0$ for one pair L, R but $RL = 0$ for all pairs L, R and (b) a semigroup S of order five in which $RL \neq 0$ for one pair L, R but $LR = 0$ for all pairs L, R — where in each case, L, R denote minimal left and right ideals, respectively, of S.

The case $M = S$ of Theorem 8.2 (when the proof is considered) determines the structure of completely simple semigroups except for the following point:

Lemma 8.2. (Rees.) *If e, f are nonzero idempotents of the simple semigroup S (with zero) then $e = ab$ for a in eSf, b in fSe. The mapping $x \to bxa$ induces an isomorphism of eSe into fSf.*

Corollary. *If S is completely simple, $f = ba$ and the subsemigroups eSe, fSf are isomorphic groups with zero.*

Proof. Since S is simple and $f \neq 0$, then $S = SfS$ (Compare the proof that (ii) of Theorem 8.2 implies (iii).) Hence e is in $eSe = (eSf)(fSe)$, so $e = ab$ for a in eSf, b in fSe. Thus $bSa \subset fSf$. If x, y are in eSe then $(bxa)(bya) = bxeya = b(xy)a$ and $a(bxa)b = exe = x$. Hence the mapping $x \to bxa$ induces an isomorphism of eSe into fSf. In particular, $bea = ba$ is a nonzero idempotent of fSf. Hence, if f is primitive, $ba = f$ and the mapping is upon. Then, by (8.3), eSe, fSf are isomorphic groups with zero. This completes the proof.

Rees has explicitly constructed all completely simple semigroups, essentially as follows: Let A, B be non-empty sets and G a group with zero. Let $[b, a]$ be a function from $B \times A$ to G such that each $b \in B$ and each $a \in A$ occurs in at least one pair b, a with $[b, a] \neq 0$. Turn $A \times G \times B$ into a semigroup S by the definition $(a, x, b)(a', x', b') = (a, x[b, a']x', b')$. The set K of all elements $(a, 0, b)$ of S is the unique minimal ideal of S, and S/K is a completely simple semigroup. Moreover, each completely simple semigroup has such a representation. — If we replace G by a group, S itself is the model of a completely simple semigroup without zero. Thus, obviously, *any group can be imbedded in a completely simple semigroup* (with or without zero).

It is an easy consequence of Theorem 8.2 that *a simple semigroup is completely simple if and only if it satisfies the minimum conditions for both right and left ideals.* The example of Baer and Levi shows that neither of these minimum conditions implies the other. The following theorem indicates a complex situation:

Theorem 8.3. *Any semigroup S can be imbedded in a simple semigroup (with or without zero) possessing an identity element.*

Proof. If S has no identity element we first adjoin one, thus imbedding S in a semigroup S'. Hence assume that S has identity element 1. We now construct a semigroup T (without zero) generated by S and two additional elements a, b, subject only to the relations $ab = 1$, $as = a$, $sb = b$ for every s in S. Then every element of T has form $b^m s a^n$ for non-negative integers m, n, provided b^0 and a^0 are deleted when they occur. The construction may be given concretely as follows:

Let N be the set of all non-negative rational integers. Turn $N \times S \times N$ into a groupoid T according to the definition

$$(m, s, n)(m', s', n') = (m + [m' - n], f(n - m'; s, s'), n' + [n - m']), \tag{8.6}$$

where, for every rational integer x,

$$[x] = x \text{ if } x \geqq 0; \quad [x] = 0 \text{ if } x < 0; \tag{8.7}$$

$$f(x; s, s') = s, ss' \text{ or } s' \text{ according as } x > 0, x = 0, x < 0. \tag{8.8}$$

We first verify the identities

$$[x] + [y - [-x]] = [x + [y]], \tag{8.9}$$

$$f(x + [y]; f(y; s, s'), s'') = f(y - [-x]; s, f(x; s', s'')) \tag{8.10}$$

and then prove easily that T is associative. From (8.6), for each n in N, the mapping $s \to (n, s, n)$ is an isomorphism of S into T. Moreover, $(0, 1, 0)$ is the identity of T. Finally, if $p = (m, s, n)$, $q = (m', s', n')$ are arbitrary elements of T, then $q = (m', s', m + 1)\, p\, (n + 1, 1, n')$. Hence T is simple. This completes the proof of Theorem 8.3. — The semigroup T (for a suitable choice of S) must occur as a subsemigroup of every simple semigroup containing a nonzero, nonprimitive idempotent. The construction can be generalized (a) by iteration and (b) by replacing N by the non-negative elements of any simply ordered abelian group. T will still be simple if S has no idempotents but is simple; then T will have no idempotents either.

If the semigroup S with zero satisfies the minimal conditions for left and right ideals it is easily deduced from Theorems 8.1, 8.2 that S has an ascending chain of ideals S_i such that $S_0 = 0$ and, for $i = 0, 1, 2, \ldots$, S_{2i+1}/S_{2i} is the (nilpotent) nil radical of S/S_{2i} and S_{2i+2}/S_{2i+1} is the union of the completely simple ideals of S/S_{2i+1}. To make the chain finite we may impose a maximal condition on ideals. — Nothing in the literature suggests a structure theory comparable to that for rings.

Green [197] attacks the structure problem in terms of principal left, right and two-sided ideals. If x is an element of the semigroup S, $(x)_L$ denotes the smallest left ideal of S containing x. A left ideal L of S is called *principal* if and only if $L = (x)_L$ for some x in S. Similarly for principal right ideals $(x)_R$ and principal ideals $(x)_F$. If S has an identity element, 1, $S \cup 1 = S$; if S has no identity element, S can be imbedded in a semigroup $S \cup 1$ with a single additional identity element 1. In either case,

$$(x)_L = (S \cup 1)\, x\,, \quad (x)_R = x(S \cup 1)\,, \quad (x)_F = (S \cup 1)\, x (S \cup 1) \tag{8.11}$$

for each x in S.

We set up in S the following equivalence relation f:

$$x \sim y\ (f) \leftrightarrow (x)_F = (y)_F. \tag{8.12}$$

If F is an f-equivalence class of S, the ideal $I = (x)_F$ is (by definition) the same for each x in F. We call I the *principal ideal of* F. Let $K = I - F$ be the complement of F in I; (i. e., K consists of the elements of I which are not in F.) If S has a zero and if $F \neq 0$, then K is certainly non-empty; but K is empty if $F = 0$. If S has no zero then K may also be empty. Assume K not empty. Then, for k in K, $(k)_F$ is a proper subset of I and $kS \cup Sk \subset (k)_F$. Hence K is an ideal of S contained in I, and I/K is an ideal of S/K. We call K the *complementary ideal* of I and I/K the *principal factor* of S *corresponding* to the f-class F (or to any element x of S.). Even when K is empty we preserve the terminology, interpreting I/K and S/K as I and S respectively.

Lemma 8.3. (Green.) *Let F be an f-class of the semigroup S, with principal ideal I, complementary ideal K. Then (except when F is the zero* *ideal of* S*) the principal factor* I/K *is a minimal ideal of* S/K *and either* $F^2 \subset I^2 \subset K$ *or* I/K *is simple. In the latter case,* $F \subset FxF$ *for each* x *in* F.

Corollary. *If* F *is a subsemigroup of* S *and* I/K *is simple, then* F *is a simple subsemigroup (without zero) of* S.

Proof. Let A be an ideal of S such that $I \supset A \supset K$ and $A \neq K$. Then A contains an element x of $I - K = F$, so $I = (x)_F \subset A$ and hence $A = I$. Since every ideal of S/K has the form A/K where A is an ideal of S containing K, we see that I/K is a minimal ideal of S/K. First let us assume that K is non-empty, so that K is the zero of S/K. Then, as shown previously, either $(I/K)^2 = 0$ or $(I/K)^2 = (I/K)$; and, in the latter case, I/K is simple. Now I/K is essentially a semigroup $F \cup K$, where we set $xy = K$ if x, y are in F and $xy \in K$ in S, and regard K as the zero of $F \cup K$. Hence $(I/K)^2 = 0$ means $F^2 \subset K$ (in S); and $(I/K)^2 = I/K$ means that $F \subset F^2$ (in S). In the latter case, the semigroup $F \cup K$ is a simple semigroup with zero K, and hence, for each x in F, $F \cup K = FxF \cup K$. Thus $F \subset FxF$ in S. In this case, if F is a subsemigroup, $FxF \subset F$ and hence $F = FxF$, so that F is simple. On the other hand, if K is empty and F is not the zero ideal of S, then S has no zero. Then the same arguments show that $F = I$ is a minimal and hence a simple ideal of S. This completes the proof of Lemma 8.3 and Corollary.

We are now ready to prove three theorems on semigroups without zero. With slight modifications, each holds for semigroups without nonzero nilpotent elements, but REES' construction for completely simple semigroups readily affords counterexamples in case there are nonzero nilpotent elements. An *idempotent* semigroup is one whose elements are all idempotents.

Theorem 8.4. (CROISOT-ANDERSEN.) *A necessary and sufficient condition that a semigroup* S *without zero should be a union of simple subsemigroups without zero is that* $x \in Sx^2S$ *for each* x *in* S. *If the condition holds, the f-equivalence class* F_x *of* x *is a simple subsemigroup of* S *for each* x *in* S *and the mapping* $x \to F_x$ *is a homomorphism of* S *upon a commutative idempotent semigroup.*

Proof. *Necessity.* If the element x is contained in a simple subsemigroup T without zero, then $x^2 \in T$ and hence $x \in T = Tx^2T \subset Sx^2S$.

Sufficiency. Let $x \in Sx^2S$ for each x in S. Then $x \in SxS$, and hence

$$(x)_F = SxS\,. \tag{8.13}$$

Since $x^2 \in (x)_F$ in all cases, and $x \in (x^2)_F$ by hypothesis, we have $(x)_F = (x^2)_F$. Then, if $y \in S$, $(xy)_F = (xyxy)_F \subset (yx)_F$ and hence, by symmetry, $(xy)_F = (yx)_F$. Thus

$$x \sim x^2 \ (f)\,, \quad xy \sim yx \ (f)\,. \tag{8.14}$$

If $s \in S$, $(xsy)_F = (syx)_F \subset (yx)_F = (xy)_F$. Hence $(x)_F(y)_F = SxS^2yS \subset (xy)_F = (xyxy)_F = Sxy \cdot xyS \subset SxS \cdot SyS = (x)_F(y)_F$, or

$$(x)_F(y)_F = (xy)_F. \tag{8.15}$$

By (8.15), if $x \sim x'$ (f) and $y \sim y'$ (f) then $(xy)_F = (x)_F (y)_F = (x')_F (y')_F = (x'y')_F$. Hence $xy \sim x'y'$ (f). That is

$$f \text{ is a congruence relation.} \tag{8.16}$$

By (8.16), if F_x is the congruence class of x, the mapping $x \to F_x$ is a homomorphism of S upon a semigroup S'. By (8.14), S' is commutative and every element of S' is idempotent. Again, by (8.15), if $y, z \in F_x$ then $yz \sim x^2$ (f), whence, by (8.14), $yz \sim x$ (f). That is, F_x is a subsemigroup of S. Now the Corollary to Lemma 8.3 shows that F_x is simple. This completes the proof of Theorem 8.4 and Corollary.

Theorem 8.4 has been proved by McLean [223] for the case that S is idempotent; in this case the condition $x \in Sx^2S$ is automatically satisfied.

Theorem 8.5. (Clifford.) *Let S be a semigroup without zero. Then the following statements are equivalent:*

(i) *S is a union of completely simple subsemigroups without zero.*

(ii) *S is a union of subgroups.*

(iii) *x is in $Sx^2 \cap x^2S$ for every x in S.*

If (iii) *holds, each f-equivalence class of S is a completely simple subsemigroup of S.*

Proof. Assume (i). Then each element of S is contained in at least one completely simple subsemigroup without zero. By Rees's construction, the latter can be represented as (A, G, B) where G is a group and $(a, x, b)(a', x', b') = (a, x[b, a']x', b')$. Thus for fixed a, b, the mapping $(a, x, b) \to x[b, a]$ is an isomorphism of (a, G, b) upon the group G. Hence each element of S is contained in a subgroup of S. Thus (i) implies (ii).

Assume (ii). Then the element x of S is contained in a subgroup G of S, so x^2 is in G and $x \in Gx^2 \cap x^2G \subset Sx^2 \cap x^2S$. Thus (ii) implies (iii).

Assume (iii). By Theorem 3.3, where A is taken to consist of a single element x, the element x uniquely determines an idempotent e of S such that the maximal subgroup, $G(e)$, of S with e as identity element, contains x. Specifically, e is the unique element of xS such that $x = xe$. Now let e be an idempotent of S and suppose that $e = ab$ for a, b in eSe. Since $a = ae$ and $e \in aS$ we conclude that $a \in G(e)$. If a^{-1} is the inverse of a in $G(e)$, then $a^{-1} = a^{-1}e = a^{-1}ab = eb = b$. Hence (iii) ensures that $G(e)$ consists of all elements of eSe which have one-sided inverses (left or right) with respect to e. Since S is a union of subgroups and since groups are simple semigroups, by Theorem 8.4 each f-class F is a simple subsemigroup. Moreover, F contains idempotents. If the idempotent e is in F, and if x is an element of Fe, then $e \in eFxFe$, so $e = axb$ where $ax \in eSe$, $eb \in eSe$. Since $e = (ax)(eb)$, then $e = (eb)(ax) \in Fx$. Hence Fe is a minimal left ideal of F. Similarly, eF is a minimal right ideal of F and hence, by Theorem 8.2, F is completely simple. Thus (iii) implies (i).

In view of Theorems 8.4, 8.5, we give a simplified form to the next theorem.

Theorem 8.6. (CROISOT-CLIFFORD.) *Let S be a semigroup without zero. Necessary and sufficient conditions that each f-equivalence class of S be a group are.* (i) $x \in Sx^2 \cap x^2S$ *for each x of S and* (ii) *the idempotents of S commute.*

Proof. *Sufficiency.* If (i) holds, each f-class F of S is a completely simple subsemigroup, by Theorem 8.5. If (ii) holds for the idempotents of F, we verify from REES' construction that F is a group.

Necessity. Let each F_x be a group. Then (i) holds, by Theorem 8.5, and the mapping $x \to F_x$ is a homomorphism of S upon a commutative idempotent semigroup, by Theorem 8.4. If e, f are idempotents of S then ef, fe, efe, fef must lie in the same group $F = F_h$, where h is the identity element of F. If a is the inverse of ef in F then $efa = aef = h$. Hence $eh = hf = h$. Similarly, by using fe, we see that $he = fh = h$. Then, since $h = efa$ and $h = efh = (ef)^2a$, we conclude that $ef = (ef)^2$. Since h is the only idempotent of F, $ef = h$. Similarly, $fe = h$, whence (ii) holds and the proof of Theorem 8.6 is complete.

Further theorems allied to the last three will be found in CLIFFORD [172]. The essential character of the principal factors is shown by the following theorem, which implies a JORDAN-HÖLDER theorem for "principal" series:

Theorem 8.7. (GREEN.) *Let A, B be ideals of the semigroup S such that $A \subset B$, $A \neq B$ and no ideal T of S satisfies $A \subset T \subset B$, $T \neq A, B$. Then, for each x of $B - A$, the identity mapping of S induces an isomorphism of B/A upon the principal factor of S corresponding to x.*

Proof. Let F be the f-equivalence class of x and set $I = (x)_F$, $K = I - F$. Then $I \cup A = B$ and $I \cap A \subset K$. If $K \neq I \cap A$ then $K \cup A \neq A$ so $K \cup A = B$ and $I = I \cap B = (I \cap K) \cup (I \cap A) = K$, a contradiction. Hence $I \cap A = K$. Thus the identity mapping of S induces an isomorphism of $B/A = (I \cup A)/A$ upon $I/(I \cap A) = I/K$. This completes the proof.

GREEN calls a semigroup (nonzero semigroup with zero) *semisimple* if the principal factor corresponding to every (nonzero) element is simple. An element a of S is *regular* if $axa = a$ for some x in S; equivalently, $(a)_L = (e)_L$ (or $(a)_R = (f)_R$) for an idempotent e (or f) of S. A semigroup is regular if every element is regular. He notes that *every simple semigroup is semisimple*, that *every regular semigroup is semisimple*, and proves that *every semisimple semigroup which satisfies the descending chain conditions for principal right and principal left ideals is regular.* He notes (as do many others) that if an element x of a semigroup generates a finite subsemigroup then some power of x is idempotent and, a fortiori, x is regular; hence a semigroup with minimum condition for one-generator subsemigroups is regular. He also proves that *in a semigroup S satisfying the descending chain condition for principal ideals, some power x^n, $n = n(x)$, of every non-nilpotent element has a simple f-class.*

Some of these results depend on the equivalences $l, r, l \cap r$ and d on S, defined as follows:

$$x \sim y\ (l) \leftrightarrow (x)_L = (y)_L;\quad x \sim y\ (r) \leftrightarrow (x)_R = (y)_R; \tag{8.17}$$

$$x \sim y\ (l \cap r) \leftrightarrow x \sim y\ (l) \text{ and } x \sim y\ (r); \tag{8.18}$$

$$x \sim y\ (d) \leftrightarrow x \sim z\ (l) \text{ and } z \sim y\ (r) \text{ for some } z \text{ in } S. \tag{8.19}$$

The equivalences l, r are *permutable* — that is, the right side of (8.19) implies that $x \sim w\,(r)$ and $w \sim y\,(l)$ for some w in S — and consequently d is an equivalence as asserted. Indeed, if $x = pz$, $z = qx = yr$, $y = zs$ for p, q, r, s in $S \cup 1$, then $x = wr$, $w = xs = py$, $y = qw$. Since also $x = pyr$, $y = qxs$, we see that *$d \subset f$ in the sense that $x \sim y\,(d)$ implies $x \sim y\,(f)$.* Green shows that $d \neq f$ in the free semigroup on six symbols and he examines the interrelations of r, l, d, f in the presence of various minimum conditions. He also gives the following form of Theorem 4.3, Corollary 2: *if $x \sim x^2\,(l \cap r)$, the $l \cap r$-class of x is a group.*

Let S be a semigroup with identity 1. If P, Q are the r-class and l-class respectively of 1 in S, then the d-class of 1 is QP Moreover: (1) P has an identity; (2) if $xz = yz$ for x, y, z in P, then $x = y$. Clifford [171] calls S *d-simple* if and only if $S = QP$. Necessary and sufficient conditions that S be d-simple and that every two idempotents of S commute are that P satisfy (1), (2) and also: (3) the intersection of any two principal left ideals of P is a principal left ideal of P. When (3) holds, P possesses an antiisomorphism $p \to p^{-1}$ upon Q. Clifford gives an invariant construction leading from any semigroup P, subject to (1), (2), (3), to a "quotient" d-simple semigroup $P^{-1} \circ P$ with identity 1, in which P is the r-class of 1.

If S is any semigroup with identity, consider the simple semigroup $T = (N, S, N)$ constructed in (8.6), (8.7), (8.8). It may be verified that T is d-simple precisely when S is and that the idempotents of T commute precisely when those of S do. The first of these facts, taken with Lemma 8.2, suggests that a d-simple semigroup with identity can be peeled like an onion.

The reader should perhaps be reminded at this point that we have purposely omitted certain aspects of ideal theory: analytic or topological theory, arithmetic theory and also the analogue of the Noether theory of representation of an ideal as an intersection of primary ideals.

9. Multiplication semigroups

With each element x of a groupoid G are associated single-valued mappings of G into G, the *left-multiplication* $L(x)$ and the *right multiplica-*

tion $R(x)$, defined by

$$yL(x) = xy\,, \quad yR(x) = yx \tag{9.1}$$

for all y in G. Each of the identical relations

$$L(x)\,L(y) = L(yx)\,, \quad L(x)\,R(y) = R(y)\,L(x)\,, \quad R(x)\,R(y) = R(xy)$$

is, in itself, necessary and sufficient that G be a semigroup.

The set of all $R(x)$, x in G, generates (under multiplication of mappings) a semigroup $\mathfrak{S}_\varrho = \mathfrak{S}_\varrho(G)$, called the *right multiplication semigroup* of G. Similarly, the set of all $L(x)$, x in G, generates the *left multiplication semigroup* $\mathfrak{S}_\lambda = \mathfrak{S}_\lambda(G)$; and the set of all $L(x)$, $R(x)$, x in G, generates the *multiplication semigroup* $\mathfrak{S} = \mathfrak{S}(G)$.

An element x of G is called *right nonsingular* (*left nonsingular*) if $R(x)$ (if $L(x)$) is one-to-one upon G. And x is called *nonsingular* if x is both right and left nonsingular. A necessary and sufficient condition that G be a quasigroup is that every element of G be nonsingular. If G is a quasigroup, the $R(x)$ and their inverses $R(x)^{-1}$ generate the *right multiplication group* $\mathfrak{M}_\varrho = \mathfrak{M}_\varrho(G)$ of G. The *left multiplication group* $\mathfrak{M}_\lambda = \mathfrak{M}_\lambda(G)$ and the *multiplication group* $\mathfrak{M} = \mathfrak{M}(G)$ are similarly defined.

T. Evans [83] pointed out that, if G is a free loop, $\mathfrak{M}_\varrho$ and $\mathfrak{M}_\lambda$ are free groups. The converse is false.

In an early paper Suškevič [110] studied quasigroups whose right multiplications $R(x)$ form a group. Such a quasigroup is isomorphic to a quasigroup (G, o) obtained from a group $(G, \cdot)$ by the definition $x o y = [(x\theta^{-1})\,y]\,\theta$ where θ is an arbitrary but fixed permutation of G. If both the right multiplications and the left multiplications of a quasigroup form groups, the quasigroup is a group. More recently Gardaschnikoff [196] has studied groupoids whose right multiplications form a semigroup.

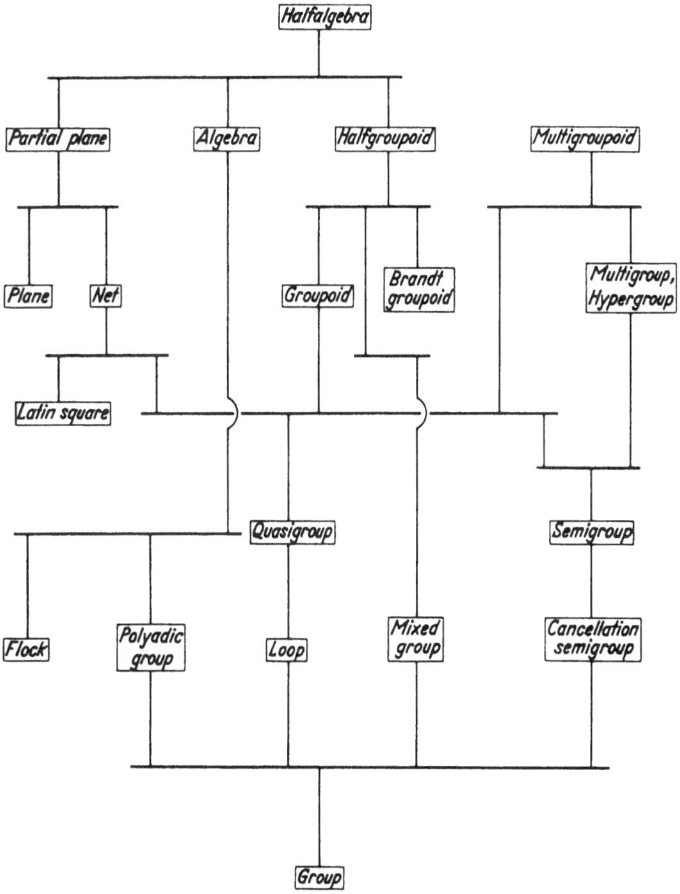
Halfalgebra
Partial plane
Algebra
Halfgroupoid
Multigroupoid
Plane
Net
Groupoid
Brandt groupoid
Multigroup, Hypergroup
Latin square
Quasigroup
Semigroup
Flock
Polyadic group
Loop
Mixed group
Cancellation semigroup
Group

III. Isotopy

1. Isotopy of groupoids

Let $(G, \cdot)$ and (H, o) be two groupoids. An ordered triple (α, β, γ) of one-to-one mappings α, β, γ of G upon H is called an *isotopism* of $(G, \cdot)$ upon (H, o), and $(G, \cdot)$ is said to be *isotopic to* or an *isotope of* (H, o), provided

$$(x\alpha) o (y\beta) = (x \cdot y)\gamma \tag{1.1}$$

for all x, y in G. Isotopy of groupoids is clearly an equivalence relation. Moreover, given the groupoid $(G, \cdot)$ and one-to-one mappings α, β, γ of G upon a set H, (1.1) defines a groupoid (H, o) isotopic to $(G, \cdot)$. The element y is right nonsingular (Chapter II, § 9) in $(G, \cdot)$ if and only if $y\beta$ is right nonsingular in (H, o); and similarly for left nonsingularity. In particular, *every isotope of a quasigroup is a quasigroup.*

The concept of isotopy seems very old. In the study of latin squares (which were known to Bachet and certainly predate Euler's problem of the 36 officers) the concept is so natural as to creep in unnoticed; and latin squares are simply the multiplication tables of finite quasigroups. It was consciously applied by Schönhart [108], Baer [348], and independently by Albert [54]. Albert earlier had borrowed the concept from topology for application to linear algebras; in the latter theory it has virtually been forgotten except for applications to the theory of projective planes.

An isotopism (α, β, I_G) of $(G, \cdot)$ upon $(G, *)$ is called a *principal* isotopism, and $(G, *)$ is called a principal isotope of $(G, \cdot)$. Principal isotopy is also an equivalence relation. Given (1.1), define $(G, *)$ by $(x\gamma\alpha^{-1}) \cdot (y\gamma\beta^{-1}) = x * y$. Then $(G, *)$ is a principal isotope of $(G, \cdot)$ and, by (1.1) with x, y replaced by $x\gamma\alpha^{-1}$, $y\gamma\beta^{-1}$ respectively, $(x\gamma) o (y\gamma) = (x * y)\gamma$. Thus *every isotope of a groupoid is isomorphic to a principal isotope of the groupoid.*

Let a be a left nonsingular and b be a right nonsingular element of $(G, \cdot)$. Then the principal isotope (G, o) of $(G, \cdot)$, defined in either of the following ways (for notation see II.9, I.2):

$$x o y = xR(b)^{-1} \cdot yL(a)^{-1}, \quad x o y = (x/b) \cdot (a\backslash y), \tag{1.2}$$

has identity element $a \cdot b$. Conversely, if (G, o) is a principal isotope defined by $x o y = (x\alpha) \cdot (y\beta)$ and if (G, o) has identity element e, set $a = e\alpha$, $b = e\beta$. Since $x = x o e = (x\alpha) \cdot b$ for all x in G, b is right nonsingular in $(G, \cdot)$ and $\alpha = R(b)^{-1}$; similarly, a is left nonsingular in

$(G, \cdot)$ and $\beta = L(a)^{-1}$. Hence *necessary and sufficient conditions that a groupoid possess an isotope with an identity element are that the groupoid have a right nonsingular element and a left nonsingular element.* In particular, *every quasigroup is isotopic to a loop.*

2. Invariants of principal isotopy

We list without formal proof some invariants of principal isotopy:

(i) *If (G, o) is an isotope with identity of the semigroup $(G, \cdot)$, then (G, o) is isomorphic to $(G, \cdot)$.* (Assuming (1.2), the mapping $L(a)\,R(b)$: $x \to a \cdot x \cdot b$, is an isomorphism of $(G, \cdot)$ upon (G, o).)

(ii) *Every isotope of a quasigroup is a quasigroup.* (Previously noted.)

(iii) *If the loop (G, o) is isotopic to the loop $(G, \cdot)$, the left, right and two-sided multiplication groups of (G, o) are respectively isomorphic.* (They are respectively identical for principal isotopes; as a containing relation in one direction, (1.2) gives $R_0(y) = R(b)^{-1}R(a\backslash y)$, $L_0(x) = L(a)^{-1}L(x/b)$.)

The above are in ALBERT [54]. The following definitions, with "associator" instead of "nucleus", are in BRUCK [70]: If $(G, \cdot)$ is a loop, the *left nucleus*, N_λ, of $(G, \cdot)$ is the set of all a in G such that $(a \cdot x) \cdot y = a \cdot (x \cdot y)$ for all x, y of G. The *middle nucleus*, N_μ, of $(G, \cdot)$ is similarly defined in terms of $(x \cdot a) \cdot y = x \cdot (a \cdot y)$; and the *right nucleus*, N_ϱ, in terms of $(x \cdot y) \cdot a = x \cdot (y \cdot a)$. The *nucleus*, N, is defined by $N = N_\lambda \cap N_\mu \cap N_\varrho$. The *centre*, Z, of $(G, \cdot)$, is the set of all a in N such that $a \cdot x = x \cdot a$ for all x in G. The four nuclei are subgroups of $(G, \cdot)$ and the centre is an abelian subgroup.

(iv) *If the loop (G, o) is isotopic to the loop $(G, \cdot)$, the left, middle and right nuclei of $(G, \cdot)$ are respectively isomorphic to those of (G, o) and the centre of $(G, \cdot)$ is isomorphic to that of (G, o).* (Assuming (1.2), the first three isomorphisms are induced respectively by $R(a)\,R(b)$, $L(a)\,R(b)$, $L(b)\,L(a)$; they coincide on the centre of $(G, \cdot)$.)

(v) *If the loop (G, o) is isotopic to the loop $(G, \cdot)$ and if θ is a homomorphism of $(G, \cdot)$ upon a loop $(H, \cdot)$, then θ induces a homomorphism of (G, o) upon a loop (H, o).* (Assuming (1.2), define (H, o) by $(x\theta)\,o\,(y\theta) = (x\theta)\,R(b\theta)^{-1} \cdot (y\theta)\,L(a\theta)^{-1}$.)

An unsolved problem is this: *Find necessary and sufficient conditions upon the loop G in order that every loop isotopic to G be isomorphic to G.* Associativity is sufficient, by (i), but is not necessary, since the multiplicative loop of any alternative division ring has the property.

In another direction, certain properties (P) are known such that isotopic loops with property (P) are isomorphic; although, in some cases, not every loop-isotope of a loop with property (P) has property (P). Associativity (by (i)) and the property of the preceding paragraph are examples. As another:

(vi) *Isotopic free loops are isomorphic.* (EVANS [83]; this follows from (v) and (i), which show that the ranks must be equal.)

A loop is called *Moufang* if it satisfies the identity $(xy)(zx) = (x(yz))x$.

(vii) *Isotopic commutative Moufang loops are isomorphic.* (This can be deduced from Chapter VII, § 5.)

A loop is called *totally symmetric* if it is commutative and satisfies the identity $x(xy) = y$. (Such loops are essentially Steiner triple systems.)

(viii) *Isotopic totally symmetric loops are isomorphic.* (BRUCK [67].)

The last two items are illuminated by the following:

(ix) *A necessary and sufficient condition that every loop isotopic to the loop G be commutative is that G be an abelian group.* (BAER [348], ALBERT [54].)

(x) *Every loop isotopic to a Moufang loop is Moufang.* (Implicit in BOL [66]; proved in BRUCK [67], [70]; see also Chapter VII.)

3. Non-invariants of principal isotopy

Certain non-invariants of principal isotopy may be illustrated in terms of the following loops $(G, \cdot)$, (G, o), where G consists of the first six natural numbers:

·	1	2	3	4	5	6
1	1	2	3	4	5	6
2	2	1	6	3	4	5
3	3	4	5	2	6	1
4	4	5	1	6	2	3
5	5	6	4	1	3	2
6	6	3	2	5	1	4

o	2	3	4	5	6	1
2	2	3	4	5	6	1
3	3	2	6	4	1	5
4	4	6	2	1	5	3
5	5	4	1	2	3	6
6	6	1	5	3	2	4
1	1	5	3	6	4	2

In terms of the multiplications of $(G, \cdot)$, $x o y = xR(3)^{-1} \cdot yL(6)^{-1}$ for all x, y of G. Hence $(G, \cdot)$, (G, o) are principal isotopes of each other.

(i) *Commutativity.* (G, o) is commutative; $(G, \cdot)$ is not.

(ii) *Number of generators.* $(G, \cdot)$ can be generated by any one of 3, 4, 5, 6. No single element generates (G, o), but any two of 3, 4, 5, 6, 1 will generate (G, o).

(iii) *Automorphism group.* The automorphism group of $(G, \cdot)$ has order 4 and is generated by the permutation (3456). That of (G, o) has order 20 and is generated by (34561) and (3465).

(iv) *Freeness.* (EVANS [83].) Let $(F, \cdot)$ be a free loop of rank 1, with free generator a. Set $b = a \cdot (a \cdot a)$ and define (F, o) by $x o y = xR(a)^{-1} \cdot yL(b)^{-1}$. Since $(G, \cdot)$ is generated by 3, there is a unique homomorphism θ of $(F, \cdot)$ upon $(G, \cdot)$ such that $a\theta = 3$. Since $b\theta = 6$, θ induces a homomorphism of (F, o) upon (G, o). If (F, o) were free it would be isomorphic to $(F, \cdot)$. But then (G, o) could be generated by a single element, a contradiction. Hence (F, o) is not free. Now let $(K, \cdot)$

be any free loop of positive rank. Without loss of generality we may assume that $(F, \cdot)$ is a subloop of $(K, \cdot)$. Then the isotope (K, o) of $(K, \cdot)$ defined by $x o y = xR(a)^{-1} \cdot yL(b)^{-1}$ has the non-free loop (F, o) as a subloop and therefore is not free. Thus *every free loop of positive rank possesses a non-free loop-isotope.* Since no loop is a free quasigroup and every quasigroup is isotopic to a loop, a like result holds for quasigroups.

Other examples, which we omit, show the following, in contrast to (iv) of § 2:

(v) *There exist isotopic loops $(G, \cdot)$, (G, o) such that the nucleus of one is not isomorphic to that of the other.*

(vi) *There exist a loop $(G, \cdot)$, a principal loop-isotope (G, o) and a homomorphism θ of $(G, \cdot)$ upon a loop such that the kernel of θ in $(G, \cdot)$ is not isotopic to the kernel of the homomorphism induced by θ in (G, o).*

Since the contrary to (vi) is stated in ALBERT [54], we should point out that (vi) may be deduced from the proof of Theorem 8H (ii), p. 311 of BRUCK [70]. For example, one kernel can be a noncommutative group and the other a commutative Moufang loop.

There seems to be no example of (v) in the literature. However, the following remarks are easily verified: There exists one and (to within an isomorphism) only one loop G of order six whose centre has order one and whose nucleus has order two. Among the isotopes of G there are four non-isomorphic loops with a nucleus of order one.

4. Isotopy and geometry

It may be verified that the concept of isotopy introduced in I.4 with respect to the quasigroups and loops defined by a 3-net is identical with the present concept. Hence a property holding for a complete class of isotopic loops can be expressed as a geometric property of a 3-net and conversely. We shall mention some of the best known examples. For more details see BLASCHKE und BOL, Geometrie der Gewebe and G. PICKERT, Projektive Ebenen.

Let G be a loop and N be the 3-net defined by G. The requirement that G (and hence all its loop-isotopes) be associative corresponds in N to the closure of a "parallelogram" known as the Reidemeister figure. The requirement that G and all its loop-isotopes be commutative (or, equivalently, that G be an abelian group) corresponds in N to the closure of a "double-triangle" known as the Thomsen triangle. The requirement that every loop (H, o) isotopic to G should satisfy the identity $(x o x) o x = x o (x o x)$ is equivalent in N to the closure of a certain "hexagon"; the corresponding nets are called *hexagonal* (in German: Sechseckgewebe). A net N is hexagonal if and only if each of its loops is *power-associative* in the sense that every element of the loop generates an abelian subgroup. The requirement that G and all its loop-isotopes have the *inverse property* or, equivalently, that G and all its loop-isotopes be *di-associative* or, equivalently, that G be *Moufang* (for definitions of these terms, see VII) corresponds in N to the closure of three Moufang figures. In each case one gets a pretty diagram by representing the three line-classes of the net as the parallels to three sides of a plane triangle, using the geometric definition of the loop operation which is indicated in I.4 and imposing the algebraic condition in question.

In the theory of projective planes it is of interest to know under what circumstances all coordinate rings of a projective plane are isomorphic. As a closely allied question: *Under what circumstances are all the loops of a 3-net isomorphic?* The latter is equivalent to a question previously raised: What loops have the property that all their loop-isotopes are isomorphic?

Since PICKERT's book is so recent, we have largely excluded geometric literature from the bibliography. A few related papers are: MOUFANG [91], BOL [66], BAER [348], BATES [63], PAIGE [96], HUGHES [352]. Many others will be found in PICKERT.

IV. Homomorphism Theory of Loops

1. Normal subloops

The *kernel*, H, of a homomorphism θ of a loop G upon a loop K with identity element 1, is the set of all h in G such that $h\theta = 1$. If x, y are in G, there exist unique elements a, b of G such that $y = xa = bx$. Then $y\theta = (x\theta)(a\theta) = (b\theta)(x\theta)$, so the equations $y\theta = x\theta$, $a\theta = 1$, $b\theta = 1$ are equivalent. The case $x\theta = 1$ shows that H is a subloop of G. Moreover $xH = Hx$ for all x in G. If x, y, z are in G, there exist unique elements p, q, r, s of G such that $z = (px)y = q(xy) = x(yr) = (xy)s$; and $z\theta = [(p\theta)(x\theta)](y\theta) = (q\theta)[(x\theta)(y\theta)] = (x\theta)[(y\theta)(r\theta)] = [(x\theta)(y\theta)](s\theta)$. Therefore the equations $z\theta = (x\theta)(y\theta)$, $p\theta = 1$, $q\theta = 1$, $r\theta = 1$, $s\theta = 1$, are equivalent. That is, $(Hx)y = H(xy) = x(Hy) = (xy)H$ for all x, y of G, and we have

$$xH = Hx\,, \quad (Hx)y = H(xy)\,, \quad y(xH) = (yx)H \tag{1.1}$$

for all x, y of G. A subloop H of G which satisfies (1.1) is called a *normal* subloop of G. From (1.1), $(xH)y = (Hx)y = H(xy) = (xy)H = x(yH) = x(Hy)$, so (1.1) implies

$$x(Hy) = (xH)y \tag{1.2}$$

for all x, y of G. If c is a centre element of G, then $cx = xc$, $(cx)y = c(xy)$, $y(xc) = (yx)c$ for all x, y of G. This gives:

Lemma 1.1. *Any subgroup of the centre of a loop G is a normal subloop of G.*

Now let H be any normal subloop of a loop G with identity element 1. Since $1 \in H$, $x \in Hx$ for every x. If $y \in Hx$, then $y = hx$ for some h in H and hence $Hy = H(hx) = (Hh)x = Hx$. That is:

$$y \in Hx \leftrightarrow Hy = Hx. \tag{1.3}$$

Consequently, the *right cosets* Hx *of* G *modulo* H partition G. By (1.2), (1.1), for all x, y in G, $(Hx)(Hy) = [(Hx)H]y = [H(Hx)]y = [(HH)x]y = (Hx)y = H(xy)$, or

$$(Hx)(Hy) = H(xy). \tag{1.4}$$

By (1.3), the mapping φ of G, defined by $x\varphi = Hx$, is a single-valued mapping of G upon the set G/H of right cosets which maps each right coset into itself. By (1.4), φ is a homomorphism of G upon a groupoid G/H. The identity element of G/H is the coset H. If $H(xy) = H(xz)$, then $x(Hy) = x(yH) = (xy)H = H(xy) = H(xz) = x(Hz)$ and hence $Hy = Hz$. If $H(yx) = H(zx)$ then $(Hy)x = H(yx) = H(zx) = (Hz)x$ and hence $Hy = Hz$. Therefore G/H is a cancellation groupoid. The equations $(Ha)(Hx) = Hb$, $(Hy)(Ha) = Hb$ have the (necessarily unique) solutions Hx, Hy defined by $ax = b$, $ya = b$. Therefore G/H is a loop, called the *quotient loop of* G *modulo* H. If H is the kernel of a homomorphism θ of G upon a loop $G\theta$, the mapping ψ of G/H, defined by $(Hx)\psi = x\theta$, is an isomorphism (*induced* by θ) of G/H upon $G\theta$. To sum up:

Theorem 1.1. *If H is a normal subloop of the loop G, then H defines a natural homomorphism $x \to Hx$ of G upon the quotient loop G/H. If H is the kernel of a homomorphism θ of G upon a loop $G\theta$, then θ induces a natural isomorphism $Hx \to x\theta$* of G/H *upon* $G\theta$.

Denote by $\mathfrak{M} = \mathfrak{M}(G)$ the multiplication group of the loop G. The *inner mapping group* $\mathfrak{J} = \mathfrak{J}(G)$ of G is the subgroup of M generated by the mappings $R(x, y)$, $L(x, y)$, $T(x)$, for all x, y of G, where

$$R(x, y) = R(x)\,R(y)\,R(xy)^{-1}, \quad L(x, y) = L(x)\,L(y)\,L(yx)^{-1}, \tag{1.5}$$
$$T(x) = R(x)\,L(x)^{-1}.$$

In view of (1.1) we see that *a subloop H of G is normal in G if and only if $H\mathfrak{J} = H$*. As an immediate consequence:

Theorem 1.2. *The intersection of any non-empty set of normal subloops of the loop G is a normal subloop of G.*

The inner mapping group has the following characterization:

Lemma 1.2. *The inner mapping group $\mathfrak{J}$ of the loop G is the set of all α in the multiplication group $\mathfrak{M} = \mathfrak{M}(G)$ such that* $1\alpha = 1$.

Proof. Since each generator β of $\mathfrak{J}$ satisfies $1\beta = 1$, then $1\mathfrak{J} = 1$. Let $\mathfrak{K}$ be the set of all α in $\mathfrak{M}$ such that $\alpha \in \mathfrak{J}R(1\alpha)$. Since $\mathfrak{J}R(x) = \mathfrak{J}T(x)\,L(x) = \mathfrak{J}L(x)$, α is in $\mathfrak{K}$ if and only if $\alpha \in \mathfrak{J}L(1\alpha)$. If α is in $\mathfrak{K}$, write $1\alpha = t$. Then $\alpha = \theta R(t)$ for θ in $\mathfrak{J}$. Let x be any element of G. Since $1\alpha R(x) = tx$ and $\alpha R(x) = \theta R(t,x)\, R(tx) \in \mathfrak{J}R(tx)$, then $\alpha R(x) \in \mathfrak{K}$. Since $1\alpha R(x)^{-1} = s$, where $sx = t$, and since $\alpha R(x)^{-1} = \theta R(t)\, R(x)^{-1} R(s)^{-1} R(s) = \theta R(s,x)^{-1} R(s) \in \mathfrak{J}R(s)$, then $\alpha R(x)^{-1} \in \mathfrak{K}$. Since $1\alpha L(x) = xt$ and $\alpha L(x) = \theta R(t)\, L(x) = \theta T(t)\, L(t,x)\, L(xt) \in \mathfrak{J}L(xt)$ then $\alpha L(x) \in \mathfrak{K}$. Since $1\alpha L(x)^{-1} = u$ where $t = xu$, and since $\alpha L(x)^{-1} = \theta R(t)\, L(x)^{-1} = \theta T(t)\, L(t)\, L(x)^{-1} L(u)^{-1} L(u) = \theta T(t) L(u,x)^{-1} L(u) \in \in \mathfrak{J}L(u)$, then $\alpha L(x)^{-1} \in \mathfrak{K}$. Therefore $\mathfrak{K}\mathfrak{M} \subset \mathfrak{K}$ and hence $\mathfrak{K} = \mathfrak{M}$. Consequently $1\alpha = 1$, for α in $\mathfrak{M}$, if and only if $\alpha \in \mathfrak{J}R(1\alpha) = \mathfrak{J}$.

Lemma 1.3. *Let θ be a homomorphism, with kernel H, of the loop G upon the loop $G\theta$. Then θ induces a homomorphism φ of $\mathfrak{M}(G)$ upon $\mathfrak{M}(G\theta)$ such that $\alpha\theta = \theta(\alpha\varphi)$ for all α in $\mathfrak{M}(G)$. In particular, if H^* is the set of all α in $\mathfrak{M} = \mathfrak{M}(G)$ such that $H(x\alpha) = Hx$ for every x in G, then $\mathfrak{M}(G/H)$ is isomorphic to $\mathfrak{M}(G)/H^*$ and $\mathfrak{J}(G/H)$ is isomorphic to $\mathfrak{J}(G)/(H^* \cap \mathfrak{J}(G))$.*

Proof. If $xy = z$ in G then $(x\theta)\,(y\theta) = z\theta$ in $G\theta$. Hence $R(y)\theta = \theta R(y\theta)$, $R(y)^{-1}\theta = \theta R(y\theta)^{-1}$, $L(x)\theta = \theta L(x\theta)$, $L(x)^{-1}\theta = \theta L(x\theta)^{-1}$. Therefore, if $\alpha = P(x_1)\, P(x_2) \ldots P(x_r)$, where $P(x_i)$ is one of $R(x_i)^{\pm 1}$, $L(x_i)^{\pm 1}$ for each i, $\alpha\theta = \theta P(x_1\theta)\, P(x_2\theta) \ldots P(x_r\theta)$. If, further, $\alpha = I_G$, then $y\theta = y\alpha\theta = y\theta P(x_1\theta) \ldots P(x_r\theta)$ for each y in G and hence $P(x_1\theta) \ldots P(x_r\theta) = I_{G\theta}$. Therefore there is a well-defined homomorphism φ of $\mathfrak{M} = \mathfrak{M}(G)$ upon $\mathfrak{M}(G\theta)$ such that $R(x)\varphi = R(x\theta)$, $L(x)\varphi = L(x\theta)$ for all x in G and $\alpha\theta = \theta(\alpha\varphi)$ for all α in $\mathfrak{M}(G)$. The kernel of φ is the set of all α in $\mathfrak{M}$ such that $\alpha\theta = \theta$; that is, $(x\alpha)\theta = x\theta$ for every x in G. In case θ is the natural homomorphism of G upon G/H, the kernel of φ is the H^* of Lemma 1.3, and the kernel of the homomorphism induced by φ on $\mathfrak{J} = \mathfrak{J}(G)$ is $\mathfrak{J} \cap H^*$. In view of Theorem 1.1, this completes the proof of Lemma 1.3.

Albert [54] sets up a correspondence between the normal subloops of a loop G and certain normal subgroups of $\mathfrak{M} = \mathfrak{M}(G)$. On the one hand, to a normal subloop H of G he makes correspond the normal subgroup H^* of $\mathfrak{M}$, defined in Lemma 1.3. On the other hand, if $\mathfrak{N}$ is any normal subgroup of $\mathfrak{M}$, the sets $x\mathfrak{N}$, $x \in G$, partition G and $(x\mathfrak{N})\,(y\mathfrak{N}) = (x\mathfrak{N})\, R(y\mathfrak{N}) = [xR(y\mathfrak{N})]\mathfrak{N} = (x \cdot y\mathfrak{N})\mathfrak{N} = y\mathfrak{N}L(x)\mathfrak{N} = yL(x)\mathfrak{N}\mathfrak{N} = (xy)\mathfrak{N}$. Hence the mapping θ defined by $x\theta = x\mathfrak{N}$ is a homomorphism of G upon a groupoid, say $G/\mathfrak{N}$. If $(xy)\mathfrak{N} = (xz)\mathfrak{N}$, then $x(y\mathfrak{N}) = y\mathfrak{N}L(x) = yL(x)\mathfrak{N} = (xy)\mathfrak{N} = (xz)\mathfrak{N} = x(z\mathfrak{N})$, so $y\mathfrak{N} = z\mathfrak{N}$. Similarly, if $(yx)\mathfrak{N} = (zx)\mathfrak{N}$ then $y\mathfrak{N} = x\mathfrak{N}$. Hence $G/\mathfrak{N}$ is a loop and the kernel, $1\mathfrak{N}$, of θ, is a normal subloop of G. Thus *every normal subgroup $\mathfrak{N}$ of $\mathfrak{M}$ determines a normal subloop $1\mathfrak{N}$ of G.* If $1\mathfrak{N} = H$ then $Hx = 1\mathfrak{N}R(x) = 1R(x)\mathfrak{N} = x\mathfrak{N}$ and $H(x\mathfrak{N}) = H(Hx) = Hx$ for every x in G, so $\mathfrak{N} \subset H^*$. Therefore we may define a *closure* operation (*) on the normal subgroups of $\mathfrak{M}$, by the definition $\mathfrak{N}^* = (1\mathfrak{N})^*$. Albert develops normality theory for G entirely in terms of $\mathfrak{M}$. He makes one slip, in dealing with isotopy, which is sufficiently indicated by III.3 (vi).

The rest of this section, and the next two, combine BAER [57], BRUCK [70].

A non-empty subset S of a loop G is called *self-conjugate* in G if $S\mathfrak{J} \subset S$ (and hence $S\mathfrak{J} = S$) where $\mathfrak{J} = \mathfrak{J}(G)$ is the inner mapping group. Each element x of G determines a self-conjugate subset $x\mathfrak{J}$, called the *conjugate class* of x in G.

Lemma 1.4. *If H is a subloop of the loop G, let K be the set of all k in H such that $k\mathfrak{J} \subset H$. Then K is the largest normal subloop of G contained in H.*

Corollary. *Every self-conjugate subset of G generates a normal subloop of G.*

Proof. Since $1 \in K$ and $(K\mathfrak{J})\mathfrak{J} \subset H$, K is a self-conjugate subset of G. Let $k \in K$, $\alpha \in \mathfrak{J}$. Then $R(k)\alpha \in \mathfrak{J}R(k\alpha)$, so $(Kk)\alpha = KR(k)\alpha \subset K\mathfrak{J}R(k\alpha) \subset H \cdot k\alpha = H$. Therefore $KK \subset K$. If $1R(k)^{-1}\alpha = s$ then $R(k)^{-1}\alpha = \beta R(s)$ for $\beta \in \mathfrak{J}$. Hence $\alpha = R(k)\beta R(s)$ and $1 = 1\alpha = (k\beta) \cdot s \in Hs$, so $s \in H$. Therefore $[KR(k)^{-1}]\alpha = K\beta \cdot s \subset H$. Hence $KR(k)^{-1} \subset K$. Similarly $KL(k)^{-1} \subset K$, so K is a subloop of G. Since $K\mathfrak{J} \subset K$, K is also normal. If N is a normal subloop of G contained in H, $N\mathfrak{J} \subset N \subset H$, so $N \subset K$. If H is generated by a self-conjugate subset S of G, then $S\mathfrak{J} \subset S \subset H$, so $S \subset K$ and hence $K = H$. This completes the proof of Lemma 1.4 and Corollary.

Theorem 1.3. *If H is a subloop generated by a non-empty set of normal subloops of the loop G then H is a normal subloop of G, called the union of the given normal subloops.*

Proof. The set-union of the given normal subloops is self-conjugate and generates H.

Lemma 1.5. *The set of all self-conjugate subsets of a loop G is a commutative semigroup under the product operation AB.*

Proof. Let A, B, C be self-conjugate subsets of G. If $a \in A$, $b \in B$, then $ab = b[aT(b)] \in bA$. Hence $AB \subset BA \subset AB$, $AB = BA$. If also $c \in C$, then $(ab)c = [aR(b,c)](bc) \in A(bc)$. Hence $(AB)C \subset A(BC)$. Similarly, $A(BC) \subset (AB)C$, so $(AB)C = A(BC)$. Finally, $(AB)\mathfrak{J} = AR(B)\mathfrak{J} \subset A\mathfrak{J}R(B\mathfrak{J}) \subset AB$.

If θ is a homomorphism of the loop G upon the loop G' and if S is a non-empty subset of G', the *inverse image*, $S\theta^{-1}$, of S is the set of all x in G such that $x\theta \in S$. In particular, $1\theta^{-1}$ is the kernel of θ.

Lemma 1.6. *Let θ be a homomorphism, with kernel K, of a loop G upon a loop G'. Then:*

(i) *If S is a non-empty subset of G, $(S\theta)\theta^{-1} = SK = KS$.*

(ii) *If H' is a (normal) subloop of G', $H'\theta^{-1}$ is a (normal) subloop of G.*

(iii) *If H is a (normal) subloop of G, $H\theta$ is a (normal) subloop of G' and $HK = KH$ is a (normal) subloop of G.*

Proof. (i) $x\theta = s\theta$ if and only if $x \in sK = Ks$.

(ii) Write $H = H'\theta^{-1}$. If $xy = z$ in G and two of x, y, z are in H then $(x\theta)(y\theta) = z\theta$ in G' and two, hence all, of $x\theta, y\theta, z\theta$ are in H'. Thus all of x, y, z are in H. Since $1 \in 1\theta^{-1} \subset H'\theta^{-1} = H$, H is a subloop of G. If H' is normal in G', $H\mathfrak{J}(G)\theta = H'\mathfrak{J}(G') \subset H'$, so $H\mathfrak{J}(G) \subset H$, H is normal in G.

(iii) Write $H' = H\theta$. If $x'y' = z'$ in G' and two of x', y', z' are in H' we pick the corresponding two of x, y, z in H to satisfy the appropriate equations from (*) $x\theta = x', y\theta = y', z\theta = z'$, and define the third by $xy = z$. Then all three of (*) hold. Since H' contains $1\theta = 1$, H' is a subloop of G'. If H is normal in G, then $H'\mathfrak{J}(G') = (H\mathfrak{J}(G))\theta \subset H\theta = H'$, so H' is normal in G'. Moreover, $H'\theta^{-1} = (H\theta)\theta^{-1} = HK = KH$, by (i), and the concluding statement follows by (ii).

Theorem 1.4. *The set of all normal subloops of the loop G is a commutative semigroup under the product operation AB.*

Proof. By Lemmas 1.5, 1.6 (iii).

Theorem 1.5. *Let H, K be subloops of the loop G such that K is normal in the subloop $\{H, K\}$ generated by H and K. Then:*

(i) $\{H, K\} = HK = KH$.

(ii) *$H \cap K$ is a normal subloop of H.*

(iii) *The identity mapping of H induces an isomorphism of HK/K upon $H/(H \cap K)$.*

Proof. Let θ be a homomorphism with kernel K of $\{H,K\}$ upon a loop. Then $(H\theta)\theta^{-1} = HK = KH$ is a subloop of $\{H, K\}$ containing H and K, so $\{H, K\} = HK$. Moreover, $(HK)\theta = H\theta$, so the identity mapping induces the isomorphism $hK \to h\theta$, $h \in H$, of $(HK)/K$ upon $H\theta$. Also θ induces a homomorphism, with kernel $H \cap K$, of H upon $H\theta$, and the identity mapping induces the isomorphism $h(H \cap K) \to h\theta$ of $H/(H \cap K)$ upon $H\theta$. Hence the mapping $hK \to h(H \cap K)$ is an isomorphism of HK/K upon $H/H \cap K$.

The inner mapping group $\mathfrak{J}(G)$ of a loop G has some of the functions of the inner automorphism group of a group — and reduces to the latter when G is a group. This suggests the definition: an *inner* automorphism of a loop G is an automorphism contained in $\mathfrak{J}(G)$. The question remains open as to when $\mathfrak{J}(G)$ consists entirely of automorphisms. A necessary condition is that G be *power-associative*; i.e., that each element generate an abelian subgroup. For this and other results see BRUCK and PAIGE [76]; also Chapter VII below. Marshall OSBORN (Ph. D. thesis, University of Chicago, 1957; not yet published) has given one of the results of BRUCK and PAIGE (loc. cit.) its definitive form by proving the following: *If G is a commutative and di-associative loop and if $\mathfrak{J}(G)$ consists of automorphisms, then G is Moufang.*

2. Loops with operators

By an operator-loop (G, Ω) we mean a pair consisting of a loop G and a set Ω (possibly empty) of endomorphisms of G. An Ω-subloop H of the operator-loop (G, Ω) is a subloop such that $H\,\Omega \subset H$.

A "universe" $\mathfrak{U}$ of operator-loops (G, Ω) may be defined recursively as follows. We start from a loop M and a set Ω of endomorphisms of M and assert the following:

(i) (M, Ω) is in $\mathfrak{U}$.

(ii) If (G, Ω) is in $\mathfrak{U}$ and if H is an Ω-subloop of G, each ω in Ω induces an endomorphism of H which we also denote by ω. Then (H, Ω) is in $\mathfrak{U}$.

(iii) If (G, Ω) is in $\mathfrak{U}$ and if N is a normal Ω-subloop of G, each ω in Ω induces an endomorphism $Nx \to N(x\omega)$ of G/N which we also denote by ω: $(Nx)\omega = N(x\omega)$. Then $(G/N, \Omega)$ is in $\mathfrak{U}$.

(iv) (G, Ω) is in $\mathfrak{U}$ only on the basis of (i), (ii), (iii).

An Ω-homomorphism θ of the operator loop (G, Ω) upon the operator-loop (H, Ω) is a homomorphism of G upon H such that $\theta\omega = \omega\theta$ for each ω in Ω. That is, $x\theta\omega = x\omega\theta$ for all x in G, ω in Ω.

Lemma 2.1. *Let θ be an Ω-homomorphism with kernel K, of the operator loop (G, Ω) upon the operator loop (G', Ω). Then:*

(i) *If H' is an (a normal) Ω-subloop of G', $H'\theta^{-1}$ is an (a normal) Ω-subloop of G.*

(ii) *If H is an (a normal) Ω-subloop of G, $H\theta$ is an (a normal) Ω-subloop of G' and HK is an (a normal) Ω-subloop of G.*

Proof. (i) If $H = H'\theta^{-1}$ then $(H\Omega)\theta = (H\theta)\Omega = H'\Omega \subset H'$, so $H\,\Omega \subset H$.

(ii) $(H\theta)\,\Omega = (H\,\Omega)\theta \subset H\theta$. In view of Lemma 1.6, this completes the proof of Lemma 2.1.

Lemma 2.2. (Dedekind's Law). *If A, B, C are subloops of the loop G with $A \subset B$, then $A(B \cap C) = B \cap AC$.*

Proof. If $x \in B \cap AC$ then $x = b = ac$ for a, b, c in A, B, C respectively. Since $A \subset B$, we deduce that $c \in B$. Thus $c \in B \cap C$ and $x = ac \in A(B \cap C)$. Hence $B \cap AC \subset A(B \cap C) \subset AB \cap AC \subset B \cap AC$, proving Lemma 2.2.

Lemma 2.3. *Let θ be an Ω-homomorphism, with kernel K, of the operator loop (G, Ω) upon the operator loop $(G\theta, \Omega)$. Let A be a normal Ω-subloop of the Ω-subloop B of G. Then:*

(i) *$A\theta$ is a normal Ω-subloop of $B\theta$.*

(ii) *$A(K \cap B)$ is a normal Ω-subloop of B.*

(iii) *θ induces an Ω-isomorphism of $B/(A(K \cap B))$ upon $B\theta/A\theta$.*

(iv) *KA is a normal Ω-subloop of the Ω-subloop KB.*

(v) *θ induces an Ω-isomorphism of $(KB)/(KA)$ upon $B\theta/A\theta$.*

(vi) *$(KB)/(KA)$ is Ω-isomorphic to $B/(A(K \cap B))$.*

Proof. (i) By Lemma 2.1 (i) with G replaced by B. (ii) Define the mapping φ of B by $b\varphi = (b\theta)(A\theta)$. Then φ is an Ω-homomorphism of B upon $B\theta/A\theta$ and $1\varphi^{-1} = B \cap (A\theta)\theta^{-1} = B \cap AK$. By Lemma 2.2, $B \cap AK = A(B \cap K)$. Hence $A(B \cap K)$ is a normal Ω-subloop of B (by Lemma 2.1 (i)). (iii) The natural isomorphism α: $(b(A(B \cap K))\alpha = b\varphi = (b\theta)(A\theta)$, is an Ω-isomorphism of $B/(A(B \cap K))$ upon $B\theta/A\theta$. (iv) By Lemma 2.1 (i), $(A\theta)\theta^{-1} = KA$ is a normal Ω-subloop of $(B\theta)\theta^{-1} = KB$. (v) Define ψ on KB by $x\psi = (x\theta)((KA)\theta) = (x\theta)(A\theta)$. Then ψ is an Ω-homomorphism of KB upon $B\theta/A\theta$ with kernel KA, and β: $(b(KA))\beta = (b\theta)(A\theta)$, is an Ω-isomorphism of $(KB)/(KA)$ upon $(B\theta)/A\theta$. (vi) α^{-1} is an Ω-isomorphism. Hence $\beta\alpha^{-1}$ is an Ω-isomorphism of $(KB)/(KA)$ upon $B/(A(K \cap B))$.

Theorem 2.1. (Zassenhaus' Lemma.) *Let (G, Ω) be an operator loop. Let K be a normal Ω-subloop of the Ω-subloop U of G and let L be a normal Ω-subloop of the Ω-subloop V of G. Then*

$$(K(U \cap V))/(K(U \cap L)) \quad \textit{and} \quad (L(U \cap V))/(L(V \cap K))$$

are Ω-isomorphic Ω-loops.

Proof. Clearly $A = U \cap L$, $B = U \cap V$ are Ω-subloops of U. The natural homomorphism of V upon V/L induces a homomorphism of B with kernel $B \cap L = A$. Therefore A is normal in B. Hence, by Lemma 2.3 (vi) with G replaced by U, $(KB)/(KA)$ is Ω-isomorphic to $B/(A(K \cap B))$. However, $K \cap B = K \cap U \cap V = V \cap K$. Hence

(i) $(KB)/(K(U \cap L))$ and $B/((U \cap L)(V \cap K))$ are Ω-isomorphic. Interchanging U and V, K and L in (i), we get

(ii) $(LB)/(L(V \cap K))$ and $B/((U \cap L)(V \cap K))$ are Ω-isomorphic.

Since the second terms in (i), (ii) are the same, the first are Ω-isomorphic. Since $B = U \cap V$, this completes the proof of Theorem 2.1.

The set $\mathfrak{E}$ of all endomorphisms of a loop G is a semigroup. $\mathfrak{E}$ has an identity I (the identity automorphism) and a *zero*, 0, defined by $x0 = 1$ for all x in G. The *centralizer*, $\mathfrak{C}$, of a subset Ω of $\mathfrak{E}$, is the set of all Ω-endomorphisms of G. If $\alpha, \beta \in \mathfrak{C}$, $\omega \in \Omega$, then $(\alpha\beta)\omega = \alpha\beta\omega = \alpha\omega\beta = \omega\alpha\beta = \omega(\alpha\beta)$, so $\mathfrak{C}$ is a subsemigroup of $\mathfrak{E}$, containing I and 0. Ω is said to be *irreducible* if the only Ω-subloops of G are G and 1. We now may state a famous lemma:

Schur's Lemma. *If Ω is an irreducible set of endomorphisms of a loop G, then the centralizer, $\mathfrak{C}$, of Ω in the semigroup of all endomorphisms of G, is a group with zero.*

Proof. Let $\alpha \in \mathfrak{C}$, $\alpha \neq 0$. Since $1\alpha^{-1}$ is an Ω-subloop of G distinct from G, $1\alpha^{-1} = 1$. Hence α is one-to-one. Since $G\alpha$ is an Ω-subloop of G distinct from 1, then $G\alpha = G$. Therefore α is an automorphism of G, and $\mathfrak{C}$ is a group with zero.

3. The refinement theorem

By an Ω-chain

$$G = G_0 \supset G_1 \supset \cdots \supset G_r = 1 \tag{3.1}$$

of an operator-loop (G, Ω) we mean a finite sequence of Ω-subloops G_i of G such that $G_0 = G$; G_i is a normal subloop of G_{i-1} for $i = 1, 2, \ldots, r$, and $G_r = 1$. The integer r is called the *length* of the chain. The chain is called a *chief* Ω-chain if each G_i is normal in G itself. The Ω-chain

$$G = H_0 \supset H_1 \supset \cdots \supset H_s = 1 \tag{3.2}$$

is a *refinement* of the Ω-chain (3.1) provided there exists a one-to-one mapping $i \to i'$ of the integers $0, 1, 2, \ldots, r$ into the integers $0, 1, 2, \ldots, s$ such that $G_i = H_{i'}$ for $i = 0, 1, \ldots, r$. The Ω-chains (3.1), (3.2) are *isomorphic* if $r = s$ and there exists a permutation $i \to i'$ of the integers $0, 1, 2, \ldots, r$ such that the quotient loops G_i/G_{i+1} and $H_{i'}/H_{i'+1}$ are Ω-isomorphic for $i = 0, \ldots, r-1$.

Theorem 3.1. (Schreier Refinement Theorem.) *Any two (chief) Ω-chains of the operator-loop (G, Ω) have isomorphic (chief) refinements.*

Proof. Given the chains (3.1), (3.2), define

$$G_{i,0} = G_{i-1}, \quad G_{i,j} = G_i(G_{i-1} \cap H_j),$$
$$H_{0,j} = G_{j-1}, \quad H_{i,j} = H_j(H_{j-1} \cap G_i)$$

for $i = 1, 2, \ldots, r$; $j = 1, 2, \ldots, s$. Then $G_{i,s} = G_i = G_{i+1,0}$, $H_{r,j} = H_j = H_{0,j+1}$ and $G_{r,s} = H_{r,s} = 1$. Also, by Theorem 2.1 with $U = G_{i-1}$, $V = H_{j-1}$, $K = G_i$, $L = H_j$; $G_{i,j-1}/G_{i,j}$ and $H_{i-1,j}/H_{i,j}$ are Ω-isomorphic operator loops. Therefore

$$G = G_{1,0} \supset G_{1,1} \supset \cdots \supset G_{1,s} \supset G_{2,1} \supset \cdots \supset G_{2,s} \supset G_{3,1} \supset \cdots \supset G_{r,s} = 1 \tag{3.3}$$

and

$$G = H_{0,1} \supset H_{1,1} \supset \cdots \supset H_{r,1} \supset H_{1,2} \supset \cdots \supset H_{r,2} \supset H_{1,3} \supset \cdots \supset H_{r,s} = 1 \tag{3.4}$$

are isomorphic refinements of (3.1), (3.2) respectively. If (3.1), (3.2) are chief, then, by Theorems 1.2, 1.4, so are (3.3), (3.4).

A strictly decreasing (chief) Ω-chain which has no strictly decreasing (chief) refinement aside from itself is called a (chief) *Ω-composition series.* Every refinement of such a chain may be made into an isomorphic chain by suitable deletions. Hence we have, as an immediate consequence of Theorem 3.1:

Theorem 3.2. (Jordan-Hölder Theorem.) *If the operator loop (G, Ω) possesses a (chief) Ω-composition series, every two (chief) Ω-composition series of (G, Ω) are isomorphic.*

The theory of (descending or ascending) transfinite chains — and, more generally, of normal systems — will be deferred until Chapter VI.

Theorems 3.1, 3.2 can be carried over to ascending transfinite chains, the proofs being essentially unchanged. (See KUROŠ, Theory of Groups.)

4. Normal endomorphisms

The set $\mathfrak{S}$ of all single-valued mappings of the loop G into itself is a semigroup under multiplication and a loop under the operation $(+)$ defined by $x(\theta + \varphi) = (x\theta)(x\varphi)$ for all x in G. The multiplicative identity, 1, is the identity mapping of G. The additive identity, 0, defined by $x0 = 1$ for all x in G, is the *zero endomorphism* of G. We have $\theta(\varphi + \psi) = \theta\varphi + \theta\psi$ for all θ, φ, ψ of $\mathfrak{S}$ and $(\varphi + \psi)\theta = \varphi\theta + \psi\theta$ when θ is an endomorphism of G. We call θ, φ *orthogonal* if $\theta\varphi = \varphi\theta = 0$. We call φ the *orthogonal complement* of θ provided $\theta + \varphi = 1$, $\theta\varphi = \varphi\theta = 0$; this implies that $\theta = \theta \cdot 1 = \theta^2 + \theta\varphi = \theta^2$. Thus a necessary (but not sufficient) condition that θ have an orthogonal complement is that θ be idempotent. If θ is an idempotent endomorphism and if $\theta + \varphi = 1$, then $\theta = \theta \cdot 1 = \theta + \theta\varphi$ and $\theta = 1 \cdot \theta = \theta + \varphi\theta$, so φ is the orthogonal complement of θ. Moreover, $\varphi = \varphi \cdot 1 = \varphi\theta + \varphi^2 = \varphi^2$, but φ may not be an endomorphism. However, if φ is also an endomorphism and if $\psi + \theta = 1$, then $\psi\theta = 0$ and $\psi\varphi = (\psi + \theta)\varphi = \varphi$, so $\psi = \psi(\theta + \varphi) = \varphi$. Thus: *if an endomorphism θ has an orthogonal complement φ which is an endomorphism, then θ is the orthogonal complement of φ.* An element α of $\mathfrak{S}$ will be called *centralizing* if $G\alpha$ is contained in the centre Z of G.

A loop word $W_n = W_n(X_1, \ldots, X_n)$ is an element of the free loop L_n with free generators $X_1, \ldots, X_n$. (Compare I.3.) A loop word W_n will be called *purely non-abelian* if, in each loop G, $W_n(c_1x_1, c_2x_2, \ldots, c_nx_n) = W_n(x_1, x_2, \ldots, x_n)$ for all x_i in G and c_i in the centre of G. And W_n will be called *normalized* if $W_n = 1$ in each case that an X_i is replaced by 1. An endomorphism θ of a loop G will be called a *normal endomorphism* of G provided

$$W_n(x_1, x_2, \ldots, x_n)\theta = W_n(x_1\theta, x_2, \ldots, x_n) = W_n(x_1, x_2\theta, \ldots, x_n) \qquad (4.1)$$
$$= \cdots = W_n(x_1, x_2, \ldots, x_n\theta)$$

for each choice of a positive integer n, a normalized purely non-abelian word W_n, and elements $x_1, x_2, \ldots, x_n$ of G. [If W_n is normalized purely non-abelian and if $F_n(X_1, \ldots, X_n) = W_n(X_{1'}, \ldots, X_{n'})$ for a permutation $i \to i'$ of $1, 2, \ldots, n$, then F_n is also normalized purely non-abelian. Hence (4.1) *can be replaced by* $W_n(x_1, \ldots, x_n)\theta = W_n(x_1\theta, \ldots, x_n)$.] By a *projection* of G we mean an idempotent normal endomorphism of G.

As a direct consequence of the definition, *every centralizing endomorphism of a loop G is normal.*

Lemma 4.1. *Let $\mathfrak{N}$ be the set of all normal endomorphisms of a loop G with commutator-associator subloop G'. Then $\mathfrak{N}$ is a multiplicative semigroup containing* 0 *and* 1. *Moreover:*

(i) *If θ is in $\mathfrak{N}$ and θ^{-1} exists, then θ^{-1} is in $\mathfrak{N}$.*

(ii) *Each θ in $\mathfrak{N}$ maps normal subloops of G upon normal subloops of G and centre elements upon centre elements. Indeed, each θ commutes with every inner mapping of G.*

(iii) *The restriction of $\mathfrak{N}$ to G' is a commutative semigroup. That is, $\mathfrak{N}$ maps G' into itself and*

$$a\theta\varphi = a\varphi\theta$$

for every a in G' and all θ, φ in $\mathfrak{N}$. Moreover,

$$a\theta^3 = a\theta$$

for every a in G', θ in $\mathfrak{N}$.

(iv) *A necessary and sufficient condition that the element θ of $\mathfrak{N}$ be centralizing is that $G'\theta = 1$.*

(v) *Let θ be in $\mathfrak{N}$ and let $\theta + \lambda = \mu$ for single-valued mappings λ, μ of G into G such that $\lambda\theta$ is centralizing. Then:*

(a) $\theta + \lambda = \lambda + \theta$

(b) *Both or neither of λ, μ is an endomorphism of G.*

(c) *Both or neither of λ, μ is in $\mathfrak{N}$.*

(vi) *Let θ be in $\mathfrak{N}$ and let λ, μ be single-valued mappings of G into G such that $\lambda\theta$ or $\mu\theta$ is centralizing. Then $(\theta + \lambda) + \mu = \theta + (\lambda + \mu)$, $(\lambda + \theta) + \mu = \lambda + (\theta + \mu)$, $(\lambda + \mu) + \theta = \lambda + (\mu + \theta)$.*

(vii) *A necessary and sufficient condition that the endomorphism θ of G have an orthogonal complement which is an endomorphism of G is that θ be a projection of G.*

Corollary 1. *The normal automorphisms of the loop G form a multiplicative group.*

Corollary 2. *The centralizing endomorphisms of the loop G form an ideal in the multiplicative semigroup $\mathfrak{N}$ of all normal endomorphisms of G and form an additive abelian group which is part of the centre of the additive loop of all single-valued mappings of G into G. They also form a ring.*

Proof. Corollary 1 is a consequence of (i); Corollary 2, of (ii)–(vii). We examine each of (i)—(vii) in turn.

(i) If W_n is normalized purely non-abelian and θ is a normal automorphism of G,

$$\begin{aligned} W_n(x_1\theta^{-1}, x_2, \ldots, x_n) &= W_n(x_1\theta^{-1}, x_2, \ldots, x_n)\theta\theta^{-1} \\ &= W_n(x_1\theta^{-1}\theta, x_2, \ldots, x_n)\theta^{-1} \\ &= W_n(x_1, x_2, \ldots, x_n)\theta^{-1} \end{aligned}$$

for all x_i in G. This is enough to establish (4.1) for θ^{-1}.

(ii) The loop words A_2, B_3, C_3, defined by

$$XY = Y[X \cdot A_2(X, Y)],$$

$$(XY)Z = [X \cdot B_3(X, Y, Z)](YZ), \quad Z(YX) = (ZY)[X \cdot C_3(X, Y, Z)],$$

are normalized and purely non-abelian. Moreover, if $T(y)$, $R(y, z)$, $L(y, z)$ are the generators, defined by (1.5), of the inner mapping group $\mathfrak{J} = \mathfrak{J}(G)$, then

$$xT(y) = x \cdot A_2(x, y),$$
$$xR(y, z) = x \cdot B_3(x, y, z), \quad xL(y, z) = x \cdot C_3(x, y, z)$$

for all x, y, z in G. If θ is in $\mathfrak{N}$, $xT(y)\theta = [x \cdot A_2(x, y)]\theta = (x\theta)[A_2(x, y)\theta] = (x\theta) \cdot A_2(x\theta, y) = x\theta T(y)$. Hence $T(y)\theta = \theta T(y)$ for every y in G. Similarly, $\alpha\theta = \theta\alpha$ for each generator α of $\mathfrak{J}$ and hence for every element α of $\mathfrak{J}$. In particular, if H is a normal subloop of G, $(H\theta)\mathfrak{J} = (H\mathfrak{J})\theta \subset H\theta$, showing that the subloop $H\theta$ is normal in G. If c is in the centre of G, $(c\theta)\mathfrak{J} = (c\mathfrak{J})\theta = c\theta$, so $c\theta$ is in the centre also.

(iii) The commutator-associator subloop G' of G (cf. I.2) is generated by the commutators (x, y) and associators (x, y, z), defined by

$$xy = (yx)(x, y), \quad (xy)z = [x(yz)](x, y, z), \tag{4.2}$$

for all x, y, z of G. Since (X, Y) and (X, Y, Z) are normalized purely non-abelian words, each θ in $\mathfrak{N}$ satisfies

$$(x, y)\theta = (x\theta, y) = (x, y\theta), \tag{4.3}$$
$$(x, y, z)\theta = (x\theta, y, z) = (x, y\theta, z) = (x, y, z\theta) \tag{4.4}$$

for all x, y, z in G. If θ, φ are in $\mathfrak{N}$, $(x, y)\theta\varphi = (x\theta, y)\varphi = (x\theta, y\varphi) = (x, y\varphi)\theta = (x, y)\varphi\theta$ and, similarly, $(x, y, z)\theta\varphi = (x\theta, y\varphi, z) = (x, y, z)\varphi\theta$. Moreover, $(x, y)\theta = (x\theta, y\theta)$, since θ is an endomorphism, and $(x\theta, y\theta) = (x, y)\theta^2$, since θ is normal. Hence $(x, y)\theta = (x, y)\theta^2 = (x, y)\theta^3$ and, similarly, $(x, y, z)\theta = (x\theta, y\theta, z\theta) = (x, y, z)\theta^3$. Since the commutators and associators generate G', we have $a\theta\varphi = a\varphi\theta$ and $a\theta = a\theta^3$ for every a in G'.

(iv) We note that an element c of G is in Z if and only if $(c, x) = (c, x, y) = (x, c, y) = (x, y, c) = 1$ for all x, y in G. Therefore, by (4.3), (4.4), the element θ of $\mathfrak{N}$ is centralizing if and only if $(x, y)\theta = (x, y, z)\theta = 1$ for all x, y, z of G; that is, if and only if $G'\theta = 1$.

(v) (a) Since θ is in $\mathfrak{N}$ and $\lambda\theta$ is centralizing, $(x\theta, x\lambda) = (x, x\lambda\theta) = 1$ for all x in G. Then $x(\theta + \lambda) = x\theta \cdot x\lambda = (x\lambda \cdot x\theta)(x\theta, x\lambda) = x\lambda \cdot x\theta = x(\lambda + \theta)$ for every x in G, so $\theta + \lambda = \lambda + \theta$.

(v) (b) For arbitrary x, y in G, set $z = (x\theta \cdot y\theta)(x\lambda \cdot y\lambda)$. Since $(x\theta, y\theta, x\lambda \cdot y\lambda) = (x, y\theta, x\lambda\theta \cdot y\lambda\theta) = 1$, $z = x\theta \cdot [y\theta \cdot (x\lambda \cdot y\lambda)]$. Since $(y\theta, x\lambda, y\lambda) = (y, x\lambda\theta, y\lambda) = 1$, $z = x\theta \cdot [(y\theta \cdot x\lambda) \cdot y\lambda]$. Since $(y\theta, x\lambda) = (y, x\lambda\theta) = 1$, $z = x\theta \cdot [(x\lambda \cdot y\theta) \cdot y\lambda]$. Since $(x\lambda, y\theta, y\lambda) = (x\lambda, y, y\lambda\theta) = 1$, $z = x\theta \cdot [x\lambda \cdot (y\theta \cdot y\lambda)]$. Since $(x\theta, x\lambda, y\theta \cdot y\lambda) = (x, x\lambda\theta, y\theta \cdot y\lambda) = 1$, $z = (x\theta \cdot x\lambda)(y\theta \cdot y\lambda)$. Comparing the first and last expressions for z, we have

$$(xy)\theta \cdot (x\lambda \cdot y\lambda) = x\mu \cdot y\mu \tag{4.5}$$

where $\mu = \theta + \lambda$. If λ is an endomorphism of G, the left hand side of (4.5) becomes $(xy)\theta \cdot (xy)\lambda = (xy)\mu$, and we deduce that μ is an endomorphism. If μ is an endomorphism of G, the right hand side of (4.5) becomes $(xy)\mu = (xy)\theta \cdot (xy)\lambda$, and we deduce that λ is an endomorphism.

(v) (c) Assume that λ, μ are endomorphisms. If W_n is a normalized purely non-abelian word, there exists a normalized purely non-abelian word $F = F_{n+1}$ such that

$$\begin{aligned} &W_n(xy, z_2, \ldots, z_n) \\ &= [W_n(x, z_2, \ldots, z_n) \cdot W_n(y, z_2, \ldots, z_n)] \cdot F(x, y, z_2, \ldots, z_n) \end{aligned}$$

for all $x, y, z_2, \ldots, z_n$ of G. Since $F(x\theta, x\lambda, z_2, \ldots, z_n) = F(x, x\lambda\theta, z_2, \ldots, z_n) = 1$, we replace x, y by $x\theta$, $x\lambda$ and deduce that

$$W_n(x\mu, z_2, \ldots, z_n) = W_n(x\theta, z_2, \ldots, z_n) \cdot W_n(x\lambda, z_2, \ldots, z_n) .$$

If one of λ, μ is normal and we write ν for the other, then, since θ is normal, we see, just as with (4.5), that $W_n(x, z_2, \ldots, z_n)\nu = W_n(x\nu, z_2, \ldots, z_n)$ for all $x, z_2, \ldots, z_n$. Hence ν is normal. That is, both or neither of λ, μ is in $\mathfrak{N}$.

(vi) Since one of $\lambda\theta$, $\mu\theta$ is centralizing, $(x\theta, x\lambda, x\mu) = (x, x\lambda\theta, x\mu)$ $= (x, x\lambda, x\mu\theta) = 1$ and thence $x[(\theta + \lambda) + \mu] = (x\theta \cdot x\lambda) \cdot x\mu$ $= x\theta \cdot (x\lambda \cdot x\mu) = x[\theta + (\lambda + \mu)]$ for all x in G. Therefore $(\theta + \lambda) + \mu$ $= \theta + (\lambda + \mu)$. Similarly for the other equations of (vi).

(vii) First let θ, φ be endomorphisms of G such that $\theta + \varphi = 1$, $\theta\varphi = \varphi\theta = 0$ and hence $\theta^2 = \theta$, $\varphi^2 = \varphi$. If W_n is any normalized purely non-abelian word,

$$\begin{aligned} W_n(x_1\theta, x_2, \ldots, x_n) &= W_n(x_1\theta, x_2, \ldots, x_n)\theta \cdot W_n(x_1\theta, x_2, \ldots, x_n)\varphi \\ &= W_n(x_1\theta, x_2\theta, \ldots, x_n\theta) \cdot W_n(1, x_2\varphi, \ldots, x_n\varphi) \\ &= W_n(x_1, x_2, \ldots, x_n)\theta . \end{aligned}$$

Hence θ is normal as well as idempotent, and therefore is a projection of G. Next let θ be a projection of G and φ a mapping defined by $\theta + \varphi = 1$. Since $\varphi\theta = 0$, φ is in $\mathfrak{N}$ by (v). This completes the proof of Lemma 4.1.

The following Lemmas are easy consequences of Lemma 4.1:

Lemma 4.2. *If $\theta_1, \theta_2, \ldots, \theta_n$ are normal endomorphisms of a loop G such that the product $\theta_i\theta_j$ is centralizing for all $i, j = 1, 2, \ldots, n$, $i \neq j$, then their sum is a normal endomorphism of G and is independent of ordering or bracketing.*

Lemma 4.3. *If θ, φ are normal endomorphisms of a loop G such that $\theta\varphi$ is centralizing and $\theta\varphi = \varphi\theta$, then, for every positive integer n, the binomial theorem is valid for the expansion of the normal endomorphism $(\theta + \varphi)^n$.*

An automorphism θ of the loop G will be called a *centre automorphism* provided that $x\theta = x \bmod Z$ for each x in G, where Z is the centre of G. Equivalently, $\theta = 1 + \lambda$ where λ is a centralizing endomorphism of G and θ is a one-to-one mapping of G upon G. In view of Lemma 4.1 (v), centre automorphisms are automatically normal.

The author will amplify the theory of normal endomorphisms in a paper to be published separately. Some of the results may be indicated briefly: (1) The present definition is equivalent to the usual one in the case of groups: an endomorphism θ of a *group* G is normal in the present sense if and only if θ commutes with every inner automorphism of G. (2) If θ is a normal endomorphism of a loop G, there exists a normal endomorphism φ of G such that $\theta + \varphi = \varphi + \theta = 1$. (3) Specht's Lemma (Wilhelm Specht, Gruppentheorie, p. 227, Satz 7) is valid for loops and has the following generalization: If α, β are normal endomorphisms of a loop G such that $x\alpha = x\beta$ for every x in the commutator-associator subloop G', there exists a centralizing endomorphism γ of G such that $\alpha = \beta + \gamma$. (4) Necessary and sufficient conditions (much more concrete than our definition) are derived for normality of an endomorphism. (5) Other (inequivalent) definitions of a normal endomorphism are explored and interrelated. — It is inconvenient to give further details concerning (4) and (5), since the theory of commutative Moufang loops (See Chapter VIII) plays a prominent rôle in the paper. In any case, the results just described will not be used in what follows.

5. Direct decomposition

The *outer direct product* (H, Ω) of a finite set of operator loops (H_i, Ω), $i = 1, 2, \ldots, n$, is defined as follows: H is the set of all ordered n-tuples $(h_1, h_2, \ldots, h_n)$, $h_i \in H_i$, with equality and multiplication defined componentwise and with $(h_1, h_2, \ldots, h_n)\,\omega = (h_1\omega, h_2\omega, \ldots, h_n\omega)$ for all h_i in H_i, ω in Ω. Thus (H, Ω) is an operator loop. If the H_i are Ω-subloops of an operator loop (G, Ω) we say that (G, Ω) is a *direct product* of the H_i and we write

$$G = H_1 \otimes H_2 \otimes \cdots \otimes H_n, \tag{5.1}$$

provided there exists an Ω-isomorphism α of (G, Ω) upon the outer direct product (H, Ω) of the H_i such that $h_i\alpha = (1, 1, \ldots, h_i, \ldots, 1)$ for each $i = 1, 2, \ldots, n$ and each h_i in H_i. If α exists, define the mappings θ_i of G upon H_i as follows: $x\theta_i = h_i$ if $x\alpha = (h_1, h_2, \ldots, h_n)$. Then $\theta_1, \theta_2, \ldots, \theta_n$ are idempotent Ω-endomorphisms of G and $\theta_i\theta_j = 0$ for all i, j $(i \neq j)$. Moreover,

$$1 = \theta_1 + \theta_2 + \cdots + \theta_n, \tag{5.2}$$

the order and association on the right of (5.2) being immaterial. Consequently, each θ_i is an Ω-projection of G. Conversely, given a set of

n pairwise orthogonal Ω-projections $\theta_1, \theta_2, \ldots, \theta_n$ of G which satisfy (5.2) in some (and hence in every) order and association, we define α by $x\alpha = (x\theta_1, x\theta_2, \ldots, x\theta_n)$ and see at once that (G, Ω) is the direct product of Ω-subloops $H_i = G\theta_i$. By Lemma 4.1 (ii), the H_i are normal subloops of G. The following lemma is easily proved:

Lemma 5.1. *Necessary and sufficient conditions that the operator loop (G, Ω) be a direct product of a set of* Ω-subloops $H_1, H_2, \ldots, H_n$ *are:* (a) H_i *is normal in* G for $i = 1, 2, \ldots, n$; (b) $H_1H_2 \ldots H_i \cap H_{i+1} = 1$ *for* $i = 1, 2, \ldots, n-1$; (c) $G = H_1H_2 \ldots H_n$.

In particular, G is a direct product of its Ω-subloops H, K if and only if H, K are normal in G, $H \cap K = 1$ and $G = HK$. This fact is often used to give a generalized definition in the following form. Let $\{H_i; i \in I\}$ be a set of Ω-subloops of a loop G, I being an arbitrary index set (finite or infinite.) For each i in I, let K_i be the subloop generated by the $H_j, j \neq i$. Then we say that G is a direct product of the subloops H_i, $i \in I$, if and only if, for each i in I, H_i and K_i are normal in G and satisfy $H_i \cap K_i = 1$, $G = H_iK_i$. Equivalently, the following conditions are to be met:

(a′) H_i is normal in G for each $i \in I$.

(b′) $H_i \cap K_i = 1$ for each $i \in I$.

(c′) G is generated by the set of subloops H_i, $i \in I$.

Conditions (a′), (c′) are of course direct generalizations of (a), (c) of Lemma 5.1. On the other hand, (b′) appears stronger than (b), as well as being more symmetric. Nevertheless, if I is finite and consists of the first n natural numbers, it is easily seen that (a′), (b′), (c′) are equivalent to (a), (b), (c).

In what follows we restrict attention to the theory of direct decomposition into finitely many factors. The material has been selected largely from Baer [58], [60, [61]. The present definition of normal endomorphisms (§ 4) allows some simplification in the proofs and also a closer parallel with classical group theory.

In the study of the direct decompositions of an operator loop (G, Ω) the centre Ω-automorphisms, and hence the centralizing Ω-endomorphisms, play an important rôle. If λ is a centralizing Ω-endomorphism, $G\lambda$ is an Ω-subloop of the centre $Z = Z(G)$. For this reason, the centre Z is displaced by the Ω-centre Z_Ω, namely the set-union of all Ω-subloops of G which are contained in Z. We note that Z_Ω is a normal Ω-subloop of G identical with the union of all Ω-subloops contained in Z. With this definition, an Ω-automorphism θ of G is a centre automorphism if and only if $x\theta \equiv x \bmod Z_\Omega$ for every x in G. It will be convenient to note that the commutator-associator subloop, G', of an operator loop (G, Ω), is an Ω-subloop; indeed, more generally, $(x, y)\theta = (x\theta, y\theta)$ and

$(x, y, z)\theta = (x\theta, y\theta, z\theta)$ for all x, y, z of G and for every homomorphism θ of G upon a loop.

Lemma 5.2. *Let (G, Ω) be an operator loop with commutator-associator subloop G', Ω-centre Z_Ω, such that either* $G = G'$ or $Z_\Omega = 1$. *Then:* (i) *the zero endomorphism is the only centralizing Ω-endomorphism of G;* (ii) *the identity mapping is the only centre Ω-automorphism of G.*

Proof. (i) Let λ be a centralizing Ω-endomorphism of G. If $G = G'$, then $G\lambda = G'\lambda = 1$. If $Z_\Omega = 1$, then $G\lambda = 1$. In either case, $\lambda = 0$. (ii) Let θ be a centre Ω-automorphism of G. Then $\theta = 1 + \lambda$ for a centralizing Ω-endomorphism λ of G. By (i), $\lambda = 0$. Hence $\theta = 1$.

If we assume the conclusion of Lemma 5.2 (ii), the theory of direct decomposition is particularly simple:

Lemma 5.3. *Let (G, Ω) be an operator loop whose only centre Ω-automorphism is the identity mapping. Then:* (a) *every two Ω-projections of G commute;* (b) *if $G = H_1 \otimes \cdots \otimes H_m = K_1 \otimes \cdots \otimes K_n$ for Ω-subloops H_i, K_j, and if $L(i, j) = H_i \cap K_j$, then $H_i = L(i, 1) \otimes \ldots \otimes L(i, n)$, $K_j = L(1, j) \otimes \ldots \otimes L(m, j)$ for $i = 1, 2, \ldots, m$; $j = 1, 2, \ldots, n$. Also G is the direct product of the mn Ω-subloops $L(i, j)$.*

Proof. (a) Let θ, φ be Ω-projections of G with orthogonal complements θ', φ' respectively. Thus $1 = \theta + \theta' = \theta' + \theta = \varphi + \varphi' = \varphi' + \varphi$ and $\theta\theta' = \theta'\theta = \varphi\varphi' = \varphi'\varphi = 0$. Set $\lambda = \theta\varphi\theta'$, $\mu = \theta\varphi'\theta'$. Then $\lambda\mu = \mu\lambda = 0$, $\lambda + \mu = \theta(\varphi + \varphi')\theta' = \theta\theta' = 0$. Moreover λ, μ are centralizing since, for example, $G'\lambda = G'\theta\varphi\theta' = G'\theta\theta'\varphi = 1$. Finally, $(1 + \lambda)(1 + \mu) = (1 + \lambda) + \mu = 1 + (\lambda + \mu) = 1$ and, similarly, $(1 + \mu)(1 + \lambda) = 1$. Thus $1 + \lambda$ is a centre Ω-automorphism, $1 + \lambda = 1$, $\lambda = 0$, $\theta\varphi\theta' = 0$. Hence $\theta\varphi = \theta\varphi(\theta + \theta') = \theta\varphi\theta$. Similarly, $\theta'\varphi\theta = 0$ and hence $\varphi\theta = (\theta + \theta')\varphi\theta = \theta\varphi\theta = \theta\varphi$.

(b) We can assume that $1 = \theta_1 + \cdots + \theta_m = \varphi_1 + \cdots + \varphi_n$ where the θ_i are pairwise orthogonal Ω-projections such that $H_i = G\theta_i$ and the φ_j are pairwise orthogonal Ω-projections such that $K_j = G\varphi_j$. By (a), the mn Ω-endomorphisms $\psi_{ij} = \theta_i\varphi_j = \varphi_j\theta_i$ are pairwise orthogonal idempotents; moreover $1 = (\Sigma\,\theta_i)\,(\Sigma\,\varphi_j) = \Sigma\,\psi_{ij}$, so each ψ_{ij} is an Ω-projection. Since $G\psi_{ij} = H_i\varphi_j = K_j\theta_i$, we have $G\psi_{ij} \subset K_j \cap H_i = L(i, j)$. On the other hand, if $x \in L(i, j)$ then $x\psi_{ij} = (x\theta_i)\varphi_j = x\varphi_j = x$, so $G\psi_{ij} = L(i,j)$. This is enough to prove (b).

Two Ω-subloops A, B of an operator loop (G, Ω) are called *centre Ω-isomorphic* if there exists a centre Ω-automorphism θ of G such that $A\theta = B$. This notion allows us to prove a weak form of "cancellation law" for direct decomposition:

Lemma 5.4. *If $G = H \otimes A = H \otimes B$ for Ω-subloops H, A, B of the operator loop (G, Ω), there exists a centre Ω-automorphism θ of G which induces the identity automorphism on H and maps A on B. In particular, A and B are centre Ω-isomorphic.*

Proof. Let $1 = \alpha + \alpha' = \beta' + \beta$ for Ω-projections $\alpha, \alpha', \beta, \beta'$ of G such that $A = G\alpha$, $B = G\beta$, $H = G\alpha' = G\beta'$. We shall show that $\theta = \alpha\beta + \alpha'\beta'$ has the properties stated. To begin with, $\alpha\alpha' = \alpha'\alpha = \beta\beta' = \beta'\beta = 0$. Since $G\alpha'\beta = H\beta = G\beta'\beta = 1$, then $\alpha'\beta = 0$. Similarly, $\beta'\alpha = 0$. We now verify readily that $\alpha\beta = \beta$, $\beta\alpha = \alpha$, $\alpha'\beta' = \alpha'$, $\beta'\alpha' = \beta'$. In particular, $\theta = \beta + \alpha'$. Since $\alpha'\beta = 0$ and β, α' are normal, we see that θ is a normal Ω-endomorphism. Similarly, $\varphi = \alpha + \beta'$ is a normal Ω-endomorphism. Moreover, $\theta\varphi = \beta\alpha + \alpha'\beta' = \alpha + \alpha' = 1$ and, similarly, $\varphi\theta = 1$, so θ is an Ω-automorphism. Again, $\alpha'\theta = \alpha'(\beta + \alpha') = \alpha'$, so θ induces the identity mapping on $H = G\alpha'$; and $\alpha\theta = \alpha(\beta + \alpha') = \beta$, so θ maps A upon B. Since $G'\beta\alpha' = G'\alpha'\beta = 1$, the normal Ω-endomorphism $\beta\alpha'$ is centralizing. Also, $\theta = \beta + \alpha' = \beta + (\beta + \beta')\alpha' = \beta + (\beta\alpha' + \beta') = (\beta + \beta') + \beta\alpha' = 1 + \beta\alpha'$, which shows that θ is a centre Ω-automorphism. This completes the proof of Lemma 5.4.

The next two lemmas can be generalized to wide classes of algebras (Jônsson and Tarski [15]):

Lemma 5.5. *If $G = A \otimes B$ for Ω-subloops A, B of the operator loop (G, Ω) and if S is an Ω-subloop of G containing A, then $S = A \otimes (B \cap S)$.*

Proof. Let $1 = \alpha + \beta$ for Ω-projections α, β such that $A = G\alpha$, $B = G\beta$. Since $A = A\alpha \subset S\alpha \subset A$, then $A = S\alpha$. If s is in S, then $s = (s\alpha)(s\beta)$ and $s\alpha \in S$, so $s\beta \in S$. Hence $S\beta \subset B \cap S = (B \cap S)\beta \subset S\beta$, so $B \cap S = S\beta$. Thus, by Lemma 5.1, $S = A \otimes (B \cap S)$.

Lemma 5.6. *Let the operator loop (G, Ω) have direct decompositions $G = H \otimes A = H \otimes B_1 \otimes \cdots \otimes B_n$ into Ω-subloops. Set $A_i = A \cap (H \otimes B_i)$. Then $A = A_1 \otimes \cdots \otimes A_n$ and $H \otimes B_i = H \otimes A_i$ for $i = 1, 2, \ldots, n$. Moreover, A_i is centre Ω-isomorphic to B_i* for $i = 1, 2, \ldots, n$.

Proof. For each $i = 1, 2, \ldots, n$, set $K_i = H \otimes B_i$ and let L_i be the direct product of the $B_j, j \neq i$. Since $G = H \otimes A$ and since K_i is an Ω-subloop of G containing H, Lemma 5.5 implies that $K_i = H \otimes (A \cap K_i) = H \otimes A_i$. Therefore $G = K_i \otimes L_i = (H \otimes L_i) \otimes B_i = (H \otimes L_i) \otimes A_i$. Hence, by Lemma 5.4, there exists, for each i, a centre Ω-automorphism θ_i of G which maps B_i on A_i and induces the identity mapping on $H \otimes L_i$. Thus θ_i induces the identity mapping on $K_j = H \otimes B_j = H \otimes A_j$ for each $j \neq i$. Consequently, the centre Ω-automorphism $\theta = \theta_1\theta_2 \ldots \theta_n$ maps $G = H \otimes B_1 \otimes \cdots \otimes B_n$ upon $G = H \otimes A_1 \otimes \cdots \otimes A_n$. Hence $G = H \otimes C$ where $C = A_1 \otimes \cdots \otimes A_n$. Since C is part of A, Lemma 5.5 yields $A = C \otimes (A \cap H)$; since $G = H \otimes A$, $A \cap H = 1$ and therefore $A = C = A_1 \otimes \cdots \otimes A_n$. This completes the proof of Lemma 5.6.

Consider two decompositions

$$(H): G = H_1 \otimes \ldots \otimes H_m, \quad (K): G = K_1 \otimes \ldots \otimes K_n$$

of the operator loop (G, Ω) into finitely many Ω-subloops. The decomposition (K) is a *refinement* of the decomposition (H) if each H_i is a direct product of (one or more of) the K_j. The decomposition (H) is *centre-isomorphic* to the decomposition (K) if $m = n$ and if there exist a centre Ω-automorphism θ of G and a permutation $i \to i^*$ of the integers $1, 2, \ldots, n$ such that θ maps H_i on K_{i^*} for $i = 1, 2, \ldots, n$. Centre-isomorphism of direct decompositions is clearly an equivalence relation. The decomposition (H) is said to be *exchange-isomorphic* to the decomposition (K) if $m = n$ and if there exists a permutation $i \to i^*$ of the integers $1, 2, \ldots, n$ such that

$$G = H_1 \otimes \cdots \otimes H_{i-1} \otimes K_{i^*} \otimes H_{i+1} \otimes \cdots \otimes H_n \tag{5.3}$$

for $i = 1, 2, \ldots, n$. In this case we say that the permutation $i \to i^*$ sets up an exchange-isomorphism of (H) upon (K). Exchange-isomorphism must be used with care, as the following examples show: (1) Let G be the (multiplicative) free abelian group without operators on three free generators u, v, w. Let H_1, H_2, H_3 be the subgroups generated by u, v, w respectively; K_1, K_2, K_3 be the subgroups generated by uvw, uv, uw respectively; L_1, L_2, L_3 be the subgroups H_2, H_3, H_1 respectively Then G is a direct product (H) of the H_i, (K) of the K_i and (L) of the L_i. The identity mapping of 1, 2, 3 induces an exchange-isomorphism of (H) upon (K) and of (K) upon (L) but (since $H_3 \subset K_1 \otimes K_2$ and $L_3 \subset H_1 \otimes H_2$) neither of (K) upon (H) nor of (H) upon (L). Thus neither the symmetric nor transitive laws of exchange-isomorphism are valid in their simplest forms. (2) Let G be the (multiplicative) free abelian group on two free generators u, v and let A, B, C, D be the subgroups generated by u, v, uv, u^2v^3 respectively. Then $G = A \otimes B = A \otimes C = B \otimes C = D \otimes C$, so each two successive direct decompositions are exchange isomorphic. However, since $D \otimes A \neq G$ and $D \otimes B \neq G$, neither of $G = A \otimes B$, $G = D \otimes C$ is exchange-isomorphic to the other. Thus there is no sense in which exchange-isomorphism is transitive for all loops. On the other hand, exchange-isomorphism always implies centre-isomorphism:

Lemma 5.7. *Let $(H): G = H_1 \otimes \cdots \otimes H_n$ and $(K): G = K_1 \otimes \cdots \otimes K_n$ be two direct decompositions of the operator loop (G, Ω) into n Ω-subloops. The decomposition (H) will be exchange-isomorphic to the decomposition (K) if (and only if) there exist a permutation $i \to i^*$ of the integers $1, 2, \ldots, n$ and n centre Ω-automorphisms $\theta_1, \theta_2, \ldots, \theta_n$ of G such that θ_i maps H_i on K_{i^*} and induces the identity mapping on H_j for $j \neq i$. If the conditions are satisfied there exists a unique centre Ω-automorphism of G which induces θ_i on H_i for $i = 1, 2, \ldots, n$. In particular, exchange isomorphism implies centre-isomorphism.*

Proof. If (H) is exchange-isomorphic to (K), so that (5.3) holds for each i, Lemma 5.4 ensures the existence of centre Ω-automorphisms θ_i with the properties stated. Conversely, if θ_i exists with the properties stated, then θ_i maps the direct decomposition (H) upon the direct decomposition (5.3). Now assume that the conditions are satisfied.

In view of (H), each x in H has a unique representation $x = h_1 h_2 \ldots h_n$, $h_i \in H_i$, in which the ordering and bracketing of factors is immaterial. If there exists an endomorphism θ which induces θ_i on H_i for each i, then, necessarily, $x\theta = (h_1\theta_1)(h_2\theta_2)\ldots(h_n\theta_n)$. Considering θ to be a mapping defined by the last equation we note, in view of (H), (K), that θ is an Ω-automorphism of G. Moreover, since $h_i\theta_i \equiv h_i \bmod Z_\Omega$ for each i, we have $x\theta \equiv h_1 h_2 \ldots h_n \equiv x \bmod Z_\Omega$. Hence θ is a centre Ω-automorphism and the proof of Lemma 5.7 is complete.

It is our purpose to examine hypotheses sufficient to ensure the truth of the following Refinement Theorem: *Every two direct decompositions* (H), (K) *of the operator loop* (G, Ω) *into finitely many direct* Ω*-factors possess refinements* (H′), (K′) *respectively such that* (H′) *is exchange-isomorphic to* (K′). Lemma 5.3 and the refinement theorem for Ω-chains (Theorem 3.1) both suggest conditions such as the following for all positive integers m, n:

$\mathfrak{E}(m, n)$. *Let* $(H): G = H_1 \otimes \cdots \otimes H_m$ *and* $(K): G = K_1 \otimes \cdots \otimes K_n$ *be two direct decompositions of the operator loop* (G, Ω) *into* Ω*-subloops. Then there exist* Ω*-subloops* $H(i, j)$, $K(j, i)$ *of* G *such that*

$$H_i = H(i, 1) \otimes \ldots \otimes H(i, n), \quad K_j = K(j, 1) \otimes \ldots \otimes K(j, m)$$

for $i = 1, 2, \ldots, m$; $j = 1, 2, \ldots, n$ *and such that, for each* $i = 1, 2, \ldots, m$, *and for each subset* J *of the integers* $1, 2, \ldots, n$, G *is the direct product of the* H_k *for* $k \neq i$, *the* $H(i, j)$ *for* j *in* J *and the* $K(j, i)$ *for* j *not in* J.

More simply, we would be content to insist that, for each ordered pair (i, j), $i = 1, 2, \ldots, m$; $j = 1, 2, \ldots, n$, G was the direct product of the H_k for $k \neq i$, the $H(i, p)$ for $p \neq j$, and $K(j, i)$. However, the stronger form is easier to handle and turns out to hold under weak conditions. That some conditions are certainly necessary has been demonstrated as follows: There exists an abelian group G (without operators) possessing direct decompositions without isomorphic refinements (Baer [B 1]). There exists an abelian operator group (G, Ω) satisfying the ascending chain condition on Ω-subgroups but possessing two non-isomorphic decompositions into direct-indecomposable Ω-subgroups (Krull [B 7]). There exists a nonabelian group satisfying the ascending chain condition on arbitrary subgroups but possessing non-isomorphic decompositions into two direct-indecomposable factors (Kuroš [B 8]).

It is an interesting open question as to whether $\mathfrak{E}(2, 2)$ implies $\mathfrak{E}(m, n)$ for all positive integers m, n. We shall now show that $\mathfrak{E}(m, n)$ is implied by a suitable use of the following strengthened form of $\mathfrak{E}(2, 2)$ introduced by Baer [61]:

$\mathfrak{B}$. If $G = A \otimes B = D \otimes E$ *are two direct decompositions of the operator loop* (G, Ω) *into two* Ω*-factors, then*

$$A = A_1 \otimes A_2, \quad B = B_1 \otimes B_2, \quad D = D_1 \otimes D_2, \quad E = E_1 \otimes E_2$$

for direct Ω-factors such that the members of the pairs (A_1, D_1), (B_1, E_1), (A_2, E_2), (B_2, D_2) are interchangeable in the following strict sense: $G = G_1 \otimes G_2$ *where*

$$G_1 = A_1 \otimes B_1 = D_1 \otimes B_1 = A_1 \otimes E_1 = D_1 \otimes E_1 , \qquad (5.4)$$

$$G_2 = A_2 \otimes B_2 = E_2 \otimes B_2 = A_2 \otimes D_2 = E_2 \otimes D_2 .$$

In $\mathfrak{B}$, G_i may be regarded as defined by any one of the products to which it is equated. To compare $\mathfrak{B}$ with $\mathfrak{E}(2, 2)$, take $H_1 = A$, $H_2 = B$, $K_1 = D$, $K_2 = E$, $H(1, 1) = A_1$, $H(2, 2) = B_1$, $K(1, 1) = D_1$, $K(2, 2) = E_1$, and so on. Thus $\mathfrak{B}$ asserts, in particular, that G can be written in 16 ways as a direct product of 4 factors consisting of one factor from each of the 4 pairs $H(i, j)$, $K(j, i)$.

Theorem 5.1. (Refinement Theorem.) *If every direct Ω-factor of an operator loop (G, Ω) satisfies $\mathfrak{B}$, then (G, Ω) (and every direct Ω-factor of (G, Ω)) satisfies $\mathfrak{E}(m, n)$ for all positive integers m, n.*

Proof. We begin by noting that $\mathfrak{E}(1, n)$ and $\mathfrak{E}(m, 1)$ are trivial. Next we consider certain implications which do not involve the use of $\mathfrak{B}$.

(i) *If $m > 1$, $\mathfrak{E}(m, n)$ implies $\mathfrak{E}(m - 1, n)$.* To see this, assume $H_m = 1$ in the statement of $\mathfrak{E}(m, n)$. Then certainly $H(m, j) = 1$ for all j. Since $\mathfrak{E}(m, n)$ implies an exchange-isomorphism, we see from Lemma 5.7 that $K(j, m) = 1$ for all j. Thus we delete the trivial direct factors H_m, $H(m, j)$, $K(j, m)$ and get $\mathfrak{E}(m - 1, n)$.

A more difficult step is to prove a converse to (i) for $m > 2$:

(ii) *If $m > 2$, $\mathfrak{E}(m - 1, n)$ implies $\mathfrak{E}(m, n)$.* Strictly speaking, we assume $\mathfrak{E}(m - 1, n)$ for every direct Ω-factor of G. By (i), since $m - 1 \geqq 2$, $\mathfrak{E}(m - 1, n)$ implies $\mathfrak{E}(2, n)$. Let $G = H_1 \otimes \cdots \otimes H_m = K_1 \otimes \cdots \otimes K_n$ for Ω-subloops H_i, K_j, and define $L_1 = H_1 \otimes \cdots \otimes H_{m-1}$, $L_2 = H_m$. Applying $\mathfrak{E}(2, n)$ to the direct decompositions $G = L_1 \otimes L_2 = K_1 \otimes \cdots \otimes K_n$, we deduce the existence of direct Ω-decompositions $L_i = L(i, 1) \otimes \cdots \otimes L(i, n)$, $K_j = M(j, 1) \otimes M(j, 2)$ for $i = 1, 2; j = 1, 2, \ldots, n$ such that:

(a) *For $i = 1$ or 2 and for each subset J of the integers $1, 2, \ldots, n$, G is the direct product of the L_k with $k \neq i$, the $L(i, j)$ for j in J and the $M(j, i)$ for j not in J.*

Quoting (a) for $i = 1$ and for J the empty set, we get

$$G = L_2 \otimes M(1, 1) \otimes \cdots \otimes M(n, 1) . \qquad (5.5)$$

Since also $G = L_2 \otimes L_1$, we deduce from (5.5) and Lemma 5.6 that $L_1 = P_1 \otimes \cdots \otimes P_n$ and that $L_2 \otimes P_j = L_2 \otimes M(j, 1)$ for $j = 1, 2, \ldots, n$, where $P_j = [L_2 \otimes M(j, 1)] \cap L_1$. Since L_1 is a direct Ω-factor of G we may apply $\mathfrak{E}(m - 1, n)$ to the direct Ω-decompositions $L_1 = H_1 \otimes \cdots \otimes H_{m-1} = P_1 \otimes \cdots \otimes P_n$. Thus there exist direct Ω-decompositions $H_i = H(i, 1) \otimes \cdots \otimes H(i, n)$, $P_j = P(j, 1) \otimes \cdots \otimes P(j, m - 1)$ for $i = 1, 2, \ldots, m - 1$; $j = 1, 2, \ldots, n$ such that

(b) *For each* $i = 1, 2, \ldots, m-1$ *and for each subset* J *of the integers* $1, 2, \ldots, n$, L_1 *is the direct product of the* K_k *with* $k \neq i$, $k < m$, *the* $H(i, j)$ *with* j *in* J *and the* $P(j, i)$ *with* j *not in* J.

Since $L_2 \otimes M(j, 1) = L_2 \otimes P_j = L_2 \otimes P(j, 1) \otimes \cdots \otimes P(j, m-1)$, we deduce from Lemma 5.6 that $M(j, 1) = K(j, 1) \otimes \cdots \otimes K(j, m-1)$ and that $L_2 \otimes P(j, i) = L_2 \otimes K(j, i)$ for $i = 1, 2, \ldots, m-1$; $j = 1, 2, \ldots, n$, where $K(j, i) = [L_2 \otimes P(j, i)] \cap M(j, 1)$. Finally, we define $K(j, m) = M(j, 2)$ and $H(m, j) = L(2, j)$ for $j = 1, 2, \ldots, n$. Then $H_i = H(i, 1) \otimes \cdots \otimes H(i, n)$, $K_j = K(j, 1) \otimes \cdots \otimes K(j, m)$ for $i = 1, 2, \ldots, m$; $j = 1, 2, \ldots, n$, and we wish to prove:

(c) *For each* $i = 1, 2, \ldots, m$ *and for each subset* J *of the integers* $1, 2, \ldots, n$, G *is the direct product of the* H_k *with* $k \neq i$, *the* $H(i, j)$ *with* j *in* J *and the* $K(j, i)$ *with* j *not in* J.

In proving (c) we may assume without loss of generality that J consists of the integers $1, 2, \ldots, k$ for some $k \leq n$. Moreover we may restrict attention to the cases $i = 1$, $i = m$.

Case 1. $i = 1$. Set $T = H_2 \otimes \cdots \otimes H_{m-1}$, so that $L_1 = H_1 \otimes T$. Since $G = L_2 \otimes L_1$, we apply (b) with $i = 1$ and get

$$G = L_2 \otimes T \otimes H(1, 1) \otimes \cdots \otimes H(1, k) \otimes P(k+1, 1) \otimes \cdots \otimes P(n, 1).$$

Since $L_2 \otimes P(j, 1) = L_2 \otimes K(j, 1)$, we may replace each $P(j, 1)$ by a $K(j, 1)$. And since $L_2 = H_m$, $L_2 \otimes T$ is the direct product of the H_p for $p \neq 1$. Thus we have (c) for $i = 1$.

Case 2. $i = m$. Applying (a) with $i = 2$, we get

$$G = L_1 \otimes L(2, 1) \otimes \cdots \otimes L(2, k) \otimes M(k+1, 2) \otimes \cdots \otimes M(n, 2).$$

Since L_1 is the product of the H_p with $p \neq m$ and since $L(2, j) = H(m, j)$, $M(j, 2) = K(j, m)$ for each j, we have proved (c) for $i = m$.

This completes the proof that $\mathfrak{E}(m, n)$ holds for G. Since direct Ω-factors of direct Ω-factors of G are thenselves direct Ω-factors of G, $\mathfrak{E}(m, n)$ holds for every direct Ω-factor of G. Hence we have proved (ii).

Now we come to the crucial use of $\mathfrak{B}$.

(iii) *If* $n > 1$, $\mathfrak{B}$ *and* $\mathfrak{E}(2, n-1)$ *imply* $\mathfrak{E}(2, n)$. As in (ii), we assume $\mathfrak{B}$ and $\mathfrak{E}(2, n-1)$ for each direct Ω-factor of G. Suppose that $G = H_1 \otimes H_2 = K_1 \otimes \cdots \otimes K_n$ for Ω-subloops H_i, K_j of G and set $A = K_1 \otimes \cdots \otimes K_{n-1}$, $B = K_n$, $D = H_1$, $E = H_2$. Applying $\mathfrak{B}$ to the direct decompositions $G = A \otimes B = D \otimes E$, we deduce the existence of direct Ω-decompositions $G = G_1 \otimes G_2$, $A = A_1 \otimes A_2$, $B = B_1 \otimes B_2$, $D = D_1 \otimes D_2$, $E = E_1 \otimes E_2$ such that (5.4) holds. Since $A = A_1 \otimes A_2 = K_1 \otimes \cdots \otimes K_{n-1}$ is a direct Ω-factor of G, we apply $\mathfrak{E}(2, n-1)$ and get direct Ω-decompositions $A_i = A(i, 1) \otimes \cdots \otimes A(i, n-1)$, $K_j = K(j, 1) \otimes K(j, 2)$, for $i = 1, 2$; $j = 1, 2, \ldots, n-1$, such that:

(a) *For* $i = 1$ *or* 2 *and for each subset* J *of the integers* $1, 2, \ldots, n-1$, A *is the direct product of the* A_k *with* $k \neq i$, *the* $A(i, j)$ *with* j *in* J *and the* $K(j, i)$ *with* j *not in* J ($j < n$).

Since $G_1 = E_1 \otimes D_1 = E_1 \otimes A_1 = E_1 \otimes A(1,1) \otimes \cdots \otimes A(1, n-1)$, we see from Lemma 5.6 that $D_1 = H(1,1) \otimes \cdots \otimes H(1, n-1)$ and that $E_1 \otimes H(1,j) = E_1 \otimes A(1,j)$ for $j < n$, where $H(1,j) = [E_1 \otimes A(1,j)] \cap D_1$. Similarly, since $G_2 = D_2 \otimes E_2 = D_2 \otimes A_2 = D_2 \otimes A(2,1) \otimes \cdots \otimes A(2, n-1)$, we have $E_2 = H(2,1) \otimes \cdots \otimes H(2, n-1)$ and $D_2 \otimes H(2,j) = D_2 \otimes A(2,j)$ for $j < n$, where $H(2,j) = [D_2 \otimes A(2,j)] \cap E_2$. In addition, we define $H(1,n) = D_2$, $H(2,n) = E_1$, $K(n,1) = B_2$, $K(n,2) = B_1$. Then $H_i = H(i,1) \otimes \cdots \otimes H(i,n)$, $K_j = K(j,1) \otimes K(j,2)$ for $i = 1, 2$; $j = 1, 2, \ldots, n$ and we are to prove:

(b) *For $i = 1$ or 2 and for each subset J of the integers $1, 2, \ldots, n$, G is the direct product of the H_k with $k \neq i$, the $H(i,j)$ with j in J and the $K(j,i)$ with j not in J.*

Clearly it will suffice to prove (b) for $i = 1$. We note that $H_2 = E$. We may assume without loss of generality that $J = J'$ or $J = J' \cup n$ where J' consists of $1, 2, \ldots, k$ for some integer k with $k < n$. Since $G = G_1 \otimes G_2 = A_1 \otimes E_1 \otimes G_2$ and since, by (a) with J replaced by J',

$$A_1 \otimes G_2 = A_1 \otimes A_2 \otimes B_2$$

$= B_2 \otimes A_2 \otimes A(1,1) \otimes \cdots \otimes A(1,k) \otimes K(k+1,1) \otimes \cdots \otimes K(n-1,1)$, we have

$G = G_2 \otimes E_1 \otimes A(1,1) \otimes \cdots \otimes A(1,k) \otimes K(k+1,1) \otimes \cdots \otimes K(n-1,1)$.
In the last equation, since $E_1 \otimes A(1,j) = E_1 \otimes H(1,j)$ for $j < n$, each $A(1,j)$ can be replaced by $H(1,j)$. Moreover, if $P = B_2$ or D_2, we have $G_2 \otimes E_1 = E_2 \otimes P \otimes E_1 = E \otimes P = H_2 \otimes P$. Hence
$G = H_2 \otimes P \otimes H(1,1) \otimes \cdots \otimes H(1,k) \otimes K(k+1,1) \otimes \cdots \otimes K(n-1,1)$.
If $J = J'$ we take $P = B_2 = K(n,1)$; if $J = J' \cup n$ we take $P = D_2 = H(1,n)$. This completes the proof of (b). Consequently, $E(2,n)$ holds for G and, as in the proof of (ii), for every direct Ω-factor of G. The proof of (iii) is complete.

Now the proof of Theorem 5.1 follows easily: Assuming $\mathfrak{B}$, we infer from (iii) that $\mathfrak{E}(2,n)$ holds for all n; thence we infer from (ii) that $\mathfrak{E}(m,n)$ holds for all m, n.

In view of Theorem 5.1 it becomes of interest to study conditions under which $\mathfrak{B}$ is valid. First we need a lemma:

Lemma 5.8. *Let $G = S \otimes T$ for Ω-subloops of the operator loop (G, Ω). Then:* (i) $G' = S' \otimes T'$; (ii) $Z_\Omega(G) = Z_\Omega(S) \otimes Z_\Omega(T)$.

Proof. Let $1 = \alpha + \beta$, $S = G\alpha$, $T = G\beta$ for Ω-projections α, β of G. Clearly $G'\alpha = S'$, $G'\beta = T'$. Hence $G' = S' \otimes T'$, proving (i). Again, from the definition of centre, $Z(G) = Z(S) \otimes Z(T)$. In particular, $Z_\Omega(S) \subset Z(S) \subset Z(G)$, so $Z_\Omega(S)$ is an Ω-subloop of $Z(G)$ and hence $Z_\Omega(S) \subset Z_\Omega(G)$. Similarly, $Z_\Omega(T) \subset Z_\Omega(G)$. On the other hand, $Z_\Omega(G)\alpha$ is an Ω-subloop of G contained in $Z(G) \cap S = Z(S)$, so $Z_\Omega(G)\alpha \subset Z_\Omega(S)$. Similarly, $Z_\Omega(G)\beta \subset Z_\Omega(T)$. Thus $Z_\Omega(G) \subset Z_\Omega(S) Z_\Omega(T) \subset Z_\Omega(G)$, so $Z_\Omega(G) = Z_\Omega(S) Z_\Omega(T) = Z_\Omega(S) \otimes Z_\Omega(T)$. This proves (ii).

Theorem 5.2. *Let (G, Ω) be an operator loop with Ω-centre Z_Ω. Then the following statements are equivalent.*

(i) *Every direct Ω-factor of G satisfies $\mathfrak{B}$.*

(ii) *For every direct Ω-factor S of G, the Ω-factor $Z_\Omega(S)$ of Z_Ω satisfies $\mathfrak{B}$.*

Corollary. *If every direct Ω-factor of Z_Ω satisfies $\mathfrak{B}$, then every direct Ω-factor of G satisfies $\mathfrak{B}$.*

Proof. That (i) implies (ii) is a direct consequence of Lemma 5.8 (ii). In view of Lemma 5.8, if we assume that $Z_\Omega = Z_\Omega(G)$ satisfies $\mathfrak{B}$ and prove that G satisfies $\mathfrak{B}$, the proof that (ii) implies (i) will be complete. Henceforth we assume that Z_Ω satisfies $\mathfrak{B}$. Let $A = G\alpha$, $B = G\beta$, $D = G\delta$, $E = G\varepsilon$ for Ω-projections α, β, δ, ε of G such that $1 = \alpha + \beta = \delta + \varepsilon$. By Lemma 5.8, $Z_\Omega = Z_\Omega(A) \otimes Z_\Omega(B) = Z_\Omega(D) \otimes Z_\Omega(E)$. Since Z_Ω satisfies $\mathfrak{B}$,

$$Z_\Omega(A) = A^* \otimes A^{**}, \quad Z_\Omega(B) = B^* \otimes B^{**}, \qquad (5.6)$$
$$Z_\Omega(D) = D^* \otimes D^{**}, \quad Z_\Omega(E) = E^* \otimes E^{**}$$

and $Z_\Omega = Z^* \otimes Z^{**}$ for Ω-subloops such that

$$Z^* = A^* \otimes B^* = D^* \otimes B^* = A^* \otimes E^* = D^* \otimes E^*, \qquad (5.7)$$
$$Z^{**} = A^{**} \otimes B^{**} = E^{**} \otimes B^{**} = A^{**} \otimes D^{**} = E^{**} \otimes D^{**}.$$

Since $Z^*\alpha = A^* = D^*\alpha$, we see that α maps D^* upon A^*. If $d^*\alpha = 1$ for d^* in D^*, then $d^* = d^*(\alpha + \beta) = d^*\beta \in D^* \cap Z^*\beta = D^* \cap B^* = 1$. Hence, and similarly:

$$\begin{aligned}
&\alpha \text{ induces isomorphisms of } D^* \text{ upon } A^* \text{ and } E^{**} \text{ upon } A^{**}, \\
&\beta \text{ induces isomorphisms of } E^* \text{ upon } B^* \text{ and } D^{**} \text{ upon } B^{**}, \qquad (5.8)\\
&\delta \text{ induces isomorphisms of } A^* \text{ upon } D^* \text{ and } B^{**} \text{ upon } D^{**}, \\
&\varepsilon \text{ induces isomorphisms of } B^* \text{ upon } E^* \text{ and } A^{**} \text{ upon } E^{**}.
\end{aligned}$$

Now we define the following Ω-subloops of G:

$$\begin{aligned}
A_1 &= A \cap (D \otimes E^*), & A_2 &= A \cap (E \otimes D^{**}), \\
B_1 &= B \cap (E \otimes D^*), & B_2 &= B \cap (D \otimes E^{**}), \qquad (5.9)\\
D_1 &= D \cap (A \otimes B^*), & D_2 &= D \cap (B \otimes A^{**}), \\
E_1 &= E \cap (B \otimes A^*), & E_2 &= E \cap (A \otimes B^{**}).
\end{aligned}$$

If a is in $A_1 \cap A_2$ then $a = (a\delta)(a\varepsilon)$ is in $E^* \otimes D^{**}$. Since $a\beta = 1$ and since, by (5.8), β induces an isomorphism of $E^* \otimes D^{**}$, then $a = 1$. Hence and similarly,

$$A_1 \cap A_2 = B_1 \cap B_2 = D_1 \cap D_2 = E_1 \cap E_2 = 1. \qquad (5.10)$$

Since $G'\alpha\delta\beta = G'\alpha\beta\delta = 1$, $\alpha\delta\beta$ is centralizing. Consequently, if a is in A, then $a\delta\beta = a\alpha\delta\beta$ is in $B \cap Z_\Omega = Z_\Omega(B) = B^* \otimes B^{**} = (E^* \otimes D^{**})\beta$. Hence $a\delta = a_1(e^* d^{**})$ for unique a_1, e^*, d^{**} in A, E^*, D^{**} respectively.

Then $1 = a\delta\varepsilon = (a_1\varepsilon)e^*$, so $a_1\varepsilon$ is in E^* and $a_1 = (a_1\delta)(a_1\varepsilon)$ is in $A \cap (D \otimes E^*) = A_1$. If $a = a_1 a_2$ then a_2 is in A and $a_1\delta \cdot a_2\delta = a\delta = (a\delta)\delta = [a_1(e^* d^{**})]\delta = (a_1\delta)d^{**}$. Hence $a_2\delta = d^{**}$ and $a_2 = (a_2\delta)(a_2\varepsilon)$ is in $A \cap (E \otimes D^{**}) = A_2$. Thus $A \subset A_1 A_2 \subset A$ so $A = A_1 A_2 = A_1 \otimes A_2$. Similarly

$$A = A_1 \otimes A_2, \quad B = B_1 \otimes B_2, \quad D = D_1 \otimes D_2, \quad E = E_1 \otimes E_2. \tag{5.11}$$

By (5.7), (5.8), we have $A^* \subset A \cap (A^* D) \subset A \cap (Z^* D) \subset A \cap [(D^* \otimes E^*)D] = A \cap (D \otimes E^*) = A_1$. If a_1 is in A_1 then $a_1 = de^*$ for d in D, e^* in E^*. Since $1 = a_1\beta = (d\beta)(e^*\beta)$ and since $e^*\beta$ is in B^*, then $d\beta$ is in B^* and $d = (d\alpha)(d\beta)$ is in $D \cap (A \otimes B^*) = D_1$. Therefore a_1 is in $D_1 \otimes E^*$. Similarly,

$$\begin{aligned}
&A^* \subset A_1 \subset D_1 \otimes E^* \subset D_1 Z^*, \quad A^{**} \subset A_2 \subset E_2 \otimes D^{**} \subset E_2 Z^{**},\\
&B^* \subset B_1 \subset E_1 \otimes D^* \subset E_1 Z^*, \quad B^{**} \subset B_2 \subset D_2 \otimes E^{**} \subset D_2 Z^{**},\\
&D^* \subset D_1 \subset A_1 \otimes B^* \subset A_1 Z^*, \quad D^{**} \subset D_2 \subset B_2 \otimes A^{**} \subset B_2 Z^{**},\\
&E^* \subset E_1 \subset B_1 \otimes A^* \subset B_1 Z^*, \quad E^{**} \subset E_2 \subset A_2 \otimes B^{**} \subset A_2 Z^{**}.
\end{aligned} \tag{5.12}$$

Since $A_1 \cap B_1 \subset A \cap B = 1$, then $A_1 \cap B_1 = 1$. Similarly, $D_1 \cap E_1 = 1$. By (5.12), $D_1 \cap B_1 = [D_1 \cap (A_1 \otimes B^*)] \cap [B_1 \cap (E_1 \otimes D^*)] = [B_1 \cap (A_1 \otimes B^*)] \cap [D_1 \cap (E_1 \otimes D^*)] = B^* \cap D^* = 1$. Similarly, $A_1 \cap E_1 = 1$. Then, by (5.12) again, $A_1 \otimes B_1 \subset D_1 Z^* B_1 \subset D_1 \otimes B_1 \subset A_1 Z^* B_1 \subset A_1 \otimes B_1 \subset A_1 E_1 Z^* \subset A_1 \otimes E_1 \subset D_1 Z^* E_1 \subset D_1 \otimes E_1 \subset A_1 Z^* B_1 Z^* \subset A_1 \otimes B_1$. Thus, and similarly,

$$\begin{aligned}
&A_1 \otimes B_1 = D_1 \otimes B_1 = A_1 \otimes E_1 = D_1 \otimes E_1,\\
&A_2 \otimes B_2 = E_2 \otimes B_2 = A_2 \otimes D_2 = E_2 \otimes D_2.
\end{aligned} \tag{5.13}$$

In view of (5.13), G satisfies $\mathfrak{B}$. And, with this, the proof of Theorem 5.2 is complete.

Theorems 5.1, 5.2 suggest that our next concern should be with the validity of $\mathfrak{B}$ in abelian operator groups. Actually we proceed with a little more generality. As before, $K(\theta)$ denotes the kernel of the endomorphism θ of the operator loop (G, Ω). The *radical*, $R(\theta)$, of θ, is defined to be the union of the kernels $K(\theta^i)$, $i = 1, 2, \ldots$. Clearly $R(\theta)$ is a normal subloop of G. If θ is an Ω-endomorphism, then $R(\theta)$ is an Ω-subloop of G. In what follows we impose ascending and descending chain conditions on the Ω-subloops of the abelian group G/G'.

Lemma 5.9. (Fitting's Lemma.) *Let (G, Ω) be an operator loop with commutator-associator subloop G', such that G/G' satisfies the ascending and descending chain conditions for Ω-subloops. If θ is a normal Ω-endomorphism of G, there exists a unique normal Ω-subloop, $C(\theta)$, of G such that θ induces an automorphism of $C(\theta)$ and*

$$G = C(\theta) \otimes R(\theta). \tag{5.14}$$

Proof. By Lemma 4.1 (iii), $a\theta = a\theta^3$ for every a in G'. If $a \in G' \cap R(\theta)$, then $a\theta = a\theta(\theta^2)^i = 1$ for a suitable positive integer i. Hence

$$G' \cap R(\theta) = G' \cap K(\theta) . \tag{5.15}$$

If $a \in G'$ and $a\theta^2 = ab$, then $b \in G'$ and $a\theta = a\theta^3 = (ab)\theta = a\theta \cdot b\theta$, so $b\theta = 1$, $b \in G' \cap K(\theta)$. Thus $G'\theta^2 \cdot K(\theta) = G' \cdot K(\theta)$. If i, j are positive integers, with $i < 2j$, then $G'K(\theta) \supset G'\theta^i \cdot K(\theta) \supset G'\theta^{2j} \cdot K(\theta) = G'\theta^2 \cdot K(\theta) = G' \cdot K(\theta)$. Hence

$$G'\theta^i \cdot K(\theta) = G'K(\theta) \supset G' \tag{5.16}$$

for every positive integer i. By (5.16), the descending chain

(D) $$GK(\theta) \supset G\theta \cdot K(\theta) \supset G\theta^2 \cdot K(\theta) \supset \cdots$$

of Ω-subloops is essentially a chain for G/G' and therefore eventually becomes constant. However, if $G\theta^s \cdot K(\theta) = G\theta^{s+1} \cdot K(\theta)$, then $G\theta^{s+1} = G\theta^{s+2}$. That is, the chain $G \supset G\theta \supset G\theta^2 \supset \cdots$, eventually becomes constant. Again, the chain

(A) $$G' \subset G'K(\theta) \subset G'K(\theta^2) \subset G'K(\theta^3) \subset \cdots$$

eventually becomes constant. However, if $G'K(\theta^t) = G'K(\theta^{t+1})$ for positive t, then, by (5.15), $K(\theta^{t+1}) = K(\theta^t)\,[G' \cap K(\theta^{t+1})] = K(\theta^t)\,[G' \cap K(\theta)] = K(\theta^t)$. That is, the chain $K(\theta) \subset K(\theta^2) \subset K(\theta^3) \subset \cdots$, eventually becomes constant. Consequently there exists a positive integer n such that $P = G\theta^n = G\theta^{n+1} = \cdots$ and $R(\theta) = K(\theta^n) = K(\theta^{n+1}) = \cdots$. Clearly $P\theta = P$ and $P \cap K(\theta) \subset P \cap R(\theta)$. If p is in $P \cap R(\theta)$ then $p = x\theta^n$ for x in G such that $x\theta^{2n} = 1$. Hence $x \in R(\theta) = K(\theta^n)$, so $p = x\theta^n = 1$. Therefore $P \cap R(\theta) = 1$ and θ induces an automorphism of P. If y is in G then $y\theta^n \in P = P\theta^n$, so $y\theta^n = p\theta^n$ for p in P and $y = pz$ where $z\theta^n = 1$, z is in $R(\theta)$. Thus $G = PR(\theta)$, $P \cap R(\theta) = 1$, so $G = P \otimes R(\theta)$. If (5.14) holds, where θ induces an automorphism of $C(\theta)$, then $P = G\theta^n = C(\theta)\theta^n = C(\theta)$, showing the uniqueness of $C(\theta)$. This proves Lemma 5.9.

Lemma 5.10. *Let (G, Ω) be an operator loop with commutator-associator subloop G' such that G/G' satisfies the ascending and descending chain conditions on Ω-subloops. If $G = S \otimes T$ for Ω-subloops S, T and if the normal Ω-endomorphism φ of G induces an isomorphism of S into S, then φ induces an automorphism of S.*

Proof. Let $S = G\alpha$, $T = G\beta$ for Ω-projections α, β such that $1 = \alpha + \beta$. The normal Ω-endomorphism $\theta = \alpha\varphi$ maps G into S and coincides with φ on S. In particular, since φ is one-to-one on S, $S \cap R(\theta) = 1$. By Lemma 5.9, $G = C(\theta) \otimes R(\theta)$ where θ induces an automorphism of $C(\theta)$. Since $C(\theta) = C(\theta)\theta \subset G\theta \subset S$, $S = C(\theta) \otimes [R(\theta) \cap S] = C(\theta)$ by Lemma 5.5. Therefore φ, with θ, induces an automorphism of S.

Lemma 5.11. *If $\alpha, \beta, \delta, \varepsilon$ are Ω-projections of the operator loop (G, Ω) such that $1 = \alpha + \beta = \delta + \varepsilon$, and if the direct Ω-decompositions $G\alpha = A_1 \otimes A_2$,*

$G\beta = B_1 \otimes B_2$, $G\delta = D_1 \otimes D_2$, $G\varepsilon = E_1 \otimes E_2$ are such that

α induces an isomorphism of D_1 upon A_1 and of E_2 upon A_2,

β induces an isomorphism of E_1 upon B_1 and of A_2 upon B_2,   (5.17)

δ induces an isomorphism of A_1 upon D_1 and of B_2 upon D_2,

ε induces an isomorphism of B_1 upon E_1 and of A_2 upon E_2,

then

$$G\alpha \otimes B_1 = G\alpha \otimes E_1, \qquad G\alpha \otimes B_2 = G\alpha \otimes D_2,$$
$$G\beta \otimes A_1 = G\beta \otimes D_1, \qquad G\beta \otimes A_2 = G\beta \otimes E_2, \qquad (5.18)$$
$$G\delta \otimes E_1 = G\delta \otimes B_1, \qquad G\delta \otimes E_2 = G\delta \otimes A_2,$$
$$G\varepsilon \otimes D_1 = G\varepsilon \otimes A_1, \qquad G\varepsilon \otimes D_2 = G\varepsilon \otimes B_2.$$

Proof. Set $P = G\alpha \otimes B_1$. Since $E_1\alpha \subset G\alpha$ and $E_1\beta \subset B_1$, then $E_1 \subset P$. If $x \in E_1 \cap G\alpha$, then $x\beta = 1$, whence, since β induces an isomorphism of E_1, $x = 1$. Therefore $Q = G\alpha \otimes E_1$ exists, and $Q \subset P$. Since $G'\beta\varepsilon\alpha = G'\beta\alpha\varepsilon = 1$, $\beta\varepsilon\alpha$ is centralizing and $-\beta\varepsilon\alpha$ exists. Since $\beta\varepsilon = \beta\varepsilon\alpha + \beta\varepsilon\beta$, then $\beta\varepsilon\beta = (-\beta\varepsilon\alpha) + \beta\varepsilon$. Inasmuch as $B_1\beta\varepsilon\beta = B_1$, $B_1(-\beta\varepsilon\alpha) \subset G\alpha$ and $B_1\beta\varepsilon = E_1$, we see that $B_1 \subset Q$. Hence $P \subset Q \subset P$, so $P = Q$. This proves the first of equations (5.18). The others are proved similarly.

Theorem 5.3. *Let (G, Ω) be an operator loop with commutator-associator subloop G' such that G/G' satisfies the ascending and descending chain conditions for Ω-subloops. Then every direct Ω-factor of G satisfies $\mathfrak{B}$ and hence satisfies $\mathfrak{E}(m, n)$ for all positive integers m, n.*

Proof. If $G = S \otimes T$ for Ω-subloops S, T, then, by Lemma 5.8 (i), $G' = S' \otimes T'$. The mapping $H \to K = H \otimes T'$ is one-to-one from the Ω-subloops H of S which contain S' into the Ω-subloops K of G which contain G'. Consequently, the chain conditions for G/G' imply those for S/S'. Accordingly, we need merely prove that G satisfies $\mathfrak{B}$. Let $G = A \otimes B = D \otimes E$ and $A = G\alpha$, $B = G\beta$, $D = G\delta$, $E = G\varepsilon$ for Ω-projections $\alpha, \beta, \delta, \varepsilon$ such that $1 = \alpha + \beta = \delta + \varepsilon$. By Lemma 5.9 with $\theta = \alpha\delta\alpha$, $G = C(\theta) \otimes R(\theta)$ where θ induces an automorphism of $C(\theta)$. Since $G\theta \subset A$, we have, by Lemma 5.5,

$$A = A_1 \otimes A_2, \quad A_1 = C(\alpha\delta\alpha), \quad A_2 = A \cap R(\alpha\delta\alpha). \qquad (5.19)$$

If $\varphi = \alpha\varepsilon\alpha$ then $\therefore = \alpha(\delta + \varepsilon)\alpha = \theta + \varphi$. Also $\alpha\theta = \theta\alpha = \theta$ and hence $\theta\varphi = \varphi\theta$. Consequently, $R(\theta)\varphi \subset R(\theta)$. Since, also $A\varphi \subset A$, we see that $A_2\varphi \subset A_2$. If $a \in A_2 \cap R(\varphi) = A \cap R(\theta) \cap R(\varphi)$ then $a\theta^s = a\varphi^t = 1$ for positive integers s, t. Since $G'\theta\varphi = G'\delta\varepsilon\alpha = 1$, $\theta\varphi = \varphi\theta$ is centralizing and, consequently, the binomial theorem is applicable to $(\theta + \varphi)^n$. Taking $n = s + t$, we deduce that $a = a(\theta + \varphi)^n = 1$. Hence φ induces an isomorphism of A_2 into A_2 and thus, by Lemma 5.10, an automorphism of A_2. Therefore

$$\alpha\delta\alpha,\ \alpha\varepsilon\alpha \text{ induce automorphisms of } A_1, A_2 \text{ respectively.} \tag{5.20}$$

Similarly, we get

$$D = D_1 \otimes D_2, \quad D_1 = C(\delta\alpha\delta), \quad D_2 = D \cap R(\delta\alpha\delta) \tag{5.21}$$

where

$$\delta\alpha\delta\ ,\ \delta\beta\delta \text{ induce automorphisms of } D_1, D_2 \text{ respectively.} \tag{5.22}$$

By the chain conditions $A_1 = G(\alpha\delta\alpha)^n$ for a suitable integer n. Then $D_1\alpha = D_1(\delta\alpha\delta)^n\alpha = D_1\delta(\alpha\delta\alpha)^n \subset A_1$. Similarly $A_1\delta \subset D_1$ and therefore $D_1 = D_1\delta\alpha\delta = (D_1\alpha)\delta \subset A_1\delta \subset D_1$. Thus $D_1 = A_1\delta$, $A_1 = D_1\alpha$, and

$$\alpha, \delta \text{ induce isomorphisms of } D_1 \text{ upon } A_1, A_1 \text{ upon } D_1, \text{ respectively.} \tag{5.23}$$

Next we define

$$B_2 = D_2\beta, \quad E_2 = A_2\varepsilon. \tag{5.24}$$

Since $\delta\beta\delta$ induces an automorphism of D_2, and similarly,

$$\beta, \delta \text{ induce isomorphisms of } D_2 \text{ upon } B_2, B_2 \text{ upon } D_2, \text{ respectively.} \tag{5.25}$$
$$\varepsilon, \alpha \text{ induce isomorphisms of } A_2 \text{ upon } E_2, E_2 \text{ upon } A_2, \text{ respectively.}$$

If x is in $B_2 \cap (D_1 \otimes E)$, then $x\delta$ is in $B_2\delta \cap D_1 = D_2 \cap D_1 = 1$. Since x is in B_2 and $x\delta = 1$, then $x = 1$. Therefore $B_2 \otimes D_1 \otimes E$ exists. Since $\delta\beta = \delta\beta\delta + \delta\beta\varepsilon$ and $\delta\beta\varepsilon$ is centralizing, $\delta\beta\delta = \delta\beta + (-\delta\beta\varepsilon)$. Therefore $D_2 = D_2\delta\beta\delta \subset (D_2\delta\beta)\,[D_2(-\delta\beta\varepsilon)] \subset B_2E = B_2 \otimes E$. Hence $G = D \otimes E = D_1 \otimes D_2 \otimes E = D_1 \otimes B_2 \otimes E$. Similarly, $G = A_1 \otimes E_2 \otimes B$. Consequently, by Lemma 5.5,

$$B = B_1 \otimes B_2, \quad E = E_1 \otimes E_2, \tag{5.26}$$

where

$$B_1 = B \cap (D_1 \otimes E), \quad E_1 = E \cap (A_1 \otimes B). \tag{5.27}$$

If b is in B_1, then, by (5.27), (5.23), $b\,\delta\alpha$ is in A_1. Since $1 = b\,\alpha = b(\delta + \varepsilon)\alpha = b\,\delta\alpha \cdot b\,\varepsilon\alpha$, $b\,\varepsilon\alpha$ is also in A_1. Then $b\varepsilon = b\,\varepsilon\alpha \cdot b\,\varepsilon\beta$ is in $E \cap (A_1 \otimes B) = E_1$. Hence $B_1\varepsilon \subset E_1$, and, similarly, $E_1\beta \subset B_1$. If b is in B_1, and $b\varepsilon = 1$, then $b = b\delta$ is in D_1 and $1 = b\alpha = (b\delta)\alpha$. Therefore, by (5.23), $b\delta = 1$ and hence $b = 1$. Thus, and similarly, ε, β induce isomorphisms of B_1, E_1 respectively. Hence $\varepsilon\beta$ induces an isomorphism of B_1 into B_1 and thence, by Lemma 5.10, an automorphism of B_1. Since $B_1 = B_1\varepsilon\beta \subset E_1\beta \subset B_1$, then $B_1 = E_1\beta$. We now may state:

$$\varepsilon, \beta \text{ induce isomorphisms of } B_1 \text{ upon } E_1, E_1 \text{ upon } B_1, \text{ respectively.} \tag{5.28}$$

Since (5.23), (5.25), (5.28) imply (5.17), we may use (5.18). In particular,

$$P = B \otimes A_1 = B \otimes D_1, \quad Q = E \otimes D_1 = E \otimes A_1.$$

Since $P \cap Q \cap B = Q \cap B = B \cap (D_1 \otimes E) = B_1$ and $P \cap Q \cap E = P \cap E = E \cap (A_1 \otimes B) = E_1$, we deduce by several uses of Lemma 5.4 that

$$P \cap Q = A_1 \otimes B_1 = D_1 \otimes B_1 = A_1 \otimes E_1 = D_1 \otimes E_1 = G_1. \qquad (5.29)$$

Similarly, if

$$D_3 = D \cap (A_2 \otimes B), \quad B_3 = B \cap (E_2 \otimes D), \qquad (5.30)$$

the equations

$$S = B \otimes A_2 = B \otimes E_2, \quad T = D \otimes E_2 = D \otimes A_2$$

yield

$$S \cap T = A_2 \otimes B_3 = E_2 \otimes B_3 = A_2 \otimes D_3 = E_2 \otimes D_3 = G_2. \qquad (5.31)$$

In particular,

$$G_2\beta = B_3 = D_3\beta\,. \qquad (5.32)$$

By (5.31), $D_3\alpha \subset G_2\alpha \subset A_2$. If $d \in D_3$, $d(\delta\alpha\delta)^{n+1} = (d\alpha)(\alpha\delta\alpha)^n\delta$ for every positive integer n. By (5.19), $A_2 \subset R(\alpha\delta\alpha)$. Therefore $D_3 \subset D \cap R(\delta\alpha\delta) = D_2$ by (5.21). Since $G = A \otimes B = G_1 \otimes A_2 \otimes B_2 = D_1 \otimes A_2 \otimes B$, then, by Lemma 5.5, $D = D_1 \otimes [D \cap (A_2 \otimes B)] = D_1 \otimes D_3$. Again, since $D_3 \subset D_2 \subset D$, $D_2 = D_3 \otimes (D_2 \cap D_1) = D_3$. Therefore $D_3 = D_2$ and, by (5.32), (5.24), $B_3 = D_3\beta = D_2\beta = B_2$. With this, the proof of Theorem 5.3 is complete. In view of Theorem 5.2 we may add:

Corollary. *If (G, Ω) is an operator loop whose Ω-centre satisfies the ascending and descending chain conditions on Ω-subloops, then every direct Ω-factor of G satisfies $\mathfrak{B}$ and hence satisfies $\mathfrak{C}(m, n)$ for all positive* integers m, n.

For an application of $\mathfrak{B}$ (under the name of "weak refinability") to the theory of direct decomposition into infinitely many factors, see Baer [B 2].

It should be clear from the foregoing treatment that the open questions in the theory of direct decomposition of operator loops are mainly open questions concerning abelian operator groups. The situation remains unchanged when we consider direct decomposition into infinitely many factors. For many topics not covered here and for other points of view (notably, that of lattice theory) see Kuroš, Theory of Groups, Specht, Gruppentheorie and the papers of Baer (cited above).

For significant and broad results in the theory of direct decomposition of general algebras see Jônsson and Tarski [15] (whose theory, however, does not say much for abelian groups) and Goldie [10].

6. Characteristic free factors

A subloop H of a loop G is called a *characteristic* subloop of G provided that $H\theta \subset H$ for every automorphism θ of G. If the inner mapping group $\mathfrak{J}(G)$ consists entirely of automorphisms (e. g., when G is a group), characteristic subloops are normal. However, there exist loops with non-normal characteristic subloops (BRUCK [69], [70], BATES [64].) This fact is particularly well illustrated by the following results of BATES [64] on free decomposition:

Theorem 6.1. *No nontrivial free factor of a loop G is normal in G.*

Theorem 6.2. *If $G = A * B$ for nontrivial subloops A, B of a loop G, then A is a characteristic free factor of G if and only if the following conditions are satisfied:* (a) *No nontrivial free factor of A is a free loop.* (b) *No nontrivial free factor of A is isomorphic to a free factor of B.*

Theorem 6.3. *Every two free decompositions, $G = \Pi^* H(v)$, $G = \Pi^* K(w)$ of a loop G into characteristic subloops $H(v)$, $K(w)$, have a common refinement, $G = \Pi^*(H(v) \cap K(w))$, where the free factors $H(v) \cap K(w)$ are also characteristic.*

Theorem 6.4. *If a loop G is a free product of loops $H(v)$, then there exists one and only one free decomposition of G into free factors $K(w)$ such that* (i) *each $K(w)$ is a characteristic subloop of G and is a free product of one or more of the $H(v)$*; (ii) *no proper refinement of the free decomposition $G = \Pi^* K(w)$ has property* (i).

The proofs of these theorems will be omitted. A free factor A of the loop G is called *nontrivial* if $A \neq 1$ and $A \neq G$. In view of Theorem 6.2, we see that if G is the free product (in the sense of loops) of any collection of non-isomorphic simple groups, each free factor is a characteristic subloop of G.

7. Hamiltonian loops

A loop is called *hamiltonian* if every subloop is normal. According to this definition, in contrast to the more usual one, abelian groups are included among the hamiltonian groups. D. A. NORTON [95] found that hamiltonian loops can have complicated structures. In order to get sharp theorems he imposed additional hypotheses. A loop is *power-associative (di-associative)* if every element generates (every two elements generate) a subgroup. A power-associative loop is a p-loop (p a prime) if every element has p-power order. NORTON's main results may be stated as follows:

Theorem 7.1. *A power-associative hamiltonian loop in which every element has finite order is a direct product of hamiltonian p-loops.*

Theorem 7.2. *A di-associative hamiltonian loop G is either an abelian group or a direct product $G = A \otimes T \otimes H$ where A is an abelian group whose elements have finite odd order, T is an abelian group of exponent* 2 *and H is a non-commutative loop with the following properties:*

(i) *The centre $Z = Z(H)$ has order two, elements $1, e$ where $e \neq 1$, $e^2 = 1$.*

(ii) *If x is a non-central element of H, $x^2 = e$.*

(iii) *If x, y are in H and $(x, y) \neq 1$, then $(x, y) = e$ and x, y generate a quaternion group.*

(iv) *If x, y, z are in H and $(x, y, z) \neq 1$, then $(x, y, z) = e$.*

NORTON showed conversely that if A, T, H are as specified in Theorem 7.2, $A \otimes T \otimes H$ is a di-associative hamiltonian loop. Furthermore, if H satisfies the additional hypothesis that any three elements x, y, z for which $(x, y, z) = 1$ are contained in a subgroup, then H is either a quaternion group or a Cayley loop. The latter is a di-associative loop of order 16 with three generators a_1, a_2, a_3 such that

$$a^4_i = 1, \quad a^2_i = a^2_j \neq 1, \quad a_i a_j = a^3_j a_i \qquad (i \neq j),$$
$$(a_i a_j) a_k = a^3_i (a_j a_k), \qquad i, j, k \text{ distinct.}$$

8. Loops with transitive automorphism groups

It will be convenient in this section to talk of additive loops instead of multiplicative loops; then the identity element, 1, is replaced by the zero element, 0, and direct products by direct sums. Let G be an additive loop, and let $\mathfrak{A}(G)$ denote the group of all automorphisms of G. Clearly $0\mathfrak{A}(G) = 0$. We say that *G has a transitive automorphism group* if $\mathfrak{A}(G)$ is transitive on the nonzero elements of G; that is, to every two nonzero elements a, b of G there corresponds at least one θ in $\mathfrak{A}(G)$ such that $a\theta = b$. In this case, $\mathfrak{A}(G)$ is clearly an irreducible set of endomorphisms of G. Hence, by SCHUR's Lemma, the centralizer, $\mathfrak{C}$, of $\mathfrak{A}(G)$ in the semigroup of all endomorphisms of G, consists of the zero endomorphism and a group $\mathfrak{C}^*$ of automorphisms of G; in particular, $\mathfrak{C}^*$ must be the centre of $\mathfrak{A}(G)$ and is, therefore, an abelian group. If, further, G is an abelian group, $\mathfrak{C}$ (as is well known) is an abelian group under addition and hence is a field. This proves one half of the following:

Theorem 8.1. *Let G be an additive abelian group. A necessary and sufficient condition that G have a transitive automorphism group is that G be the additive group of a vector space over a field.*

Proof. Necessity has been proved above, so let G be the additive group of a vector space over a field F of right operators. If a is a nonzero element of G, aF is a minimal F-subgroup and $\mathfrak{A}(G)$ is clearly transitive on aF. If the element b of G is not in aF we observe that $S = aF + bF = aF \oplus bF$. By use of Zorn's Lemma we deduce the existence of an F-subgroup M of G such that $G = S \oplus M$. Then it is clear that there exists a unique F-automorphism of G which interchanges a, b and induces the identity mapping on M. — The details concerning ZORN's Lemma will be indicated in the proof of the next theorem.

Theorem 8.2. *Let G be an additive loop which is not an abelian group and which has a transitive automorphism group. Then the following statements are equivalent:* (i) *G satisfies the ascending chain condition on normal subloops.* (ii) *G contains a normal subloop N which is minimal in the property that $N \neq 0$.* (iii) *The only normal subloops of G are G and 0; that is, G is simple.*

Proof. For each nonzero element a of G, let $N(a)$ denote the smallest normal subloop of G which contains a. First assume (i) and let K be a nonzero G-normal subloop of $N(a)$. Then K contains a nonzero element b and $N(b) \subset K \subset N(a)$. There exists an automorphism θ of G such that $b\theta = a$. Hence $N(b)\theta = N(b\theta) = N(a)$ and we have an ascending chain, $N(b) \subset N(b)\theta \subset N(b)\theta^2 \subset \ldots$, of normal subloops of G. By (i), there exists a non-negative integer n such that $N(b)\theta^n = N(b)\theta^{n+1}$; hence, since θ is an automorphism, $N(b) = N(b)\theta = N(a)$. Therefore $K = N(a)$, showing that $N(a)$ is a minimal nonzero normal subloop of G. Thus (i) implies (ii).

Next assume (ii) and let b be a nonzero element of N. By the minimality of N, $N(b) = N$. If a is a nonzero element of G and θ is an automorphism mapping b on a, then $N(a) = N(b)\theta = N\theta$. Hence $N(a)$ is a minimal nonzero normal subloop of G for each nonzero a. By ZORN's Lemma, there exists at least one normal subloop M of G which is maximal in the property that $N(a) \cap M = 0$. Hence $L = N(a) + M = N(a) \oplus M$ is a normal subloop of G. If there exists an element c in G but not in L, then $L \cap N(c) \neq N(c)$ and hence, by the minimality of $N(c)$, $L \cap N(c) = 0$. But then $L + N(c) = L \oplus N(c) = N(a) \oplus [M \oplus N(c)]$, whence $M \neq M \oplus N(c)$ and $N(a) \cap [M \oplus N(c)] = 0$, a contradiction. Therefore $G = N(a) \oplus M$. If $M \neq 0$, there exists a nonzero element m in M. The element $s = a + m$ is neither in $N(a)$ nor in M, so $N(s) \cap N(a) = N(s) \cap M = 0$. Moreover, $G = N(s) \oplus K$ for at least one normal subloop K. Since the centre $Z = Z(G)$ is characteristic and since $G \neq Z$, then $Z = Z\mathfrak{A}(G) = 0$. Hence by Lemma 5.3,

$$N(s) = [N(s) \cap N(a)] \oplus [N(s) \cap M] = 0\,,$$

a contradiction. Therefore $M = 0$, $G = N(a)$. Thus (ii) implies (iii). Since (iii) obviously implies (i), the proof of Theorem 8.2 is complete.

Weaker forms of Theorems 8.1, 8.2 are proved in BRUCK [73]. Examples of loops with transitive automorphism groups are given in PAIGE [96], BRUCK [73], ARTZY [56], BRUCK [75], HUGHES [353]. If a group G with transitive automorphism group has a nonzero element of finite order p, then p is a prime and every nonzero element has order p. If, in addition, G is not an abelian group, then G must coincide with its commutator subgroup and hence cannot be finite. This may mean — in case BURNSIDE's conjecture is true — that G cannot be finitely generated either. For groups without nonzero elements of finite order the situation is better known:

Theorem 8.3. *If G is an additive group in which every nonzero element has infinite order, then G can be imbedded in an additive group G^* (of countable order if G is countable) such that every two nonzero elements of G^* are conjugate in G^*.*

Proof. See NEUMANN [B 12]; in particular, p. 539, corollaries 19.7, 19.8.

9. Homomorphism theory of quasigroups

The homomorphism theory of quasigroups was studied for the finite case by GARRISON [84], [85] and in general by KIOKEMEISTER [89]. (See also ŠIK [357].) Another approach was used by T. EVANS [5] and others (see I, § 5), imbedding the theory in a theory of general algebras. We shall use the notation of KIOKEMEISTER.

Each homomorphism θ of a quasigroup G upon a quasigroup $G\theta$ uniquely determines a normal equivalence relation α on G by the requirement that, for all a, b in G, $a\alpha b$ if and only if $a\theta = b\theta$. An equivalence relation α on G is normal if and only if: (i) $ca\alpha cb$ implies $a\alpha b$; (ii) $ac\alpha bc$ implies $a\alpha b$; (iii) $a\alpha b$ and $c\alpha d$ imply $ac\alpha bd$. The equivalence class $R(a, \alpha)$ of the element a with respect to the normal relation α is the set of all x in G such that $x\alpha a$. The equivalence classes $R(a, \alpha)$, $a \in G$, form a quasigroup G/α with multiplication defined by $R(a, \alpha)R(b, \alpha) = R(ab, \alpha)$. If the homomorphism θ of G upon a quasigroup $G\theta$ determines the normal relation α as above, then the mapping $R(a, \alpha) \rightarrow a\theta$ is an isomorphism of G/α upon $G\theta$. The intersection, $\alpha \cap \beta$, and union, $\alpha \cup \beta$, of two normal relations α, β are defined as usual: $a(\alpha \cap \beta)b$ if and only if $a\alpha b$ and $a\beta b$; $a(\alpha \cup \beta)b$ if and only if there exist elements $a_0, a_1, \ldots, a_{2n}$ of G such that $a_0 = a$, $a_{2n} = b$ and $a_{2i}\alpha a_{2i+1}$, $a_{2i+1}\beta a_{2i+2}$ for $i = 0, 1, \ldots, n-1$. The system $(L, \cap, \cup)$ consisting of the set $L = L(G)$ of all normal relations α on G under the defined operations $\cap$, $\cup$ is a modular lattice. In particular, $R(a, \alpha \cap \beta) = R(a, \alpha) \cap R(a, \beta)$ and $R(ab, \alpha \cup \beta) = R(a, \alpha)R(b, \beta)$ for all a, b in G, α, β in L.

A subquasigroup H of the quasigroup G is called a normal divisor of G if and only if $H = R(a, \alpha)$ for some a in G, α in L; when this occurs, α is unique. For each a in G, the subset L_a of L consisting of all α in L such that $R(a, \alpha)$ is a normal divisor is either empty or a modular lattice. In particular, $R(a, \alpha \cup \beta) = R(a, \alpha)R(a, \beta)$ for all α, β in L_a and there exists a Jordan-Hölder Theorem for series of normal divisors containing a. If $e = e^2$ is an idempotent element of G, $L_e = L$. The stronger type of JORDAN-HÖLDER Theorem for composition series, in which each term is a maximal normal divisor of the preceding term (but not necessarily of G), is established by GARRISON [85] for finite quasigroups and series containing a specified idempotent e; but, as GARRISON shows, a change of idempotent can effect even the length of a composition series.

BRUCK [67] showed the existence of non-trivial quasigroups H, K such that the only subquasigroup of the direct product $G = H \otimes K$ is G itself. KIOKEMEISTER [89] began a theory of direct products limited to quasigroups containing idempotents. A much more complete theory of this sort, for general algebras containing neutral elements, has been given by JÓNSSON and TARSKI [15].

10. Homomorphism theory of other systems

SMILEY [109] used lattice-theoretic methods, based upon a fundamental lemma of ORE, to develop the homomorphism theory for loops. Later KIOKEMEISTER [89] applied similar methods to quasigroups (§ 9).

SMILEY [51] extended his methods to left division systems with left identity elements. A groupoid G is a left division system according to SMILEY, if, for each ordered pair a, b of elements of G, there is one and only one x in G such that $xa = b$. Attention is restricted to homomorphisms of left division systems with left identity elements (here called systems) upon like systems and to the kernels of such homomorphisms. A kernel is called a normal subsystem and is characterized internally. The basic lemma is the following analogue of Zassenhaus' Lemma (§ 2): *If A, B, A_1 and B_1 are subsystems of G, if A_1, B_1 are normal subsystems of A, B respectively, then $A_1(A \cap B_1)$ is a normal subsystem of $A_1(A \cap B)$, $B_1(B \cap A_1)$ is a normal subsystem of B_1 $(B \cap A)$ and the identity mapping of G induces an isomorphism of $A_1(A \cap B)/A_1(A_1 \cap B)$ upon $B_1(B \cap A)/B_1(B_1 \cap A)$.*

In his University of Wisconsin thesis (1954), WAYNE COWELL defined a *loop image* M to be a groupoid with identity element such that $aM = Ma = M$ for every a in M. By a theorem of BATES and KIOKEMEISTER (cf. I, Theorem 3.2) a groupoid M is a loop image if and only if there exists a loop G and a multiplicative homomorphism of G upon M. Let G be a loop, let the subloop K of G be the kernel of a multiplicative homomorphism θ of G upon a loop image and let K^* be the semigroup generated (under multiplication only) by the right and left mappings $R(k)$, $L(k)$ of G as k ranges over K. Then K is called a quasinormal kernel if, for each x in G, xK^* is the set of all y in G such that $y\theta = x\theta$. Equivalently, K is a subloop of G such that $(xK^*)(yK^*) \subset (xy)K^*$ for all x, y in G. The set Q of all quasinormal kernels is a commutative semigroup under the operation $P(H, K)$ defined by $P(H, K) = 1H^*K^*$ for all quasinormal kernels H, K. More generally, the congruence set up by a quasinormal kernel of G permutes with every multiplicative congruence of G. An example is given to show that the intersection of two quasinormal subloops need not be quasinormal. It is shown by construction that to each loop image M there corresponds at least one loop G with quasinormal kernel K such that G/K is isomorphic to M and $xK^* = K(Kx)$ for all x in G. Moreover, if M contains elements a, e, f, distinct from each other and the identity element, such that $ea = a = af$, there exists a loop H and a homomorphism θ of H upon M such that the kernel of θ in H is not quasinormal. These two constructions make plain the fact that loop images cannot be characterized by hypotheses on kernels alone. Two unsolved problems suggest themselves: (i) *What loops G have the property that every image of G under a multiplicative homomorphism is also a loop?* (ii) *What loops G have the property that each multiplicative homomorphism of G has a quasinormal kernel?* — Some classes of loops with property (i) are discussed in VII.

(Since the preceding paragraph was written, COWELL [349] has published some of his results and added others.)

V. Lagrange's Theorem for Loops

1. Coset expansion

Several authors (HAUSMANN and ORE [87], GRIFFIN [86], MURDOCH [92], [93] and others) have discussed coset expansion for quasigroups. We shall treat only loops. The loop G is said to have a *right coset expansion modulo* its subloop H provided the right cosets Hx partition G. Since each x of G lies in at least one right coset, namely Hx, the condition for a right coset expansion is: If $y \in Hx$, then $Hy = Hx$. Equivalently,

$$H(hx) = Hx \tag{1.1}$$

for every h in H, x in G. Similarly for left coset expansion. The usual proof of Lagrange's Theorem for groups gives:

Lemma 1.1. *If the finite loop G has a right (or left) coset expansion with respect to the subloop H, then the order of H divides the order of G. In particular, the order of every normal subloop of G divides the order of G.*

If H is a subloop of the left or middle nucleus of G, then (1.1) holds. If H is a subloop of the right nucleus of G, the left-right dual of (1.1) holds. Hence:

Theorem 1.1. *If G is a finite loop and if H is a subloop of any one of the nuclei of G, then the order of H divides the order of G.*

If G is a di-associative loop and if H is a cyclic subgroup of G, then (1.1) holds. Hence:

Theorem 1.2. *If G is a finite di-associative loop, the order of every element of G divides the order of G.* (Applies to Moufang loops.)

Lagrange's theorem fails for loops; for example, there exists a loop of order 5 with subloops of order 2. On the other hand, there exists a finite loop G with a subloop H of order dividing that of G such that G has neither a right nor a left coset expansion modulo H. For various examples see BRUCK [67], [69], [70].

2. Lagrange theorems

Consider the following properties $\mathfrak{L}$, $\mathfrak{L}'$ of the loop G:

$\mathfrak{L}$. *The order of every subloop of the finite loop G divides the order of G.*

$\mathfrak{L}'$. *If H is a subloop of a subloop K of the finite loop G, the order of H divides the order of K. In other words, every subloop of G has property $\mathfrak{L}$.*

Property $\mathfrak{L}$ is weaker than property $\mathfrak{L}'$ (BRUCK [70]). Lagrange's theorem for groups becomes: *Every finite group has property $\mathfrak{L}'$.* We note that $\mathfrak{L}, \mathfrak{L}'$ are preserved under homomorphisms into loops and that $\mathfrak{L}'$ is inherited by subloops.

Lemma 2.1. *Let H be a normal subloop of the subloop K of the loop G.* (i) *If H and K/H have property $\mathfrak{L}$, so has K.* (ii) *If H and K/H have property $\mathfrak{L}'$, so has K.*

Proof. Let S be a subloop of K. Then $S \cap H$ is a normal subloop of S and the loops $S/(S \cap H)$, SH/H are isomorphic.

(i) Let the common order of $S/(S \cap H)$, SH/H be n. Since SH/H is a subloop of K/H and the latter has property $\mathfrak{L}$, the order of K/H is np for some integer p. Let $S \cap H$ have order q, so that S has order nq. Since $S \cap H$ is a subloop of H and since H has property $\mathfrak{L}$, the order of H is qr for some integer r. Since K/H and H have orders np and qr respectively, the order of K is $npqr$. Hence the order, nq, of S divides the order, $npqr$, of K. Thus K has property $\mathfrak{L}$.

(ii) Since H has property $\mathfrak{L}'$, $S \cap H$ has property $\mathfrak{L}$. Since K/H has property $\mathfrak{L}'$, SH/H and $S/(S \cap H)$ have property $\mathfrak{L}$. By (i), since $S \cap H$ and $S/(S \cap H)$ have property $\mathfrak{L}$, then S has property $\mathfrak{L}$. Thus K has property $\mathfrak{L}'$.

Theorem 2.1. *Each finite loop G has a normal subloop N uniquely characterized by the conditions:* (a) *N has property $\mathfrak{L}'$.* (b) *If K is a normal subloop of G with property $\mathfrak{L}'$, then $K \subset N$.*

Proof. Let N, K be normal subloops of G with property $\mathfrak{L}'$, N being maximal with respect to $\mathfrak{L}'$. Since K, $N/(N \cap K)$, NK/K have property $\mathfrak{L}'$, then, by Lemma 2.1 (ii), so has NK. Therefore $NK = N$, $K \subset N$. If K is also maximal with respect to $\mathfrak{L}'$, then $N \subset K$ so $N = K$.

Theorem 2.2. *A necessary and sufficient condition that a finite loop G have property $\mathfrak{L}'$ is that G possess a normal chain*

$$G = G_0 \supset G_1 \supset G_2 \supset \cdots \supset G_n = 1 \tag{2.1}$$

such that G_{i-1}/G_i has property $\mathfrak{L}'$ for $i = 1, 2, \ldots, n$.

Corollary. *If* (2.1) *holds and each G_{i-1}/G_i is either a group or a loop without proper subloops, then G has property $\mathfrak{L}'$.* (Applies to loops satisfying certain nilpotency conditions.)

Proof. Necessity is obvious. As for sufficiency, let i be the least nonnegative integer such that G_i has property $\mathfrak{L}'$. If $i > 0$, we deduce from Lemma 2.1 that G_{i-1} has property $\mathfrak{L}'$, a contradiction. Hence $G = G_0$ has property $\mathfrak{L}'$. The Corollary is immediate.

It is natural to enquire whether Sylow's First Theorem (for finite groups) has an analogue for finite loops with property $\mathfrak{L}'$. However, GRIFFIN [86] has constructed finite loops of arbitrary odd order exceeding 5 without proper subloops. (For a correction to Griffin's construction see BRUCK [69].)

VI. Nilpotency of Loops

1. A general theory of nilpotency

BRUCK [70] contains a theory of nilpotency of loops which embodies some of the basic features of nilpotency for groups. We shall sketch a less general approach which seems conceptually simpler. Let $\mathfrak{L}$ be any class of loops such that:

(a) *Every subloop of a member of* $\mathfrak{L}$ *is in* $\mathfrak{L}$.

(b) *Every loop which is a homomorphic image of a member of* $\mathfrak{L}$ *is in* $\mathfrak{L}$.

For example, $\mathfrak{L}$ may consist of all loops or of all loops satisfying a prescribed set of identities. By a *nilpotency function*, f, *for* $\mathfrak{L}$ we mean a function f with the following properties:

(i) *If* G *is in* $\mathfrak{L}$, $f(G)$ *is a uniquely defined subloop of* G.

(ii) *If* G *is in* $\mathfrak{L}$ *and if* H *is a subloop of* G, *then* $H \cap f(G) \subset f(H)$.

(iii) *If* G *is in* $\mathfrak{L}$ *and if* θ *is a homomorphism of* G *upon a loop, then* $f(G)\theta \subset f(G\theta)$.

(iv) *If* G *is in* $\mathfrak{L}$, *if* N *is a normal subloop of* G *and if* A *is the intersection of all normal subloops* K *of* G *such that* NK/K *is a subloop of* $f(G/K)$, *then* NA/A *is a subloop of* $f(G/A)$.

When $\mathfrak{L}$ is the class of all loops, the most important example of a nilpotency function is obtained by defining $f(G)$ to be the centre of G; this leads to the notion of *central nilpotency*. For *nuclear nilpotency* we define $f(G)$ to be the nucleus of G. Similarly, we could take $f(G)$ to be the left, middle or right nucleus of G. Again, if $f(G)$ is the *Moufang centre* of G; that is, the set of all a in G such that $(aa)(xy) = (ax)(ay)$ for all x, y in G, then (i) holds for every loop G. The proof of (i) is tricky but (ii)—(iv) give no trouble.

If $\mathfrak{L}$ is the set of all loops with the *inverse property* (see VII) we may define $f(G)$ to be the *Moufang nucleus* of G; that is, the set of all a in G such that $a[(xy)a] = (ax)(ya)$ for all x, y in G.

If $\mathfrak{L}$ is the set of all Moufang loops (see VII) we may define $f(G)$ to be the set of all a in G such that $ax = xa$ for every x in G; then $f(G)$ coincides with the Moufang centre of G.

If $\mathfrak{L}$ is the class of all commutative Moufang loops (see VIII) we may take $f(G)$ to be the *distributor* of G, namely the set of all a in G such that $(xy, z, a) = (x, z, a)(y, z, a)$ for all x, y, z in G.

If $\mathfrak{L}$ is the class of all groups, we may take $f(G)$ to be the n-centre of G (BAER [B 14]).

These examples suggest another approach. Let $\mathfrak{C}$ be a non-empty set of loop words W_n such that $W_n(1, X_2, \ldots, X_n) = 1$; e. g., a set of loop words which are normalized purely non-abelian in the sense of

IV.4. For every loop G, let $f(G)$ be the set of all a in G such that $W_n(a, x_2, \ldots, x)_n = 1$ for all $x_2, \ldots, x_n$ in G and each W_n in $\mathfrak{C}$. Finally, define $\mathfrak{L}$ to be the class of all loops G for which $f(G)$ is a subloop. Then $\mathfrak{L}$ satisfies (a), (b) and f is a nilpotency function for $\mathfrak{L}$.

Now let $\mathfrak{L}$ be a class of loops subject to (a), (b) and let f be any nilpotency function for $\mathfrak{L}$. For G in $\mathfrak{L}$ we define the *f-centre*, $Z_f(G)$, to be the union of all normal subloops of G which are contained in $f(G)$. For any normal subloop N of G, we define $(N, G)_f$ to be the subloop A whose existence is guaranteed in (iv). Clearly $Z_f(G)$ and $(N, G)_f$ are normal subloops of G; moreover, $(N, G)_f$ is a subloop of N. The (transfinite) *lower f-series*, $\{G_\alpha\}$, of G is defined inductively as follows: (i) $G_0 = G$; (ii) for any ordinal α, $G_{\alpha+1} = (G_\alpha, G)_f$; (iii) if α is a limit ordinal, G_α is the intersection of all G_β with $\beta < \alpha$. Clearly G_α is a normal subloop of G for every α; and, if $\alpha < \gamma$, G_γ is a subloop of G_α. Moreover, if α, β are ordinals such that $\alpha < \beta$ and $G_\alpha = G_{\alpha+1}$, then $G_\alpha = G_\beta$. Consequently there exists a least ordinal σ such that $G_\sigma = G_\tau$ for every ordinal τ with $\sigma < \tau$. The loop G is *transfinitely f-nilpotent* of class σ if $G_\sigma = 1$ and is *f-nilpotent* if, in addition, σ is a finite ordinal. The (transfinite) *upper f-series*, $\{Z_\alpha\}$, of G is defined inductively as follows: (i) $Z_0 = 1$; (ii) for any ordinal α, $Z_{\alpha+1}$ is the unique subloop of G containing Z_α such that $Z_{\alpha+1}/Z_\alpha$ is the f-centre, $Z_f(G/Z_\alpha)$, of G/Z_α; (iii) if α is a limit ordinal, Z_α is the union of all Z_β with $\beta < \alpha$. Again, Z_α is a normal subloop of G for each α, but this time Z_α is a subloop of Z_γ for $\alpha < \gamma$. As before, there exists a least ordinal λ such that $Z_\lambda = Z_\mu$ for every ordinal μ such that $\lambda < \mu$. The loop G is called *transfinitely upper-f-nilpotent* of class λ if $Z_\lambda = G$ and *upper-f-nilpotent* if, in addition, λ is finite.

In the rest of this section it is to be understood that we are dealing with loops of a class $\mathfrak{L}$ satisfying (a), (b) and with nilpotency functions for $\mathfrak{L}$.

Lemma 1.1. *Let G be a loop with lower f-series $\{G_\alpha\}$. (i) If H is a subloop of G, then $H_\alpha \subset G_\alpha$ for every ordinal α. (ii) If θ is a homomorphism of G upon a loop K, then $(G_\alpha)\theta \subset K_\alpha$ for every ordinal α and $(G_n)\theta = K_n$ for every non-negative integer n.*

Proof. (i) Certainly $H_0 = H \subset G = G_0$. Let α be an ordinal and assume inductively that $H_\beta \subset G_\beta$ for every $\beta < \alpha$. If $\alpha = \gamma + 1$ for some $\gamma < \alpha$, then $H_\gamma \subset G_\gamma$, so $H_\gamma G_\alpha / G_\alpha$ is a subloop of G_γ / G_α and hence of $f(G/G_\alpha)$. Therefore, by property (ii) of f, $H_\gamma G_\alpha / G_\alpha$ is a subloop of $f(HG_\alpha / G_\alpha)$. The natural isomorphism of HG_α / G_α upon $H/(H \cap G_\alpha)$ maps $H_\gamma G_\alpha / G_\alpha$ upon $H_\gamma (H \cap G_\alpha)/(H \cap G_\alpha)$. Hence, by properties (iii), (ii) of f, $H_\gamma(H \cap G_\alpha)/(H \cap G_\alpha)$ is contained in $f(H/(H \cap G_\alpha))$. Consequently $H_\alpha = H_{\gamma+1}$ is contained in $H \cap G_\alpha$ and hence in G_α. If α is a limit ordinal, H_α is the intersection of the H_β with $\beta < \alpha$ and hence is contained in the intersection of the G_β with $\beta < \alpha$, that is, in G_α. This proves (i).

It will be convenient to note at this point that if φ is an isomorphism of a member M of $\mathfrak{L}$, property (ii) of f yields $f(M)\varphi = f(M\varphi)$.

(ii) Certainly $(G_0)\theta = K_0$. Let α be an ordinal and assume inductively that $(G_\beta)\theta \subset K_\beta$ for every $\beta < \alpha$. If α is a limit ordinal it follows at once that $(G_\alpha)\theta \subset K_\alpha$. If $\alpha = \gamma + 1$ for some ordinal $\gamma < \alpha$, let $H = (K_\alpha)\theta^{-1}$ be the inverse image of K_α. Then θ induces a natural isomorphism, say φ, of G/H upon K/K_α. Since $(G_\gamma H)\theta = (G_\gamma)\theta \subset K_\gamma$, by our inductive hypothesis, and since K_γ/K_α lies in $f(K/K_\alpha)$, then $G_\gamma H/H$ lies in $f(K/K_\alpha)\varphi^{-1} = f(G/H)$. Therefore $G_\alpha = G_{\gamma+1}$ is contained in H and $(G_\alpha)\theta$ is contained in $H\theta = K_\alpha$. This proves the first part of (ii). In particular, $(G_n)\theta \subset K_n$ for every positive integer n. Next we consider a positive integer n and assume inductively that $(G_{n-1})\theta = K_{n-1}$. Certainly θ induces a natural homomorphism, say ψ, of G/G_n upon $K/(G_n)\theta$. Since G_{n-1}/G_n is in $f(G/G_n)$, then $K_{n-1}/(G_n)\theta = (G_{n-1}/G_n)\psi$ is in $f(G/G_n)\psi$ and hence, by property (iii) of f, in $f(K/(G_n)\theta)$. Thus $K_n \subset (G_n)\theta$ and consequently $(G_n)\theta = K_n$. This proves (ii) and completes the proof of Lemma 1.1.

Theorem 1.1. *If a loop G is transfinitely f-nilpotent of class σ, then every subloop H of G is transfinitely f-nilpotent of class at most σ. If, further, σ is finite, every homomorphic loop-image of G is f-nilpotent of class at most σ.*

Proof. This is a corollary of Lemma 1.1.

Lemma 1.2. *Let the loop G have upper and lower f-series $\{Z_\alpha\}$, $\{G_\alpha\}$ and assume that $G_{\beta+1} \subset Z_{\gamma+1}$ for some ordinals β, γ. Then $G_{\beta+2} \subset Z_\gamma$ and $G_\beta \subset Z_{\gamma+2}$.*

Proof. Since $G_{\beta+1}Z_\gamma/Z_\gamma$ is contained in $Z_{\gamma+1}/Z_\gamma = Z_f(G/Z_\gamma)$ and the latter is part of $f(G/Z_\gamma)$, then Z_γ contains $(G_{\beta+1}, G)_f = G_{\beta+2}$. Since $(G_\beta, G)_f = G_{\beta+1}$ is contained in $Z_{\gamma+1}$, then, by property (iii) of f, $G_\beta Z_{\gamma+1}/Z_{\gamma+1}$ is contained in $f(G/Z_{\gamma+1})$ and hence, being normal, in $Z_f(G/Z_{\gamma+1}) = Z_{\gamma+2}/Z_{\gamma+1}$. Therefore G_β is contained in $Z_{\gamma+2}$. This proves Lemma 1.2.

Theorem 1.2. *A necessary and sufficient condition that the loop G be f-nilpotent of (finite) class n is that G be upper-f-nilpotent of class n.*

Proof. By the arguments of Lemma 1.2, the inequalities $G_n \subset Z_0$ and $G_0 \subset Z_n$ are equivalent. — Known facts about nilpotency of groups show that the theorem fails for the transfinite case.

The following lemma suggests an order relation among nilpotency functions:

Lemma 1.3. *Let f, h be nilpotency functions for a class $\mathfrak{L}$ such that $f(G) \subset h(G)$ for every G in $\mathfrak{L}$. Then, for all normal subloops of a member, G, of $\mathfrak{L}$,*

$$\text{if } K \subset L \text{ then } (K, G)_h \subset L, G)_f\,. \tag{1.1}$$

Proof. If $K \subset L$ then $K(L, G)_f/(L, G)_f \subset L/(L, G)_f \subset f(G/(L, G)_f) \subset \subset h(G/(L, G)_f)$ and therefore $(L, G)_f$ contains $(K, G)_h$.

Theorem 1.3. *Under the hypothesis of Lemma 1.3, if G is transfinitely f-nilpotent of class σ then G is transfinitely h-nilpotent of class at most σ.*

Proof. If $\{G_\alpha^{(f)}\}$, $\{G_\alpha^{(h)}\}$ are the lower series corresponding to f, h respectively, we verify that $G_\alpha^{(h)} \subset G_\alpha^{(f)}$ for every ordinal α by transfinite induction based on (1.1). Consequently, $G_\sigma^{(h)} = 1$.

Among the many unanswered questions in connection with nilpotency, we will mention only the following: *Is every free loop transfinitely centrally nilpotent?*

2. The Frattini subloop

If $S, T, \ldots,$ are subsets or elements of a loop G, let $\{S, T, \ldots\}$ denote the subloop of G generated by $S, T, \ldots$. An element x of a loop G is a *non-generator* of G (NEUMANN [B 11], BRUCK [70]) if, for every subset S of G, $\{x, S\} = G$ implies $\{S\} = G$. It is easy to see that the non-generators of G form a subloop, $\varphi(G)$, of G. This is the *Frattini subloop.*

Lemma 2.1. *If θ is a homomorphism of the loop G upon a loop, then $\varphi(G)\theta \subset \varphi(G\theta)$.*

Proof. Let a be an element of $\varphi(G)$ and let S be a subset of $G\theta$ such that $\{a\theta, S\} = G\theta$. If $T = S\theta^{-1}$, then $H = \{a, T\}$ is a subloop of G containing the kernel of θ. Since, also, $H\theta = G\theta$, then $H = G$. Therefore $\{T\} = G$ and hence $\{S\} = \{T\}\theta = G\theta$, showing that $a\theta$ is in $\varphi(G\theta)$.

Theorem 2.1. *Let G be any loop. If G has at least one maximal proper subloop, then $\varphi(G)$ is the intersection of all maximal proper subloops of G. In the contrary case, $\varphi(G) = G$.*

Proof. If G has maximal proper subloops, let H be the intersection of all such subloops; in the contrary case, let $H = G$. First assume $H \neq G$ and let x be in G but not in H. Then there exists a maximal proper subloop M of G which does not contain x. Hence $\{x, M\} = G$ but $\{M\} = M \neq G$, so x is not in $\varphi(G)$. Therefore $\varphi(G) \subset H$; and this is also true when $H = G$. Next assume $\varphi(G) \neq G$ and let y be in G but not in $\varphi(G)$. Then there exists a subset S of G such that $\{y, S\} = G$ but $\{S\} \neq G$. By Zorn's Lemma, the set of all subloops of G which contain $\{S\}$ but not y has at least one maximal element, say K. If L is a subloop of G containing K properly, then L contains both y and S, so $L = G$. Therefore K is a maximal proper subloop of G, so $H \subset K$ and, in particular, y is not in H. Hence $H \subset \varphi(G)$; and this is also true when $\varphi(G) = G$. Consequently $\varphi(G) = H$, proving Theorem 2.1. NEUMANN has pointed out (correcting a misstatement in BRUCK [70]) that $\varphi(G) = G$ if G is the additive group of rationals but $\varphi(G) \neq G$ if G is finitely generated.

By Lemma 2.1, $\varphi(G)$ is a characteristic subloop of G for every loop G. Thus $\varphi(G)$ is normal in G if G is a group but (see BRUCK [69] for examples) there exist loops with non-normal Frattini subloops.

Theorem 2.2. *Let the loop G be transfinitely upper centrally nilpotent. Then $\varphi(G)$ is normal in G and $G/\varphi(G)$ is an abelian group. Hence either $\varphi(G) = G$ or $G/\varphi(G)$ is a subdirect product of cyclic groups of prime order.*

Proof. We need only consider the case that $\varphi(G) \neq G$. Then G has at least one maximal proper subloop M and $\varphi(G)$ is the intersection of all such subloops. First suppose that $MZ = G$ where Z is the centre of G; then it is easy to see that M is normal in G. We note that, for central nilpotency, $f(G)$ is the centre (a normal subloop) of G. Hence, if $\{Z_\alpha\}$ is the upper central series of G, $Z_{\alpha+1}/Z_\alpha$ is the centre of G/Z_α for every ordinal α. By hypothesis, $Z_\sigma = G$ for some ordinal σ. Thus the set of all ordinals α such that $MZ_\alpha = G$ is non-empty and therefore contains a least ordinal, say λ. Then $MZ_\lambda = G$. Since $MZ_0 = M \neq G$, λ is positive. If $\beta < \lambda$ then $MZ_\beta \neq G$ and hence, by the maximality of M, $Z_\beta \subset M$. If λ is a limit ordinal we deduce that $Z_\lambda \subset M$, a contradiction. Hence $\lambda = \mu + 1$ for an ordinal μ and $Z_\mu \subset M \cap Z_\lambda$. Then $(M/Z_\mu)(Z_\lambda/Z_\mu) = G/Z_\mu$ and Z_λ/Z_μ is the centre of G/Z_μ. Therefore, by the special case considered above, M/Z_μ is normal in G/Z_μ. Hence M is normal in G. Since $G = MZ_\lambda$, G/M is isomorphic to $Z_\lambda/(M \cap Z_\lambda)$. Since $M \cap Z_\lambda$ contains Z_μ, G/M is a homomorphic image of the abelian group Z_λ/Z_μ. Therefore M contains the commutator-associator subloop G' of G. Consequently, $\varphi(G)$ is a normal subloop of G containing G'; and $G/\varphi(G)$ is an abelian group, say A. Since A is an abelian group such that $\varphi(A) = 1$, then A is a subdirect product of cyclic groups of prime order. This completes the proof of Theorem 2.2.

A nontrivial abelian group G is (centrally) nilpotent of class 1. If G is the additive group of rationals, $\varphi(G) = G$, whereas, if G is a subdirect product of cyclic groups of prime order, $\varphi(G) = 1$. Hence we can expect little more from the hypothesis of Theorem 2.2. Moreover (Bruck [70]) if n is a positive integer and if $A_1, A_2, \ldots, A_n$ are nontrivial abelian groups such that A_n has at least three elements, there exists a loop G with upper central series $1 = Z_0 \subset Z_1 \subset \ldots \subset Z_n = G$ such that Z_i/Z_{i-1} is isomorphic to A_i for $i = 1, 2, \ldots, n$. Again, there exist finite, centrally nilpotent loops which are not direct products of loops of prime-power order and there exist finite loops of prime-power order which are not centrally nilpotent.

Lemma 2.2. *Let G be a finite loop with multiplication group $\mathfrak{M}(G)$ and let p be a prime. Then the following statements are equivalent:* (i) *G is centrally nilpotent and has order a power of p.* (ii) *$\mathfrak{M}(G)$ has order a power of p.*

Proof. We use the easily proved fact that the mapping $z \to R(z)$ is an isomorphism of the centre Z of G upon the centre $\mathfrak{Z}$ of $\mathfrak{M}(G)$. According to Albert (see IV.1) $\mathfrak{M}(G/Z)$ is isomorphic to $\mathfrak{M}(G)/Z^*$ where Z^* is the set of all θ in $\mathfrak{M}(G)$ such that $(x\theta)Z = xZ$ for every x in G. First assume (ii). Then $\mathfrak{Z}$, and hence Z, is a nontrivial group of order a power of p. If $\{Z_a\}$ is the upper central series of G, we deduce that the groups Z_n/Z_{n-1}, $n = 1, 2, \ldots$, have p-power order and are nontrivial for all n with $G \neq Z_{n-1}$. Hence (ii) implies (i). Next assume (i). If θ is in Z^*, the function f, defined by $x\theta = xf(x)$ for all x in G, has values in Z. If φ is in Z^* and $x\varphi = xg(x)$, then $x(\theta\varphi) = xR(f(x))\varphi = x\varphi R(f(x)) = x[f(x)g(x)]$. Hence Z^* is isomorphic to a multiplicative group of single-valued mappings of G into Z. Since Z, as a normal subloop of G, has p-power order, then Z^* has p-power order also. If G has central class $c > 0$, then G/Z has central class $c - 1$. Hence we assume inductively that $\mathfrak{M}(G)/Z^*$ has p-power order and deduce that $\mathfrak{M}(G)$ has p-power order. Thus (i) implies (ii) and the proof of Lemma 2.2 is complete.

Theorem 2.3. *Let G be a finite centrally nilpotent loop of primepower order p^n, $n > 0$. Then:* (i) *Every subloop of G has p-power order.* (ii) *$G/\varphi(G)$ is an abelian group of order $p^d > 1$ and exponent p.* (iii) *The elements $x_1, x_2, \ldots, x_t$ of G form a minimal set of generators of G if and only if $t = d$ and $x_1\varphi(G), x_2\varphi(G), \ldots, x_d\varphi(G)$ generate $G/\varphi(G)$.* (iv) *The group of all automorphisms of G has order dividing $p^k(p^d - 1)(p^{d-1} - 1)\cdots(p - 1)$ where $2k = d(2n - 1 - d)$.*

Proof. The lower central series of G satisfies the condition of V, Theorem 2.2. Hence (i) holds. Since G is finite, $\varphi(G) \neq G$; hence (ii) follows from Theorem 2.2. The rest of the proof is the same as for p-groups (PHILIP HALL [B 6].) However, in contrast with the case for p-groups, G can have a single generator ($d = 1$) without being a cyclic group (BRUCK [70].)

3. Hall's enumeration principle

PHILIP HALL [B 6] seems to have been the first to apply to group theory in a systematic fashion the well known MÖBIUS formulas from number theory. We shall give our own interpretation of the principles involved. (See also Birkhoff's Lattice Theory and the references cited therein.)

Let L be a finite partially ordered set consisting of a non-empty set L together with a binary relation ($<$) such that, for all elements a, b, c of L: (i) If $a < b$ and $b < c$, then $a < c$; (ii) At most one of the following holds: $a < b$, $a = b$, $b < a$; (iii) L has a unique (maximal) element e such that $a \leqq e$ for each a in L. Let F be the vector space of all single-valued functions $a \to f(a)$ from L to the ring of rational integers, where we define $(f + g)(a) = f(a) + g(a)$, $(nf)(a) = n \cdot f(a)$ for all f, g in F and all integers n. With each a in L associate the following function h_a: $h_a(b) = 1$ if $a \leqq b$ and $h_a(c) = 0$ if $a \nleqq c$. We first show that the h_a form a basis for F over the integers. To see this, we observe that if m is a minimal element of L (an element such that $a \leqq m$ implies $a = m$) then $a \neq m$ implies $h_a(m) = 0$. If f is in F, the function g, defined by $g(a)$

$= f(a) - f(m)h_m(a)$, has the property that $g(m) = 0$. Hence g and the h_a for $a \neq m$ may be thought of as defined for the subsystem L' obtained from L by deleting m. Therefore we may assume inductively that g is a unique linear combination with integer coefficients of the h_a for $a \neq m$; and then we see that f is a like linear combination of the h_a, $a \in L$. If L has n elements we enumerate the elements of L in some arbitrary fashion as $a(1), a(2), \ldots, a(n)$ and form a matrix H such that the element of H in the ith row, jth column is the number $h_{a(i)}(a(j))$. Since the h_a form a basis of F, H is nonsingular. In particular, the submatrix of $n-1$ rows, n columns, obtained from H by deleting the row corresponding to h_e, has rank $n-1$. Equivalently, there is, aside from a rational multiple, exactly one nonzero rational-valued function φ on L such that

$$\sum \varphi(a) f(a) = 0 \tag{3.1}$$

if and only if the element f of F is a linear combination of the h_a for $a \neq e$. We normalize φ by the requirement that $\varphi(e) = 1$. Then, by substituting in turn for f each h_a with $a \neq e$, we get the following complete description of φ:

$$\varphi(e) = 1; \quad \varphi(a) = -\sum_{a<b} \varphi(b) \qquad \text{if } a < e. \tag{3.2}$$

As a familiar example of the above, let e be a fixed positive integer, $e \geqq 2$, and let L be the set of all positive integral divisors of e, where $a \leqq b$ in L means that a is a divisor of b. Then $\varphi(a) = \mu(e/a)$ for each a in L, where μ is the familiar Möbius function of number theory.

Next suppose that the partially ordered set L consists of finitely many non-empty subsets of a set G, including G itself, where $a < b$ means that a is a proper subset of b. Here $e = G$ is the unique maximal element of L. Now consider some finite collection S of non-empty subsets s of G. For each s in S define an *incidence* function f_s on L as follows: $f_s(a) = 1$ or 0 according as s is or is not a subset of a. Furthermore, let $f = f_S$ be the sum of the f_s for s in S. Since f, along with each f_s, is an element of F, then f is a linear combination of the h_a. We are interested in the case that f is a linear combination of the h_a for $a \neq e$, so that f satisfies (3.1). With this in mind, we note that $f_s(a) = 1$, $a < b$ implies $f_s(b) = 1$. Hence the simplest non-trivial hypothesis is that to each s in S there corresponds an a in L, $a < e$, such that $f_s = h_a$. Equivalently, for each s in S: (I) There exists at least one a in L such that $s \subset a$, $a < e$. (II) There exists a unique a in L which is minimal in the property that $s \subset a$.

Now we come to the case by Philip Hall for the enumeration theory of finite p-groups. Let G be a finite centrally nilpotent loop of p-power order, where p is a prime, and let L be the set of all subloops M of G which contain the Frattini subloop $D = \varphi(G)$. Here, if D has index p^d, L is essentially the lattice of all subgroups of the elementary abelian group G/D of order p^d. Consequently, as HALL shows, the function φ of (3.1), (3.2) is given by $\varphi(M) = (-1)^i p^{i(i-1)/2}$ if the element M of L is a subloop of index p^i in G. If we consider a collection S of non-empty subsets s of G such that each s is contained in at least one maximal proper subloop of G, then (I) of the preceding paragraph is satisfied. Moreover, (II) holds since L is closed under intersection. Therefore, if $f = f_S$, (3.1) becomes

$$\sum_{i=0}^{d} (-1)^i p^{i(i-1)/2} \sum f(M_i) = 0 \tag{3.3}$$

where the inner sum is over all subloops M_i of G which contain D and have index p^i in G. The simplest application of (3.3) is as follows:

Theorem 3.1. *Let G be a centrally nilpotent loop of finite primepower order p^n. Then, for each integer m in the range $0 \leqq m \leqq n$, the number of subloops of G of order p^m is congruent to 1 modulo p.*

Proof. If $m = n$, the number in question is one, so we assume $0 \leqq m < n$. Let S be the set of all subloops of G of order p^m. Then, if $f = f_S$, $f(M)$ is the number of such subloops contained in M; and the total number is $f(G)$. Certainly every member of S is contained in a maximal proper subloop of G. Hence we may use (3.3) and, by taking congruences modulo p, we get $f(G) \equiv \sum f(M_1) \bmod p$. Each M_1 has order p^{n-1}, so we assume inductively that $f(M_1) \equiv 1 \bmod p$. Then $f(G) \equiv t \bmod p$ where t is the number of maximal proper subloops of G. Equivalently, t is the number of subgroups of order p^{d-1} in an elementary abelian p-group of order p^d. By direct computation or otherwise, $t \equiv 1 \bmod p$. This completes the proof of Theorem 3.1.

Aside from trivialities, the above proof is that of HALL for p-groups. Similarly:

Theorem 3.2. *Let G be a di-associative, non-cyclic, centrally nilpotent loop of odd prime-power order p^n. Then:*

(a) *For $0 < m < n$, the number of subloops of G of order p^m is congruent to $1 + p$ modulo p^2* (KULAKOFF).

(b) *For $1 < m < n$, the number of cyclic subloops of G of order p^m is congruent to 0 modulo p* (MILLER).

Proof. See P. HALL [B 5], Theorems 1.52, 1.53. Note that, for (a), the case that $D = \varphi(G)$ has index p^2 is critical. Here the hypothesis of di-associativity saves us by ensuring that G is a group.

Anticipating some of the material of later chapters, we may remark that the hypotheses of Theorem 3.2 are fairly natural for Moufang loops. By MOUFANG's Theorem (VII. 4) every Moufang loop is di-associative. By VIII, Theorem 10.1, every finite (or finitely generated) commutative Moufang loop is centrally nilpotent. From this it is an easy exercise to deduce that every finite commutative MOUFANG loop is a direct product of an abelian group of order prime to 3 and a centrally nilpotent commutative Moufang loop G of order a power of 3. And Theorem 3.2 applies to G.

4. Local properties

If X is a non-empty subset of a loop G, let $N(X; G)$ denote the *smallest normal subloop* of G containing X. If X is empty, let $N(X; G) = 1$. If $X \subset Y$, clearly $N(X; G) \subset N(Y; G)$. If H, K are subloops of G such that $X \subset H \subset K$, then $H \cap N(X; K)$ is a normal subloop of H containing X, so $N(X; H) \subset N(X; K)$. Combining these facts, we see that

$$H \subset K \to N(X \cap H; H) \subset N(X \cap K; K) \qquad (4.1)$$

for all subloops H, K, not necessarily containing the subset X.

Lemma 4.1. *Let X be a subset of a loop G. Then $N(X; G)$ is the set union of the subloops $N(X \cap H; H)$ where H ranges over all finitely generated subloops of G.*

Proof. We may assume that X is non-empty, since otherwise the proof is trivial. Let M be the set union described in the lemma. If H, K are finitely generated subloops of G, there exists a finitely generated subloop L of G containing both H and K. Therefore, by (4.1), $N(X \cap L; L)$ contains both $N(X \cap H; H)$ and $N(X \cap K; K)$. Consequently, M is a subloop of G. By (4.1) again, $N(X \cap H; H) \subset N(X; G)$ for each subloop H, so $M \subset N(X; G)$. On the other hand, $N(X; G)$ is generated by the set of all elements $x\theta$ where x ranges over X and θ ranges over the inner mapping group, $\mathfrak{I}(G)$, of G. Consider a fixed pair x, θ. By the definition of $\mathfrak{I}(G)$, there exists at least one finite subset S of G such that θ lies in the subgroup of $\mathfrak{I}(G)$ generated by inner mappings of form $R(s, t)$, $L(s, t)$, $T(s)$, where s, t range over S. Let H be the subloop of G generated by the finite set S and the element x of G. Since S is in H, θ induces an inner mapping of H. Hence θ maps the normal subloop $N(X \cap H; H)$ of H into itself. Since x is in $X \cap H$, $x\theta$ is in $N(X \cap H; H)$ and hence in M. Therefore M contains a set of generators of $N(X; G)$, so $M = N(X; G)$.

If (P) is a property of loops, a loop G is said to have property (P) *locally* if every finitely generated subloop of G has property (P). And (P) is a *local* property of loops if every loop which has (P) locally also has (P). In the discussion of properties of loops, questions of separation (the element a is or is not in the subloop B) sometimes arise. Some of these questions can be handled by use of Lemma 4.1. We shall now give a fairly general example.

Let $\mathfrak{C}$ be a given non-empty set of loop words W_n. Let $\mathfrak{L}$ be a given class of loops containing the subloops of each of its members. For each loop G in $\mathfrak{L}$ and each non-empty subset X of G, let $(X; G; \mathfrak{C})$ be the subloop of G constructed as follows: $(X; G; \mathfrak{C}) = N(Y; G)$, Y being the set of all elements $W_n(p, z_2, \ldots, z_n)$ where W_n ranges over the words in $\mathfrak{C}$, p over the elements of $N(X; G)$ and $z_2, \ldots, z_n$ over the elements of G. Here $\mathfrak{C}$ may be finite or infinite.

Lemma 4.2. *If a finite subset S of the loop G is contained in $(X; G; \mathfrak{C})$, then there exists a finitely generated subloop H_0 of G such that S is contained in $(X \cap H; H; \mathfrak{C})$ for every finitely generated subloop H which contains H_0.*

Proof. Let Y be the set mentioned in the definition of $(X; G; \mathfrak{C})$. By Lemma 4.1, each element of S is in some $N(Y \cap K; K)$ where K is finitely generated. Since S is finite, S is in some $N(Y \cap L; L)$ where L is finitely generated. Hence there exists a finite subset T of $Y \cap L$ such that S is in $N(T; L)$. Each t in T has the form $W_n(p, z_2, \ldots, z_n)$ for some W_n in $\mathfrak{C}$, p in $N(X; G)$ and $z_2, \ldots, z_n$ in G. Thus there exists a finite subset P of $N(X; G)$ and a finite subset Z of G such that each t in T has form $W_n(p; z_2, \ldots, z_n)$ for some W_n in $\mathfrak{C}$, p in P, $z_2, \ldots, z_n$ in Z. Since P is a finite subset of $N(X, G)$, then, by Lemma 4.1, P is in $N(X \cap H; H)$ for any suitably large finitely generated subloop H. If we choose H to contain Z and the finitely generated subloop L we will have $S \subset N(T; L) \subset N(T; H) \subset (X \cap H; H, \mathfrak{C})$. This proves Lemma 4.2.

Next let $\mathfrak{L}$ be a class of loops satisfying (a), (b) of § 1. A nilpotency function f for $\mathfrak{L}$ will be said to have *word type* if there exists at least one non-empty set $\mathfrak{C}$ of loop words such that, for G in $\mathfrak{L}$, $f(G)$ is characterized as follows: The element a of G is in $f(G)$ if and only if $W_n(a, x_2, x_3, \ldots, x_n) = 1$ for every W_n in $\mathfrak{C}$ and all $x_2, x_3, \ldots, x_n$ in G. It is readily verified that every nilpotency function discussed in § 1 has word type. For example, a suitable set of words for central nilpotency consists of U_2, V_3, W_3 where U_2 is the commutator word and V_3, W_3 are two forms of the associator word: $U_2(X_1, X_2) = (X_1, X_2)$; $V_2(X_1, X_2, X_3) = (X_1, X_2, X_3)$; $W_3(X_1, X_2, X_3) = (X_3, X_2, X_1)$. If f has word type and if K is a normal subloop of the loop G in $\mathfrak{L}$, then the natural homomorphism $x \to xK$ maps $f(G)$ into a subloop of $f(G/K)$; consequently, the set of all elements of form $W_n(k, x_2, \ldots, x_n)$, where W_n is in $\mathfrak{C}$, k is in K and $x_2, \ldots, x_n$ are in G, must be a subset of K. Considering next the homomorphism $x \to x(K, G)_f$, we see that the smallest normal subloop of G containing the set of elements just described must be $(K, G)_f$.

Thus the following lemma is a corollary of Lemma 4.2:

Lemma 4.3. *Let f be a nilpotency function of word type for a class $\mathfrak{L}$ satisfying* (a), (b) *of* § 1. *If G is in $\mathfrak{L}$, if X is a non-empty subset of G and if S is a finite subset of $(N(X; G), G)_f$, then there exists a finitely generated subloop H_0 of G such that $S \subset (N(X \cap H; H), H)_f$ for every finitely generated subloop H which contains H_0.*

In order to make effective use of Lemma 4.3 we need the concept of a linear system $\mathfrak{S}$ of a loop G (Kuroš, Theory of Groups, vol. 2). First let $\mathfrak{S}$ be a non-empty set whose elements are subloops of G and which is simply (or linearly) ordered by inclusion: if $H, K \in \mathfrak{S}$ then either $H \subset K$ or $K \subset H$. An *extension*, $\mathfrak{T}$, of $\mathfrak{S}$ is a set of subloops of G, simply ordered by inclusion, such that $\mathfrak{S} \subset \mathfrak{T}$; and T is a *proper* extension of $\mathfrak{S}$ if $\mathfrak{S} \neq \mathfrak{T}$. By ZORN's Lemma, applied to all the extensions of $\mathfrak{S}$, $\mathfrak{S}$ must have at least one maximal extension, say $\mathfrak{M}$. That is, $\mathfrak{M}$ is an extension of $\mathfrak{S}$ such that $\mathfrak{M}$ has no proper extension. Among the processes which yield extensions of a simply ordered set $\mathfrak{S}$, we are interested in the following: (1) If G is not an element of $\mathfrak{S}$, adjoin G to $\mathfrak{S}$. (2) If the identity subloop, 1, is not an element of $\mathfrak{S}$, adjoin 1 to $\mathfrak{S}$. (3) Consider some non-empty subset $\mathfrak{T}$ of $\mathfrak{S}$ and let H be the intersection of all subloops of G which are elements of $\mathfrak{T}$. Then H is a subloop. If $\mathfrak{T}'$ is the set consisting of all elements K of $\mathfrak{S}$ such that $H \subset K$, then $\mathfrak{T} \subset \mathfrak{T}'$ and hence H is the intersection of the elements of $\mathfrak{T}'$. In particular, if L is an element of $\mathfrak{S}$ which is not in $\mathfrak{T}'$, then $L \subset K$ for every K in $\mathfrak{T}'$ and hence $L \subset H$. Therefore H may be adjoined to $\mathfrak{S}$. We note that if the elements of $\mathfrak{S}$ are normal subloops of G, then H is normal in G. Similarly, if, for some set, Ω, of endomorphisms of G, all elements of $\mathfrak{S}$ are Ω-subloops, then so is H. (4) Consider some non-empty subset $\mathfrak{T}$ of $\mathfrak{S}$ and let H be the union (i. e., the set of elements of G contained in one or more) of the elements of $\mathfrak{T}$. Since $\mathfrak{T}$, like $\mathfrak{S}$, is simply ordered, H is a subloop of G. If $\mathfrak{T}'$ is the set of all elements K of $\mathfrak{S}$ such that $K \subset H$, then $\mathfrak{T} \subset \mathfrak{T}'$ and hence H is the union of the elements of $\mathfrak{T}'$. In particular, if L is an element of $\mathfrak{S}$ which is not in $\mathfrak{T}'$, then $K \subset L$ for every K in $\mathfrak{T}'$ and hence $H \subset L$. Therefore H may be adjoined to $\mathfrak{S}$. As before, H is a normal subloop or an Ω-subloop if the same is true for each element of $\mathfrak{S}$. We note that processes (1)—(4) may be combined as follows: We partition $\mathfrak{S}$ into an *upper section* $\mathfrak{U}$ and a *lower section* $\mathfrak{L}$, one of which may be empty. That is, each element of $\mathfrak{S}$ is in exactly one of $\mathfrak{L}$, $\mathfrak{U}$, and $A \subset B$ for every A in $\mathfrak{L}$, B in $\mathfrak{U}$. If $\mathfrak{L}$ is empty, define $H_1 = 1$; otherwise let H_1 be the union of the elements of $\mathfrak{L}$. If $\mathfrak{U}$ is empty, define $H_2 = G$; otherwise let H_2 be the intersection of the elements of $\mathfrak{U}$. Then $A \subset H_1 \subset H_2 \subset B$ for every A in $\mathfrak{L}$, B in $\mathfrak{U}$. If $H_1 \neq H_2$ then H_1 is either a (unique) maximal element of $\mathfrak{L}$ or not in $\mathfrak{S}$; similarly, H_2 is either a (unique) minimal element of $\mathfrak{U}$ or not in $\mathfrak{S}$. Moreover, if x is an element of H_2 which is not in H_1 then H_2 is the intersection of all elements of $\mathfrak{S}$ containing x and H_1 is the union of all elements of $\mathfrak{S}$ not containing x. If $H_1 = H_2 = H$, then H is either a (unique) maximal element of $\mathfrak{L}$ or a (unique) minimal element of $\mathfrak{U}$ or not an element of $\mathfrak{S}$. And in any case, H_1, H_2 can be adjoined to $\mathfrak{S}$ if they are not in $\mathfrak{S}$.

By a *linear system* $\mathfrak{S}$ of a loop G we mean a (non-empty) set of subloops of G, simply ordered by inclusion, such that none of the above processes yields a proper extension of $\mathfrak{S}$. That is: (i) $\mathfrak{S}$ contains 1 and G. (ii) The intersection of any non-empty set of elements of $\mathfrak{S}$ is an element of $\mathfrak{S}$. (iii) The union of any non-empty set of elements of $\mathfrak{S}$ is an element of $\mathfrak{S}$. By a *jump* of a linear system $\mathfrak{S}$ of G we mean an ordered pair H, K of elements of $\mathfrak{S}$ such that $H \subset K$, $H \neq K$ and such that there exists no element M of $\mathfrak{S}$ satisfying $H \subset M \subset K$, $M \neq H$, $M \neq K$. If x is an element of K which is not in H, so that, in particular, $x \neq 1$, then K is the minimal element of $\mathfrak{S}$ which contains x and H is the maximal element of $\mathfrak{S}$ which does not contain x. Conversely, each element $x \neq 1$ determines a unique jump H, K of $\mathfrak{S}$ in just this manner. A linear system $\mathfrak{S}$ is called a *chief* (or *principal*) system of G if each element of $\mathfrak{S}$ is normal in G; and $\mathfrak{S}$ is called an Ω-system of G if each element of $\mathfrak{S}$ is an Ω-subloop of G. Again, $\mathfrak{S}$ is called a *normal system* if, for each jump H, K of $\mathfrak{S}$, H is normal in K. A *composition chief* system is a chief system which has no proper chief refinement. By Zorn's Lemma, every chief system can be refined to a composition chief system. Similarly for composition normal systems, composition chief Ω-systems and so on. We shall not discuss the problems connected with analogues of the Jordan-Hölder Theorem or the Schreier-Zassenhaus Refinement Theorem (IV.3); for these see Kuroš, loc. cit. supra.

A linear system $\mathfrak{S}$ of G is called *descending (ascending)* if every non-empty subset of S has a maximal (minimal) element. These two types were considered in § 1 in connection with transfinitely f-nilpotent loops and transfinitely upper f-nilpotent loops respectively.

Now let $\mathfrak{L}$ be any class of loops satisfying (a), (b) of § 1 and let f be any nilpotency function for $\mathfrak{L}$. By an *f-system*, $\mathfrak{S}$, of a loop G in $\mathfrak{L}$ we mean a chief system of G such that, for each jump H, K of $\mathfrak{S}$, $H \supset (K, G)_f$. Equivalently, $K/H \subset Z_f(G/H)$. A loop G in $\mathfrak{L}$ will be called a *Z_f-loop* if it has at least one f-system. In particular, when $f(G)$ is the centre of G for each G in $\mathfrak{L}$, we speak of *central* systems and of *Z-loops*. Malcev (see Kuroš, loc. cit.) has proved a local theorem for Z-groups. In order to generalize this to Z_f-loops we shall assume that f is of word type.

Theorem 4.1. *Let f be a nilpotency function of word type. Then every local Z_f-loop is a Z_f-loop.*

Proof. Let G be a local Z_f-loop. That is, every finitely generated subloop of G is a Z_f-loop. We begin with the following proposition:

(a) *If $F \neq 1$ is a finitely generated subloop of G and if F^* is the smallest normal subloop of G containing F, then F is not contained in $(F^*, G)_f$.*

By hypothesis, there exists a finite subset X of G which generates F. Since $F \neq 1$, we may assume without loss of generality that X does not contain 1. In our previous notation, $F^* = N(F; G) = N(X; G)$. Suppose, in contradiction to (a), that $F \subset (F^*, G)_f$. Then, in particular, $X \subset (F^*, G)_f = (N(X; G), G)_f$. By Lemma 4.3, since X is finite, there exists a finitely generated subloop H of G, containing X, such that $X \subset (N(X; H), H)_f$. We select some f-system $\mathfrak{S}$ of H; one exists by hypothesis. Since X does not contain the identity element, each x in X uniquely determines a jump P_x, Q_x of $\mathfrak{S}$ such that x is in Q_x but not in P_x. By the definition of an f-system, $(Q_x, H)_f \subset P_x$ for each x. Since X is finite and $\mathfrak{S}$ is linearly ordered, $X \subset Q_x$ for some x in X. Therefore

$N(X; H) \subset Q_x$ and hence $X \subset (N(X; H), H)_f \subset (Q_x, H)_f \subset P_x$, contradicting the fact that x is in X but not in P_x. Therefore (a) must be true.

Before proceeding, we note that (a) implies that *all minimal normal subloops of a local Z_f-loop are contained in the f-centre.* Consequently, *if a loop G has the property that every quotient loop of G is a local Z_f-loop then every composition chief system of G is an f-system.* The present weaker hypotheses require a longer argument; the fact that free groups are Z-groups but have quotients which are not Z-groups shows that we must single out, in some way, an f-system for G. We use a method of MALCEV to prove the following:

(b) *With each x of G there may be associated a normal subloop $H(x)$ of G such that* (i) $H(1) = 1$; (ii) *if $x \neq 1$ then $H(x)$ contains $(N(x; G), G)_f$ but not x;* (iii) *the set of all subloops $H(x)$, $x \in G$, is linearly ordered by inclusion.*

Suppose (b) has been proved, and let $\mathfrak{S}$ be the linear system of G obtained from the set of all $H(x)$ by forming unions and intersections and, if necessary, adjoining G. Let A, B be a jump of $\mathfrak{S}$. If x is in B but not in A, then $x \neq 1$; whence, by (ii), $H(x) \subset A$. Therefore A contains $(N(x; G), G)_f$ for every x of B which is not in A and hence, in fact, for every x of B. Consequently, $(B, G)_f \subset A$. This shows that $\mathfrak{S}$ is an f-system.

In order to prove (b), we consider the set of all finitely generated subloops of G; F will be a generic notation for such a subloop. In each F we select a fixed f-system for F. If $x \neq 1$ is in F, let F_x denote the largest element of the f-system for F which does not contain x. In addition, define $F_1 = 1$. For any fixed F, the subloops F_x, $x \in F$, are simply ordered. We wish to associate with every finite non-empty set X of elements of G a set, $\mathfrak{L}(X)$, of finitely generated subloops of G such that: (I) for each F in $\mathfrak{L}(X)$, X is a subset of F and the subloops F_x of F, $x \in X$, have a fixed ordering by inclusion, independent of F; (II) for each finite subset Y of G, there is an F in $\mathfrak{L}(X)$ which contains Y; (III) $\mathfrak{L}(X \cup Y) \subset \mathfrak{L}(X) \cap \mathfrak{L}(Y)$ for all finite non-empty subsets X, Y of G. In order to show that this can be done, we strengthen our requirements as follows: (1) If X consists of a single element x, $\mathfrak{L}(X) = \mathfrak{L}(x)$ shall consist of all F containing x; for such an X, requirements (I), (II) are clearly met. (2) If X consists of n elements, $n > 1$, then, for some one fixed enumeration (out of $n!$ possible ones) of the elements of X, say $x(1), x(2), \ldots, x(n)$, $\mathfrak{L}(X)$ shall consist of all F which contain X and satisfy the inequalities $F_{x(1)} \subset F_{x(2)} \subset \cdots \subset F_{x(n)}$. Here equalities are not excluded. Thus, with a set X of n elements we can associate $n!$ sets $\mathfrak{L}(X)$, each of which (although possibly empty) satisfies (I). When we speak of a set $\mathfrak{L}(X)$, we shall have such a set in mind. Then (III) simply means that the ordering in $X \cup Y$ is to be consistent with those in X and Y. Now consider a pair (M, θ) where M is a non-empty (possibly infinite) subset of G and θ is a single-valued mapping which assigns, to each finite non-empty subset X of M, a set $X\theta = \mathfrak{L}(X)$ such that (I), (II), (III) hold for all finite non-empty subsets X, Y of M. Order such pairs as follows: $(M, \theta) \subset (N, \varphi)$ if $M \subset N$ and φ induces θ on M. By (1), there exists a pair (M, θ) where M consists of a single element x. Therefore, by Zorn's Lemma, there exists a maximal pair (M, θ); one such that $(M, \theta) \subset (N, \varphi)$ implies $M = N$. We shall show that $M = G$. Suppose, on the contrary, that there exists an element p of G which is not in M. Consider some subset X of M, with $n \geqq 1$ elements, and let $x(1), x(2), \ldots, x(n)$ be the elements of X arranged so that $F_{x(1)} \subset F_{x(2)} \subset \cdots \subset F_{x(n)}$ for each F in $X\theta = \mathfrak{L}(X)$. For every F in $\mathfrak{L}(X)$ containing p, F_p must fit into this ordering in some one of $n+1$ possible positions, which we can indicate by inserting p in the ordered set $x(1), x(2), \ldots, x(n)$. We wish to show that $(X \cup p)\varphi = \mathfrak{L}(X \cup p)$ can be defined in at least one of the $n+1$ ways so as to satisfy (II). If not, then for each of these ways there must exist a finite set S such that no F in $L(X)$ can contain p, S and satisfy the ordering. Let T be the union of $n+1$ "contradictory" sets so chosen. By (II), there exists an F in $\mathfrak{L}(X)$ which contains T and p; but by construction, none of the $n+1$ orderings is satisfied for F. This is a contradiction. Thus, for every finite X, there is at least one way of defining $\mathfrak{L}(X \cup p)$ so that (II) holds; and there are at most $n+1$ ways if X has n elements. Next assume, for a fixed X with n elements, that for every way of defining $\mathfrak{L}(X \cup p)$ so that (II) holds, there exists a finite non-empty subset Y of M so that no choice of $\mathfrak{L}(X \cup p \cup Y)$ and/or $L(Y \cup p)$ can be made to satisfy (II). Choose such a Y for each of the (at most) $n+1$ ways and let Z be the union of X and all the Y's. Then, by construction, there is no way of choosing $L(Z \cup p)$ to satisfy (II). But, since Z is finite, this contradicts our earlier conclusion. Consequently, a pair $(M \cup p, \varphi)$ can be defined, and this contradicts the maximality of (M, θ). Therefore there exists at least one collection of sets $\mathfrak{L}(X)$ satisfying (I), (II), (III).

Now we assume a fixed choice of the sets $\mathfrak{L}(X)$; and we define $H(x)$, for each x in G, as follows: An element p of G is in $H(x)$ if and only if there exist finite non-empty subsets Y, Z of G such that $p \in F_x$ for every F of $\mathfrak{L}(Y)$ which contains Z. Here we may always assume that Z contains x, p. If p is in $H(x)$, according to a pair Y, Z, and if θ is an inner mapping of G, there exists a finite subset W of G such that θ induces an inner mapping of every F which contains W. There exists an F in $\mathfrak{L}(Y)$ containing Z, W; and, since F_x is normal in F, $p\theta \in (F_x)\theta \subset F_x$ for every such F. Thus $H(x)$ is a self-conjugate subset of G. If q is in $H(x)$ according to Y' and Z', then for every F in $\mathfrak{L}(Y \cup Y')$ which contains $Z \cup Z', p, q$ will both be in F_x. Hence $H(x)$ is a subloop. In particular, since $F_1 = 1$ for each F, $H(1) = 1$. If $x \neq 1$ and if Y is a finite subset of $(N(x; G), G)_f$ then, for every F containing a certain F', as in the proof of (a) above, we will have $Y \subset (N(x; F), F)_f \subset F_x$. Taking for Z a generating set of F', we deduce in particular that $Y \subset F_x$ for every F in $\mathfrak{L}(Y)$ which contains Z. Thus $Y \subset H(x)$. In other words, $(N(x; G), G)_f \subset H(x)$. At the same time, since $x \neq 1$ is never in $F_x, H(x)$ does not contain x. Next consider a pair x, y and suppose that the order in $\mathfrak{L}(x \cup y)$ gives $F_x \subset F_y$ for every F in $\mathfrak{L}(x \cup y)$. If p is in $H(x)$ then, for some finite sets S, T, $p \in F_x$ for every F in $\mathfrak{L}(S)$ which contains T. Then, for every F in $\mathfrak{L}(x \cup y \cup S)$ which contains T, $p \in F_x \subset F_y$. Thus $p \in H(y)$, showing that $H(x) \subset H(y)$. This completes the proof of (b) and hence the proof of Theorem 4.1.

For any nilpotency function f of a class $\mathfrak{L}$ of loops, we define a loop G to be an SN_f-loop if it has at least one normal system such that, at each jump H, K of the system, $(K, K)_f \subset H$. A similar definition may be given for SI_f-loops in terms of a chief system. The *local theorem for* SI_f-loops *holds for f of word type*. With some simplification of the work, we alter Lemma 4.3 as follows: *If S is a finite subset of* $(N(X; H), N(X; H))_f$ *then* $S \subset (N(X \cap H; H), N(X \cap H; H))_f$ *for every finitely generated subloop H containing some H_0.* The pattern of Theorem 4.1 requires little modification. It is not clear whether the local theorem holds for SN_f-loops; compare the proof of MALCEV for SN-groups (KUROŠ, loc. cit.) A loop may be called an *SD_f-loop* if it has a *descending* normal system such that, at each jump H, K, $(K, K)_f \subset H$. The following theorem is almost immediate:

Theorem 4.2. *If f is any nilpotency function (not necessarily of word type) for the class of all loops, such that $f(G)$ always contains the centre of G, then every free loop is an SD_f-loop.*

Proof. Let G be a free loop and define the f-derived series $\{G^\alpha\}$ of G inductively: $G^0 = G$; $G^{\alpha+1} = (G^\alpha, G^\alpha)_f$; if α is a limit ordinal, G^α is the intersection of the G^β for all $\beta < \alpha$. Let F be the ultimate term of this series. If $F \neq 1$, then F is a free loop of positive rank. Hence, if F' is the commutator-associator subloop of F, F/F' is a free abelian group of positive rank. In particular, $f(F/F') = F/F'$, so $(F, F)_f \subset F' \neq F$. This is a contradiction, so $F = 1$. We also note from this proof that the f-derived series of a free loop must terminate at a limit ordinal.

The f-derived series, although characteristic, need not be chief. Thus SD_f-loops may not always be SI_f-loops, although they are SN_f-loops.

5. Ordered loops

A (multiplicative) loop G is said to be *simply ordered* by a binary relation $(<)$ provided that, for all a, b, c in G: (i) exactly one of the following holds: $a < b$, $a = b$, $b < a$; (ii) if $a < b$ and $b < c$, then $a < c$; (iii) if $a < b$ then $ac < bc$ and $ca < cb$. The relation $a > b$ is interpreted, as usual, to mean $b < a$. If 1 is the identity element of G, an element a is called *positive* if $a > 1$, *negative* if $a < 1$. A loop of order one can be simply ordered in a trivial manner. If $G \neq 1$ and G is simply ordered, and if the elements a, b of G satisfy $ab = 1$, $a \neq 1$, we see from (iii) that exactly one of a, b is positive.

Lemma 5.1. *A loop $G \neq 1$ can be simply ordered if and only if G contains two non-empty subsets P, N such that* (a) *if $x \in G$ then exactly one of the following holds: $x = 1$, $x \in P$, $x \in N$;* (b) *$PP \subset P$ and $NN \subset N$;* (c) *P (and hence N) is a self-conjugate subset of G.* (Compare NEUMANN [B 13].)

Proof. (I) Let G be simply ordered and let P, N consist of the positive and negative elements of G respectively. Then (a), (b) follow immediately. If $a, b, x \in G$ and if $a < b$ then $aR(x) < bR(x)$, $aL(x) < bL(x)$ by (iii); from this we deduce readily that $a\theta < b\theta$ for every θ in the multiplication group $\mathfrak{M}(G)$ of G. In particular, the inner mapping group of G maps P into P, N into N. This proves (c).

(II) Let G be a loop with non-empty subsets P, N satisfying (a), (b), (c). From the fact that N is self-conjugate in G we deduce readily that $Nx = xN$, $N(xy) = (Nx)y = x(Ny)$ for all x, y in G. We define $x < y$ if and only if $x \in yN$. If $x < y$ and $y < z$, then $x \in yN \subset (zN)N \subset z(NN) \subset zN$, so $x < z$; that is, (ii) holds. If $x = yz$, then z is uniquely determined by x, y. Hence, by (a), exactly one of $z \in N$, $z = 1$, $z \in P$ must be true. If $z \in P$ and if $y = xw$ then $w \neq 1$. If w is also in P then $x = yz = (xw)z \in (xP)P = x(PP) \subset xP$, so $x \in xP$ and hence $1 \in P$, a contradiction. Hence $x \in yP$ if and only if $y \in xN$. This proves (i). Finally, suppose $x \in yN$. Then, for any z, $xz \in (yN)z = (yz)N$ and $zx \in z(yN) = (zy)N$. This proves (iii) and completes the proof of Lemma 4.1.

A subloop H of a simply ordered loop G is called *convex* (or *isolated*) provided H is normal in G and contains every element x of G such that $h < x < h'$ for some elements h, h' of H. If H is a convex subloop of G, the quotient group G/H can be simply ordered by defining $xH \leqq yH$ if and only if there exist elements a, b of G such that $a \in xH$, $b \in yH$ and $a \leqq b$. This order of G/H is said to be *induced* by the order of G. The kernel of an order-preserving homomorphism of G upon a loop is necessarily convex.

Theorem 5.1. *If G is a Z-loop with a central system $\mathfrak{S}$ such that, for each jump H, K of $\mathfrak{S}$, K/H is simply ordered, then G can be simply ordered so as to preserve the ordering in each K/H.* (Compare NEUMANN [B 13].)

Proof. We need only determine the P, N of Lemma 5.1. If $x \neq 1$, let H, K be the jump determined by x. We assign x to P or N according as the coset xH is positive or negative in K/H. Since K/H is part of the centre of G/H, each inner mapping of G induces the identity mapping on K/H. Hence P, N are self-conjugate subsets of G. If y is in H, then $xy \equiv yx \equiv x \bmod H$, so xy, yx, x are all in P or all in N. If y is not in K, then the rôles of x, y are simply interchanged. If y is in K but not in H, and if x, y are both in P, then so is xy. Hence $PP \subset P$, and, similarly, $NN \subset N$. This completes the proof. Note that, for each jump H, K, the homomorphism $x \to xH$ of G upon G/H is order-preserving. Hence the "lower term" H of every jump H, K is convex in the ordering of G.

If G is a free group, it is known that $G_\omega = 1$ (where ω is the first limit ordinal) and that $G_\alpha/G_{\alpha+1}$ is a free abelian group. Thus Theorem 5.1 gives one way (indeed, the original way) of proving that every free group may be simply ordered. The comparable facts are not available for free loops. Nevertheless, it is possible to show, by a tedious induction over the extension chains of Chapter I, that every free loop F can be simply ordered. The method yields no information about convex subloops and it is conceivable that F may be simply ordered so that it is order-simple. Here we define a simply ordered loop G to be *order-simple* if the only convex subloops of G are G and 1. The following theorem is due to ZELINSKY:

Theorem 5.2. *If G is a simply ordered loop whose centre Z has finite index, then G is centrally nilpotent. If some subgroup of Z is order-simple and has finite index in G, then G possesses an order-preserving isomorphism into the additive group of real numbers.*

For the proof of Theorem 5.2, see ZELINSKY [116]. ZELINSKY also shows how to construct simply ordered loops with centre of finite index which are centrally nilpotent of arbitrary class. His paper deals as well with topological loops; for a more recent study of the latter, see MALCEV [90].

We shall not attempt to discuss the literature of ordered groups. See the bibliographies of the papers cited; also BIRKHOFF, Lattice Theory.

VII. Moufang Loops

1. Groupoids with the inverse property

We saw in Chapter I that there exist quasigroups (for example, the free quasigroups) possessing (multiplicative) homomorphisms upon systems which are not quasigroups — and similarily for loops. In order to avoid this possibility it is of interest (see I.5, IV. 10) to consider a class, closed under homomorphism, of groupoids all of which are quasigroups. We may define such a class of groupoids G as follows. Let G be any groupoid with multiplication semigroup $\mathfrak{S}$ such that for each x in G there exist θ, φ in $\mathfrak{S}$ with $R(x)\theta = I$, $L(x)\varphi = I$. Clearly every homomorphic image of such a groupoid has the same property. Since $\mathfrak{S}$ is generated by the set of all $R(x)$, $L(x)$ for x in G, every α in $\mathfrak{S}$ has a β in $\mathfrak{S}$ such that $\alpha\beta = I$; consequently, $\mathfrak{S}$ is a group of permutations of G. Therefore G is a quasigroup.

We may single out a subclass in terms of the *inverse property*. A groupoid G is said to have the *left inverse property* if for each x in G there is at least one a in G such that (i) $a(xy) = y$ for every y in G. That is: $L(x)L(a) = I$. Consequently, the semigroup generated by the left multiplications of G is a group and, for given x, y in G, there is one and only one z in G such that $xz = y$. But G need not be a quasigroup, as witness the case that $xy = y$ for all x, y in G. Similarly, G has the *right inverse property* if, for each x in G, there is at least one b in G such that (ii) $(yx)b = y$ for every y in G. When (i), (ii) both hold, G is said to have the *inverse property*. In this case, G is a quasigroup, and there exist one-to-one mappings λ, μ of G upon G such that

$$L(x)^{-1} = L(x\lambda), \quad R(x)^{-1} = R(x\mu) \tag{1.1}$$

for all x in G. We note from (1.1) that

$$\lambda^2 = \mu^2 = I. \tag{1.2}$$

Clearly (1.1) can be restated as

$$(x\lambda)(xy) = y = (yx)(x\mu) \tag{1.3}$$

for all x, y in G. If $xy = z$, then, by (1.3), $(x\lambda)z = y$, $y(z\mu) = x\lambda$, $(y\lambda)(x\lambda) = z\mu$. Thus, and similarly,

$$(xy)\mu = (y\lambda)(x\lambda), \quad (xy)\lambda = (y\mu)(x\mu) \tag{1.4}$$

for all x, y in G. Examples may be given (cf. BRUCK [67]) to show that the mappings λ, μ are more or less independent of each other. Moreover, there exists an inverse property quasigroup G such that no loop isotopic to G has the inverse property.

In the case of a *loop with the inverse property* the situation is simpler. For, if G has identity element 1, equations (1.2), (1.3) give $x(x\lambda) = (x\lambda^2)(x\lambda \cdot 1) = 1 = (1 \cdot x)(x\mu) = x(x\mu)$ for each x, whence $\lambda = \mu$. Defining $x^{-1} = x\lambda$, we may now replace the above equations by

$$L(x)^{-1} = L(x^{-1}), \quad R(x)^{-1} = R(x^{-1}), \tag{1.5}$$

$$(x^{-1})^{-1} = x, \tag{1.6}$$

$$x^{-1}(xy) = y = (yx)x^{-1}, \tag{1.7}$$

$$(xy)^{-1} = y^{-1}x^{-1}, \tag{1.8}$$

holding for all x, y of G.

Another special class deserves mention: the *crossed-inverse* groupoids (ARTZY [56], BRUCK [75]). Here we assume that to each x in G there correspond a, b in G such that $(xy)a = y = b(yx)$ for every y in G. Thus $L(x)R(a) = I = R(x)L(b)$. Again G is a quasigroup and there exists a one-to-one mapping $\varkappa$ of G upon G such that

$$L(x)^{-1} = R(x\varkappa), \quad R(x)^{-1} = L(x\varkappa^{-1}) \tag{1.9}$$

for all x in G. Equivalently,

$$(xy)(x\varkappa) = y = (x\varkappa^{-1})(yx) \tag{1.10}$$

for all x, y in G. By (1.10), the equation $xy = z$ implies $z(x\varkappa) = y$, $y(z\varkappa) = x\varkappa$, $(x\varkappa)(y\varkappa) = z\varkappa$. Thus $\varkappa$ is an automorphism of G:

$$(xy)\varkappa = (x\varkappa)(y\varkappa) \tag{1.11}$$

for all x, y of G. As ARTZY shows, $\varkappa$ may have finite or infinite order. CROSSED-inverse loops deserve further study, but we shall say no more about them here.

Also see BAER [348, II] and COWELL [349].

2. Moufang elements

Let $(G, \cdot)$ be any loop and consider the principal isotope (G, o) defined by

$$x o y = xR(v)^{-1} \cdot yL(u)^{-1} \tag{2.1}$$

for all x, y in G. A necessary and sufficient condition that $(G, \cdot)$ be isomorphic to (G, o) is that there exist a permutation W of G such that $xW o yW = (xy)W$ for all x, y of G. Equivalently,

$$xU \cdot yV = (xy)W \tag{2.2}$$

for all x, y of G, where the permutations U, V are defined by $U = WR(v)^{-1}$, $V = WL(u)^{-1}$. An ordered triple (U, V, W) of permutations of a loop G is called an *autotopism* of G if (2.2) holds for all x, y. The autotopisms of G form a group under componentwise multiplication: $(U, V, W)(U', V', W') = (UU', VV', WW')$. If (U, V, W) is an autotopism of G and if

$$u = 1U, \quad v = 1V, \quad w = 1W, \tag{2.3}$$

then, from (2.2) with $y = 1$, $x = 1$ in turn, we get

$$UR(v) = VL(u) = W, \quad uv = w. \tag{2.4}$$

Consequently there is a one-to-one correspondence between autotopisms of G and isomorphisms of G upon principal isotopes.

Henceforth let G be an inverse property loop and let J be the inverse mapping of G, defined by

$$xJ = x^{-1}. \tag{2.5}$$

By (1.6), (1.8), (1.5) we have

$$J^2 = I, \quad JL(x)J = R(x^{-1}) = R(x)^{-1}, \quad JR(x)J = L(x^{-1}) = L(x)^{-1}. \tag{2.6}$$

Lemma 2.1. *If (U, V, W) is an autotopism of the inverse property loop G, then (W, JVJ, U) and (JUJ, W, V) are autotopisms of G. Moreover, if $u = 1U$, $v = 1V$, then $(U, V, W) = (S, SR(c), SR(c))(L(u), R(u),$*

$L(u)\,R(u)$) where S is a permutation of G and $c = v u^{-1}$ is an element of G such that

$$(xS)\,(yS \cdot c) = (xy)\,S \cdot c \tag{2.7}$$

for all x, y of G, and where u is an element of G such that

$$(ux)\,(yu) = [u(xy)]\,u \tag{2.8}$$

for all x, y of G.

Proof. From (2.2), $xU = (xy)\,W \cdot (yV)^{-1} = (xy)\,W \cdot yVJ$. Replacing x by xy, y by $y^{-1} = yJ$, we get $(xy)\,U = xW \cdot yJVJ$ for all x, y. Hence (W, JVJ, U) is an autotopism. From (2.2), $yV = xUJ \cdot (xy)W$. Replacing y by xy, x by xJ, we get $(xy)\,V = xJUJ \cdot yW$ for all x, y. Hence (JUJ, W, V) is an autotopism. Therefore, by (2.3), if $P^{-1} = U^{-1}JUJ$, we see that G has the autotopism $(U, V, W)^{-1}(JUJ, W, V) = (P^{-1}, L(u), L(u)^{-1})$ and hence, by (2.6), the autotopism $(L(u)^{-1}, JL(u)J, P^{-1})^{-1} = (L(u), R(u), P)$. Then $xP = (x \cdot 1)\,P = (xL(u))\,(1R(u)) = (xL(u))\,u = xL(u)\,R(u)$ for all x, so $P = L(u)\,R(u)$. Hence we have (2.8). Moreover, G has the autotopism $(U, V, W)\,(L(u), R(u), P)^{-1} = (S, C, D)$, say, where $S = UL(u)^{-1}$, so $1S = 1$. Since $xS \cdot yC = (xy)\,D$ for all x, y, and since $1S = 1$, we deduce that $SR(c) = C = D$ where $c = 1C$. Thus we have (2.7). Since $C = VR(u)^{-1}$, $c = vu^{-1}$. This completes the proof of Lemma 2.1.

A permutation S of a loop G is called a *pseudo-automorphism* of G provided there exists at least one element c of G, called a *companion* of S, such that (2.7) holds for all x, y of G. The pseudo-automorphisms form a group under composition. From (2.7) with $x = 1$ we deduce that

$$1S = 1\,. \tag{2.9}$$

Furthermore, by (2.7),

$$[x(yx)]\,S \cdot c = xS \cdot [(yx)\,S \cdot c] = xS \cdot [(yS)\,(xS \cdot c)] \tag{2.10}$$

for all x, y of G. If G has the inverse property, we set $y = x^{-1} = xJ$ in (2.7) and get $xS \cdot (xJS \cdot c) = 1S \cdot c = c$, whence $xJS \cdot c = (xS)^{-1} \cdot c = xSJ \cdot c$ for all x and thus

$$SJ = JS\,. \tag{2.11}$$

An element u of an inverse property loop G is called a *Moufang* element of G provided (2.8) holds for all x, y of G. By Lemma 2.1, if (U, V, W) is an autotopism of G, each of the elements $u = 1U$, $v = 1V$, $w = 1W$ is Moufang. In particular, the companion c of the pseudo-automorphism S of G is Moufang.

Lemma 2.2. *The set M of all Moufang elements of the inverse property loop G is a subloop of G. If p, q are in M then:* (i) *$R(p, q) = L(p^{-1}, q^{-1})$ is a pseudo-automorphism with companion (p, q).* (ii) *$T(p)$ is a pseudo-automorphism with companion p^{-3}.*

Proof. Define $A(p) = (L(p), R(p), L(p)R(p))$ for each p in M. Since $A(p)^{-1}$ is an autotopism, $1L(p)^{-1} = p^{-1}$ is in M. If q is also in M, $A(q)A(p)$ is an autotopism, so $1L(q)L(p) = pq$ is in M. Hence M is a subloop of G. If p, q are in M, $A(p^{-1})A(q^{-1})A(q^{-1}p^{-1})^{-1} = (S, T, X)$ is an autotopism where, in particular,

$$S = L(p^{-1})L(q^{-1})L(q^{-1}p^{-1})^{-1} = L(p^{-1}, q^{-1})$$

and $T = R(p^{-1})R(q^{-1})R(q^{-1}p^{-1})^{-1}$. Since $1S = 1$, we have $T = X = SR(c)$ where $c = 1T = (p^{-1}q^{-1})(pq)$. Since $(qp)c = pq$, then $c = (p, q)$. Thus S is a pseudo-automorphism with companion (p, q). By (2.11), $S = JSJ$. Hence, by (2.6), $S = R(p)R(q)R(pq)^{-1} = R(p, q)$. This proves (i). Since $A(p)$ is an autotopism, so is $B(p) = (JL(p)J, L(p)R(p), R(p))$ and hence so is $B(p)^{-1}A(p)^{-1} = (T(p), U, V)$. Here $U = R(p)^{-1}L(p)^{-1}R(p)^{-1}$, $V = R(p)^{-2}L(p)^{-1}$. Since $1T(p) = 1$, then $U = V$ and $T(p)$ is a pseudo-automorphism with companion k where $k = 1U = (p^{-1})^2p^{-1}$ and $k = 1V = p^{-1}(p^{-1})^2$. Thus $k = (p^{-1})^3 = p^{-3}$.

Theorem 2.1. *If G is an inverse property loop, the nuclei N_λ, N_μ, N_ϱ of G coincide with the nucleus N of G. Every pseudo-automorphism of G induces an automorphism of N; in particular, N is a characteristic normal subloop of the subloop M of Moufang elements of G.*

Proof. The element a of G is in N_λ if and only if

$$(ax)y = a(xy) \tag{2.12}$$

for all x, y of G. If (2.12) holds, we take inverses of both sides and see that a^{-1} (and hence a) is in N_ϱ. By symmetry, $N_\lambda = N_\varrho$. From (2.12) again, $y = (ax)^{-1}[a(xy)]$ for all x, y. Setting $x = a^{-1}p^{-1} = (pa)^{-1}$, $y = (pa)q$, we find that $(pa)q = p(aq)$ for all p, q. The converse also holds, so $N_\varrho = N_\lambda = N_\mu = N$. By (2.12), $(L(a), I, L(a))$ is an autotopism, so a is Moufang. Thus $N \subset M$. Let S be a pseudo-automorphism with companion c, and let w denote the equal elements in (2.12). Since $w = (ax)y$, $wS \cdot c = (ax)S \cdot (yS \cdot c)$. Since $w = a(xy)$, $wS \cdot c = aS \cdot [xS \cdot (yS \cdot c)]$. Comparing the two results, and replacing $yS \cdot c$ by y, we get

$$(ax)S \cdot y = aS \cdot (xS \cdot y) \tag{2.13}$$

for all x, y in G. From (2.13) with $y = 1$,

$$(ax)S = aS \cdot xS \tag{2.14}$$

for all x in G. By (2.14) in (2.13), aS is in N. Similarly, aS^{-1} is in N. Therefore, by (2.14), S induces an automorphism of N. By Lemma 2.2, every inner mapping of M is induced by a pseudo-automorphism of G. Consequently, N is normal in M. This completes the proof of Theorem 2.1.

Theorem 2.2. *If G is a commutative inverse property loop, the nuclei of G all coincide with the centre Z of G. Moreover, every pseudo-automorphism of G is an automorphism of G.*

Proof. The first statement follows from Theorem 2.1. From (2.7) with x, y interchanged, $xS \cdot (yS \cdot c) = yS \cdot (xS \cdot c)$ for all x, y and, consequently, $x(cy) = (xc)y$ for all x, y of G. Hence c is in Z and therefore may be deleted from both sides of (2.7). This completes the proof of Theorem 2.2.

Theorem 2.3. *Let G be an inverse property loop and let (G, o) be the principal isotope defined by $x o y = xR(v)^{-1} \cdot yL(u)^{-1}$. A necessary and sufficient condition that (G, o) have the left (right) inverse property is that u (that v) be a Moufang element of G. If u, v are Moufang elements of G, a necessary and sufficient condition that (G, o) be isomorphic to G is that G possess a pseudo-automorphism with companion vu^{-1}.*

Proof. Since (G, o) is a loop, (G, o) can have the left inverse property if and only if there exists a permutation λ of G such that $x\lambda o (x o y) = y$ for all x, y. Equivalently, $(x o y) L(u)^{-1} = x\lambda R(v)^{-1} J \cdot y$. Replacing x, y by $xR(v)$, $yL(u)$ respectively and setting $P = R(v)\lambda R(v)^{-1} J$, we see that (G, o) has the left inverse property if and only if G has an autotopism of form $(P, L(u), L(u)^{-1})$. By Lemma 2.1, this will be true if and only if u is Moufang. Similarly for the right inverse property. By the discussion at the beginning of this section, G will be isomorphic to (G, o) if (and only if) G has an autotopism (U, V, W) with $1U = u$, $1V = v$. Thus u, v must be Moufang; moreover, by Lemma 2.1, G must have a pseudo-automorphism S with companion $c = vu^{-1}$. Conversely, if u is Moufang and if G has a pseudo-automorphism S with companion $c = vu^{-1}$, set $U = SL(u)$, $V = SR(c)R(u)$, $W = SR(c)L(u)R(u)$; then (U, V, W) is an autotopism with $1U = u$, $1V = cu = v$. This completes the proof of Theorem 2.3.

3. Moufang loops

We begin with a lemma.

Lemma 3.1. *If the loop G satisfies any one of the following (Moufang) identities, then G has the inverse property and satisfies all three:*

$$(xy)(zx) = [x(yz)]x\,, \tag{3.1}$$

$$[(xy)z]y = x[y(zy)]\,, \tag{3.2}$$

$$x[y(xz)] = [(xy)x]z\,. \tag{3.3}$$

Moreover, G satisfies the identities

$$(xx)y = x(xy)\,, \quad (xy)x = x(yx)\,, \quad (yx)x = y(xx)\,. \tag{3.4}$$

Proof. If (3.1) holds, define x^{-1} by $x^{-1}x = 1$. From (3.1) with $x = y^{-1}$, $zx = (y^{-1}(yz))x$ or $z = y^{-1}(yz)$. Hence G has the left inverse property In particular, $y^{-1} = y^{-1}(yy^{-1})$, so $yy^{-1} = 1$, and thus $(x^{-1})^{-1} = x$. From

(3.1) with $y = 1$, $x(zx) = (xz)x$. Hence, from (3.1) with $z = x^{-1}$, $xy = [x(yx^{-1})]x = x[(yx^{-1})x]$ or $y = (yx^{-1})x$, so G has the inverse property. Then, since $(L(x), R(x), L(x)R(x))$ is an autotopism of G, so is $(JL(x)J, L(x)R(x), R(x))$. Thus, for all x, y, z of G, $y(xzx) = (yx)JL(x)J \cdot zL(x)R(x) = (yx \cdot z)R(x) = [(yx)z]x$. Interchanging y and x, we get (3.2).

If (3.2) holds, define x^{-1} as before and set $z = y^{-1}$. Then $[(xy)y^{-1}]y = xy$, so G has the right inverse property and moreover $xx^{-1} = 1$. Now set $x = y^{-1}$ in (3.2) to get $zy = y^{-1}[y(zy)]$. Taking $z = py^{-1}$ we get $p = y^{-1}(yp)$. Hence G has the inverse property. Taking inverses of both sides of (3.2) we get an identity equivalent to (3.3). Similarly, (3.3) implies (3.2). From (3.2) with x replaced by xy^{-1}, $(xz)R(y) = xR(y^{-1}) \cdot zL(y)R(y)$. Hence $y = 1R(y)$ is a Moufang element for all y, so (3.1) must hold. From (3.3) with $y = 1$, $(xx)z = x(xz)$. Similarly, (3.2) and (3.1) yield the other identities of (3.4). This completes the proof of Lemma 3.1.

A loop satisfying (3.1) is called a Moufang loop. Such loops were first studied by RUTH MOUFANG [91] under the name of "quasigroup". MOUFANG assumed (3.1), (3.2). BOL [66] showed that (3.2) implies (3.1), and BRUCK [70] showed the converse. BOL pointed out the existence of loops which satisfy the identity $[(xy)z]y = x[(yz)y]$ but fail to satisfy (3.2); such loops have the right inverse property but not the left inverse property.

A single-valued mapping θ of the Moufang loop G into itself will be called a *semi-endomorphism* of G provided that

$$(xyx)\theta = (x\theta)(y\theta)(x\theta) \tag{3.5}$$

for all x, y of G and, moreover,

$$1\theta = 1\,. \tag{3.6}$$

From (3.5) with $y = x^{-1}$, $x\theta = (x\theta)(x^{-1}\theta)(x\theta)$ and hence

$$x^{-1}\theta = (x\theta)^{-1} \tag{3.7}$$

for all x in G. We note as follows that (3.5) does not imply (3.6), thus correcting BRUCK [74]: First of all, in an abelian group A of exponent 2, every single-valued mapping of A into itself satisfies (3.5). In the general case, if θ satisfies (3.5) and if t is any element of G such that $t^2 = 1$ and $t(x\theta) = (x\theta)t$ for every x in G, define φ by $x\varphi = (x\theta)t$ and set $X = x\theta$, $Y = y\theta$. Then, by (3.1), (3.4), (3.2), $(x\varphi)(y\varphi)(x\varphi) = (tX)(Yt)(tX) = [(tX)Y][t(tX)] = [(tX)Y](t^2X) = [(tX)Y]X = t(XYX) = t(x\theta \cdot y\theta \cdot x\theta) = t[(xyx)\theta] = (xyx)\varphi$. Hence φ also satisfies (3.5). The element 1θ has the properties of t; therefore *every mapping θ which satisfies* (3.5) *can be "normalized" to the semi-endomorphism $x \to (x\theta)(1\theta)$.* — The meaning of the term *semi-automorphism* should be clear.

Lemma 3.2. *If G is a* Moufang *loop, every inner mapping of G is a pseudo-automorphism of G and every pseudo-automorphism of G is a semi-automorphism of G.*

Proof. The first statement follows from Lemma 2.2; the second, from (2.9), (2.10) and (3.3).

Lemma 3.3. *If G is a commutative* Moufang *loop, every inner mapping and every pseudo-automorphism of G is an automorphism of G. Moreover, for every semi-endomorphism θ of G:* (i) *the mapping $x \to (x\theta)^2$ is an endomorphism of G;* (ii) *if G has no elements of order 2, θ is an endomorphism of G.*

Proof. The first statement follows from Lemma 3.2 and Theorem 2.2. From (3.5) with $y = 1$, by (3.6), $x^2\theta = (x\theta)^2$. Moreover, $xyx = x(yx) = x(xy) = x^2y$ and $(xy)^2 = (xy)(yx) = xy^2x = x^2y^2$. Hence (3.5), with y replaced by y^2, gives $[(xy)\theta]^2 = (x\theta)^2(y\theta)^2$. This proves (i). Now define z by $(xy)\theta = [(x\theta)(y\theta)]z$ and set $a = (xy)\theta$, $b = (x\theta)(y\theta)$. Then $a^2 = b^2$ by (i) and, also, $a^2 = (bz)^2 = b^2z^2$; so $z^2 = 1$. Hence, if G has no elements of order 2, $z = 1$ and we have (ii). The case of the abelian groups of exponent 2 shows that (ii) requires some restriction on elements of order 2. Another suitable hypothesis would be that every element of G was a square.

Lemma 3.4. *Let G be a* Moufang *loop, $\mathfrak{S}$ be a non-empty set of semi-endomorphisms of G, F be the set of all elements of G left fixed by every element of $\mathfrak{S}$ and M be the set of all elements m of G such that $mF \subset F$. Then* (i) $1 \in M \subset F$; (ii) $F^{-1} = F$ *and* $fFf = F$ *for every f in F;* (iii) *M is a subloop of G.*

Corollary. *If G is commutative and $\mathfrak{S}$ consists of pseudo-automorphisms, $M = F$.*

Proof. (i) Certainly 1 is in M. By (3.6), 1 is in F and hence $M = M \cdot 1 \subset F$. (ii) This follows by (3.5), (3,7). (iii) Let m, m' be in M, f be in F. Then, by (ii), F contains $m[m'(mf^{-1})^{-1}]m = (mm')[(fm^{-1})m] = (mm')f$. Therefore $MM \subset M$. Again, F contains $f(mf)^{-1}f = f(f^{-1}m^{-1})f = (ff^{-1})(m^{-1}f) = m^{-1}f$. Therefore $M^{-1} \subset M$. Now we see that M is a subloop of G. The Corollary follows directly from the first sentence of Lemma 3.3.

4. Moufang's Theorem

We shall prove in generalized form the following theorem:

Moufang's Theorem. *Every Moufang loop G is di-associative. More generally, if a, b, c are elements of a Moufang loop G such that $(ab)c = a(bc)$, then a, b, c generate an associative subloop.*

Since (3.4) holds identically in every Moufang loop, the second statement of the theorem implies the first. The theorem was proved by Moufang [91] simultaneously for Moufang loops and alternative division rings, using mathematical induction. The present methods are those of BRUCK [72,] [74].

Lemma 4.1. *In a Moufang loop G, the equation* $(a, b, c) = 1$ *implies each of the equations obtained by permuting* a, b, c *or replacing any of these elements by their inverses.*

Proof. Assume (*) $(a, b, c) = 1$; that is, $(ab)c = a(bc)$. Clearly (*) is equivalent to each of $aU = a$, $bV = b$, $cW = c$ for suitable inner mappings U, V, W; for example, $U = R(b, c)$. Since $a^{-1}U = (aU)^{-1} = a^{-1}$, (*) implies $(a^{-1}, b, c) = 1$; similarly, (*) implies $(a, b^{-1}, c) = (a, b, c^{-1}) = 1$. Taking inverses in (*), we get $(c^{-1}b^{-1})a^{-1} = c^{-1}(b^{-1}a^{-1})$. Thus (*) implies $(c^{-1}, b^{-1}, a^{-1}) = 1 = (c, b, a)$. Again, (*) implies $(bc)a^{-1} = a^{-1}[a(bc)]a^{-1} = a^{-1}[(ab)c]a^{-1} = b(ca^{-1})$ or $(b, c, a^{-1}) = 1 = (b, c, a)$. This completes the proof of Lemma 4.1.

Lemma 4.2. *Let* a, b, c, d *be four elements of the Moufang loop G each three of which associate (satisfy* $(x, y, z) = 1$*). Then the following equations are equivalent:* (i) $(ab, c, d) = 1$; (ii) $((ab)^2, c, d) = 1$; (iii) $((a, b), c, d) = 1$; (iv) $(cd, a, b) = 1$; (v) $(bc, d, a) = 1$. *Hence* (i) *is equivalent to each of the equations obtained by permuting the elements* a, b, c, d *and replacing any of these elements by their inverses.*

Proof. The equation $(x, c, d) = 1$ is equivalent to $xU = x$ for an inner mapping U of G. By Lemma 3.2, we may apply Lemma 3.4. First, (i) implies (ii). Moreover, if $p = (a, b)$, we have $ab = (ba)p$, $aba = ((ba)p)a = b(apa)$ and $b(apa)b = (aba)(a^{-1}. ab) = (ab)(aa^{-1})(ab) = (ab)^2$. Hence (ii) implies (iii). Again, the equation $(x, a, b) = 1$ is equivalent to $xV = x$ where $V = R(a, b)$. By Lemma 2.2, V is a pseudo-automorphism with companion $(a, b) = p$, so $(cd)V \cdot p = cV \cdot (dV \cdot p) = c(dp)$. By (iii) and Lemma 4.1, $c(dp) = (cd)p$, so (iii) implies (iv). In particular, (i) implies (iv), whence, by symmetry, (iv) implies (i). Together, (i) and (iv) imply $[a(bc)]d = [(ab)c]d = (ab)(cd) = a[b(cd)] = a[(bc)d]$, or (v). Thus (i) implies (v) and, by repetition, (v) implies (i). In view of Lemma 4.1, the proof of Lemma 4.2 is now complete.

A (non-empty) subset A of a Moufang loop G is called *associative* if $(a, b, c) = 1$ for all a, b, c in A. By (3.4) and Lemma 4.1, a subset consisting of three elements a, b, c is associative if and only if $(ab)c = a(bc)$. An associative subset (subloop) A of G is called a *maximal* associative subset (subloop) of G if A is contained in no associative subset (subloop) distinct from A. By Zorn's Lemma, every associative subset (subloop) is contained in a maximal associative subset (subloop).

Theorem 4.1. *If G is a Moufang loop with nucleus N such that G/N is commutative, every maximal associative subset A of G is a subloop of G.*

Proof. The hypothesis on G/N simply means that every commutator of G is in N. Thus G satisfies the identity $((w, x), y, z) = 1$. If the element x of G satisfies $(x, a, b) = 1$ for all a, b in A, then, by (3.4) and Lemma 4.1, x is in A. In particular, $A^{-1} = A$. Moreover, since (iii) of

Lemma 4.2 is true, (i) also holds for all u, b, c, d in A; so $AA \subset A$. Therefore A is a subloop of G.

We note that Theorem 4.1 proves Moufang's Theorem for a class of Moufang loops containing the commutative ones. However, according to a private communication from M. F. SMILEY, there exist Moufang loops for which the conclusion of Theorem 4.1 is false.

If A, B, C are subsets of G, AB denotes the set of all products ab, a in A, b in B and, for the purposes of this section alone, (A, B, C) denotes the set of all associators (a, b, c), a in A, b in B, c in C. The *adjoint* of *A in G is the subset, A'*, consisting of all x in G such that $(A, x, G) = 1$. The *closure of A in G* is the subset $A^* = (A')'$.

Lemma 4.3. *The adjoint A' and closure A^* of a non-empty subset A of a Moufang loop G are subloops of G. Moreover, $(A, A, G) = 1$ implies $(A^*, A^*, G) = 1$.*

Proof. Let $B = A'$. By Lemma 4.1, $B^{-1} = B$. For a in A, b, b' in B, x in G we have, by three uses of the definition and two uses of (3.2), $[(a \cdot b'b)x]b = [(ab' \cdot b)x]b = (ab')(bxb) = a(b' \cdot bxb) = a[(b'b \cdot x)b] = [a(b'b \cdot x)]b$. Therefore $(a \cdot b'b)x = a(b'b \cdot x)$, $(A, BB, G) = 1$, $BB \subset B$. Thus B is a subloop. Hence $A^* = B'$ is also a subloop. If $(A, A, G) = 1$, $A \subset A'$. Hence $(A, A^*, G) = 1$, $A^* \subset A'$. Thus finally, $(A^*, A^*, G) = 1$.

Theorem 4.2. *Let A, B, C be non-empty subsets of the Moufang loop G such that $(A, A, G) = (B, B, G) = (C, C, G) = (A, B, C) = 1$. Then the subset $D = A \cup B \cup C$ is contained in an associative subloop H of G.*

Proof. Let F be the set of all elements x in G such that $(D, D, x) = (AB, C, x) = 1$, and let M be the set of all elements m in G such that $mF \subset F$. By Lemma 3.4, M is a subloop of G and $M \subset F$. In view of Lemmas 4.1, 4.2, A, B, C play symmetrical rôles in the definition of F. Since $(A, A, G) = 1$, then $(A, A, D) = (A, A, F) = (A, A, DF) = 1$ and hence, by Lemmas 4.1, 4.2, $(AA, D, F) = (AD, A, F) = (DA, A, F) = 1$. From this and $(AB, C, F) = 1$, by symmetry, $(DD, D, F) = 1$. In particular, $(DD, A, F) = 1$. Similarly, since $(D, D, D) = (D, D, F) = (DD, D, F) = 1$, then $(D, D, DF) = 1$. In particular, $(D, D, AF) = (D, A, AF) = 1$. Since $(A, A, DD) = (A, A, F) = (DD, A, F) = (A, A, (DD)F) = 1$, then $(DD, A, AF) = 1$. And, since $(D, D, A) = (D, D, AF) = (D, A, AF) = (DD, A, AF) = 1$, also $(AD, D, AF) = 1$. In particular, $(AB, C, AF) = 1$. Thus $(D, D, AF) = (AB, C, AF) = 1$, $A \subset M$. By symmetry, $D \subset M$ and, since $(D, D, M) = 1$, we may take H to be the closure of D in M. The special case $A = a$, $B = b$, $C = c$ gives Moufang's Theorem as a corollary.

For a similar proof of the following theorem, see BRUCK [74]:

Theorem 4.3. *Let A be an associative subloop of the Moufang loop G and let B be a non-empty subset of G such that $(A, A, B) = (B, B, G) = 1$. Then the subset $A \cup B$ is contained in an associative subloop of G.*

Corollary. *Every maximal associative subloop of a Moufang loop G is a maximal associative subset of G.*

Since Moufang loops are power-associative, we deduce from (3.5), (3.6) that

$$x^n\theta = (x\theta)^n \tag{4.1}$$

for every semi-endomorphism θ of the Moufang loop G, each x in G and every integer n.

5. The core

By Theorem 2.3, a necessary and sufficient condition that every loop isotopic to an inverse property loop G be an inverse property loop is that G be Moufang. As a consequence, *every loop isotopic to a Moufang loop is Moufang*. It is therefore of interest to discover invariants of classes of isotopic Moufang loops. Such an invariant is the core. The *core* of a Moufang loop G is the groupoid $(G, +)$ consisting of the elements of G under the operation $(+)$ defined by

$$x + y = xy^{-1}x \tag{5.1}$$

where y^{-1} is the inverse of y in G. From (5.1), $xz + yz = (xz)(z^{-1}y^{-1})(xz) = (xzz^{-1})(y^{-1} \cdot xz) = x(y^{-1} \cdot xz) = (xy^{-1}x)z = (x + y)z$. Thus, and similarly,

$$(x + y)z = xz + yz, \quad z(x + y) = zx + zy, \quad (x + y)^{-1} = x^{-1} + y^{-1} \tag{5.2}$$

for all x, y, z of G. It may be verified that $(G, +)$ is associative if and only if G is an abelian group of exponent 2.

The definition (5.1) can be given a simple geometric interpretation. Let G be an arbitrary loop (not necessarily Moufang) defining a 3-net N (cf. I.4, III.4). If L, M are 1-lines of N and if P is a point of M, we define the reflection, $L + M$, of M in L relative to P, as follows: Let the 2-line and 3-line through P meet L in points Q, R respectively. Let the 3-line through Q and the 2-line through R meet in a point P'. Then $L + M$ is the 1-line through P'. The geometric condition that, for all 1-lines L, M, the line $L + M$ shall be independent of the choice of P on M, is one of the three Moufang configurations (for which see Bol [66]); the other two correspond to the like conditions for reflection of 2-lines and 3-lines. Therefore, by Bol [66], the net N will satisfy all three reflection conditions if and only if G is Moufang. If G is Moufang and if L, M have equations $x = a$, $x = b$ respectively, it is easily verified that $L + M$ has equation $x = a + b$ where $a + b$ is given by (5.1).

Theorem 5.1. *Let $G = (G, \cdot)$ be a Moufang loop with multiplication group $\mathfrak{M}$, core $(G, +)$. Then*

(i) *$\mathfrak{M}$ is a group of automorphisms of $(G, +)$.*

(ii) *The semi-endomorphisms of G are the endomorphisms of $(G, +)$ which leave the identity element of G fixed.*

(iii) $(G, +)$ *is isomorphic to three subgroupoids of* $(\mathfrak{M}, +)$; *indeed*

$$L(xyx) = L(x)\,L(y)\,L(x)\,, \quad R(xyx) = R(x)\,R(y)\,R(x)\,, \tag{5.3}$$
$$P(xyx) = P(x)\,P(y)\,P(x)\,,$$

where

$$P(x) = L(x)\,R(x) = R(x)\,L(x)\,. \tag{5.4}$$

(iv) *If H is a principal isotope of G, the identity mapping of G is an isomorphism of* $(G, +)$ *upon* $(H, +)$.

Proof. (i) follows from (5.2). If θ is a semi-endomorphism of G, then, by (3.5), (3.7), (5.1), θ is an endomorphism of $(G, +)$. If θ is an endomorphism of $(G, +)$ such that $1\theta = 1$, then $y^{-1}\theta = (1 + y)\theta = 1 + y\theta = (y\theta)^{-1}$ and hence $(xyx)\theta = (x + y^{-1})\theta = x\theta + (y\theta)^{-1} = (x\theta)\,(y\theta)\,(x\theta)$. Hence θ satisfies (3.5), (3.6); so θ is a semi-endomorphism of G. This proves (ii). Since $(xy)\,x = x\,(yx)$, we see that the definition (5.4) makes sense. From (3.1), $L(x)\,L(y)\,L(x) = L(xyx)$ and, similarly, $R(x)\,R(y)\,R(x) = R(xyx)$. Since $A(x) = (L(x), R(x), P(x))$ is an autotopism of G for each x, and since $A(x)\,A(y)\,A(x) = (L(xyx), R(xyx), P(x)\,P(y)\,P(x))$, then $P(x)\,P(y)\,P(x) = L(xyx)\;R(xyx) = P(xyx)$. This proves (5.3). Since $L(y^{-1}) = L(y)^{-1}$, we have $L(x + y) = L(x) + L(y)$ in terms of the "core" addition of mappings. Similarly with L replaced by R or P. This proves (iii). Now let $H = (G, o)$ be defined by $x o y = xR(v)^{-1}\cdot yL(u)^{-1} = (xa)\,(by)$ where $a = v^{-1}, b = u^{-1}$. If $y^{(-1)} = p$ is the inverse of y in H, then, for all x, y of G, $x o y^{(-1)} o x = (x o p) o x = \{[(xa)\,(bp)]\,a\}\,(bx) = \{x\,[a\,(bp)\,a]\}\,(bx) = xqx$ where $q = [a\,(bp)\,a]\,b$ is independent of x. To determine q, we set $x = y$ and get $y = yqy$ or $q = y^{-1}$. This proves (iv) and completes the proof of Theorem 5.1.

Although, by Theorem 5.1, isotopic Moufang loops have isomorphic cores, subsequent theorems will show that the converse is false.

Theorem 5.2. *If G is a Moufang loop, a necessary and sufficient condition that the core* $(G, +)$ *be a quasigroup is that the mapping* $x \to x^2$ *be a permutation of G. When the condition holds,* $(G, +)$ *is isotopic to a loop* $G(^1/_2)$ *with the same elements as G and with operation* $(*)$ *defined by*

$$x * y = xR^+(1)^{-1} + yL^+(1)^{-1} = x^{1/2}y\,x^{1/2}. \tag{5.5}$$

The loop $G(^1/_2)$ *is an isotopic invariant, is a power-associative loop with the same identity element as G and satisfies*

$$x^{-1} * (x * y) = y\,, \quad (x * y)^{-1} = x^{-1} * y^{-1}, \tag{5.6}$$

$$x * [y * (x * z)] = [x * (y * x)] * z \tag{5.7}$$

for all x, y, z *of G, where* $x \to x^{-1}$ *is the inverse mapping of G. The endomorphisms of* $G(^1/_2)$ *are the semi-endomorphisms of G.*

Proof. A necessary and sufficient condition that $(G, +)$ be a quasigroup is that the mappings $R^+(a)$, $L^+(a)$, defined by $xR^+(a) = x + a$, $xL^+(a) = a + x$, be permutations of G. In view of Theorem 5.1 (i), only the element $a = 1$ need be considered. Since $xR^+(1) = x^2$ and $xL^+(1) = x^{-1}$, $(G, +)$ will be a quasigroup if and only if the mapping $x \to x^2$ is a permutation of G. When the condition is satisfied, we may write $xR^+(1)^{-1} = x^{1/2}$, and this gives (5.5). Now assume that H is a Moufang loop (for example, a principal isotope of G), such that the identity mapping of G is an isomorphism of $(G, +)$ upon $(H, +)$. If $(H, +) = (G, +)$ is a quasigroup and if H has identity element e, $H\,(^1/_2)$ will be a loop with operation (π) defined by $x \,\pi\, y = xR^+(e)^{-1} + yL^+(e)^{-1}$. However, since by (5.2), $R(e)$ is an automorphism of $(G, +)$, $(x * y)R(e) = [xR^+(1)^{-1} + yL^+(1)^{-1}]R(e) = xR(e)R^+(e)^{-1} + yR(e)L^+(e)^{-1} = [x\,R\,(e)]\,\pi\,[yR\,(e)]$. Hence the mapping $x \to xe$ is an isomorphism of $G(^1/_2)$ upon $H(^1/_2)$. If H is a non-principal isotope of G, $G(^1/_2)$ and $H(^1/_2)$ will still be isomorphic. By Moufang's Theorem, G is di-associative. For every integer n, $[(x^{1/2})^n]^2 = x^n = [(x^n)^{1/2}]^2$, so $(x^{1/2})^n = (x^n)^{1/2}$. Hence, for all integers m, n, $x^m * x^n = (x^{1/2})^m (x^{1/2})^{2n} (x^{1/2})^m = x^{m+n}$, showing that $G\,(^1/_2)$ is power-associative. The identities (5.6) follow immediately from (5.5). Finally, from (5.5), (5.4), (5.3), we get, for all x, y, z of G, $x * [y * (x * z)] = z\,P(x^{1/2})\,P(y^{1/2})\,P(x^{1/2}) = z\,P(a) = a^2 * z$ where $a = x^{1/2}y^{1/2}x^{1/2}$ is independent of z. Taking $z = 1$, we find $x * (y * x) = a^2$. This proves (5.7) and completes the proof of Theorem 5.2, aside from the obvious final statement.

Theorem 5.3. *If G is a Moufang loop, a necessary and sufficient condition that the core $(G, +)$ be isotopic to a Moufang loop is that the mapping $x \to x^2$ be a semi-automorphism of G. When the condition holds, the loop $G(^1/_2)$ defined by (5.5) is a commutative Moufang loop, call it H, and the identity mapping of G is an isomorphism of $(G, +)$ upon $(H, +)$.*

Proof. In view of Theorem 5.2, we shall assume that $x \to x^2$ is a permutation of G. Consider the following statements: (a) $G(^1/_2)$ is Moufang; (b) $G(^1/_2)$ has the inverse property; (c) $G(^1/_2)$ is commutative. Since every loop-isotope of a Moufang loop is Moufang, we see that $(G, +)$ is isotopic to a Moufang loop if and only if (a) holds. If (a) holds, so does (b). If (b) holds, then, in view of (5.6), $(x * y)^{-1} = (y * x)^{-1}$ for all x, y, so (c) holds. If (c) holds, then $x * (y * x) = (x * y) * x$ and (5.7) can be turned into the Moufang identity corresponding to (3.3), so (a) holds. Therefore (a), (b), (c) are equivalent. Moreover, (c) is equivalent to the identity $x(y^2 * x^2)x = x(x^2 * y^2)x$ or $(xyx)^2 = x^2y^2x^2$, which states that the permutation $x \to x^2$ is a semi-automorphism of G. Finally, (c) implies that $x * y^{-1} * x = (x * x) * y^{-1} = x^2 * y^{-1} = xy^{-1}x$. This completes the proof of Theorem 5.3. Theorem 5.3 focuses attention on identity (i) of the following lemma.

Lemma 5.1. *A di-associative loop G satisfies all or none of the following identities:* (i) $(xyx)^2 = x^2y^2x^2$; (ii) $x(y^{-1}xy) = (y^{-1}xy)x$; (iii) $((x, y), x) = 1$; (iv) $(x^n, y) = (x, y)^n$ *for all integers n;* (v) $(xy)^n = x^n y^n (x, y)^{-n(n-1)/2}$ *for all integers n;* (vi) $(xyx)^n = x^n y^n x^n$ *for all integers n.*

Proof. By definition, $xy = (yx)(x, y)$, whence

$$(x, y) = x^{-1}y^{-1}xy = (y, x)^{-1} \tag{5.8}$$

and

$$y^{-1}xy = x(x, y)\,. \tag{5.9}$$

For all integers n, by (5.9), $x^n(x^n, y) = y^{-1}x^n y = (y^{-1}xy)^n$, or

$$x^n(x^n, y) = [x(x, y)]^n\,. \tag{5.10}$$

If (i) holds, $xyx^2yx = x^2y^2x^2$ or $(yx)(xy) = (xy)(yx)$, whence, with x replaced by $y^{-1}x$, we get (ii). If (ii) holds, (5.9) yields $x^2(x,y) = x(x,y)x$ or (iii). If (iii) holds, (5.10) gives (iv). If (iv) holds with $n = 2$ or $n = -1$, (5.10) yields (iii). From (iii), (iv) and (5.8) we get (v) by a straightforward mathematical induction. Conversely, (v) for $n = 2$ yields (iii). We note that $(xy, x) = (x, y)^{-1}$ in any di-associative loop. Hence, by (iii), (v), with $k = n(n-1)/2$, $(xyx)^n = (xy)^n x^n (xy, x)^{-k} = x^n y^n (x, y)^{-k} x^n (x, y)^k = x^n y^n x^n$. This is (vi), and (vi) with $n = 2$ is (i). This completes the proof of Lemma 5.1.

Groups satisfying (ii) of Lemma 5.1 have been studied by Burnside [B 3], [B 4], Levi and van der Waerden [B 10], Bruck [70], Levi [B 9]. The connection between Burnside's two papers may be given as follows:

Lemma 5.2. *Let G be a di-associative loop and let n be an integer. If a, b are elements of G such that $(ab)^n = a^n b^n$, then $(ba)^{n-1} = a^{n-1}b^{n-1}$ and $(ba)^{1-n} = b^{1-n}a^{1-n}$. As a consequence, if the mapping $x \to x^n$ is an endomorphism of G, then the mapping $x \to x^{1-n}$ is also an endomorphism of G and the mapping $x \to x^{n-1}$ is a semi-endomorphism of G.* (Baer [B 14].)

Proof. $(ba)^{n-1} = [a^{-1}(ab)a]^{n-1} = a^{-1}(ab)^{n-1}a = a^{-1}(ab)^n b^{-1} = a^{-1}a^n b^n b^{-1} = a^{n-1}b^{n-1}$; whence, by taking inverses, $(ba)^{1-n} = b^{1-n}a^{1-n}$. The second sentence of the lemma should then be clear. Taking $n = 3$, we see that if $x \to x^3$ is an endomorphism, then $x \to x^2$ is a semi-endomorphism.

A function $f(x_1, \ldots, x_n)$ from a di-associative loop G to G is called *skew-symmetric* if interchange of any two of $x_1, \ldots, x_n$ replaces it by its inverse. For example, the commutator (x, y) is skew-symmetric.

Lemma 5.3. *Let G be a group. Then the identities of Lemma 5.1 are equivalent to* (vii) $((x, y), z) = ((z, y), x)^{-1}$ *and imply* (viii) $((x,y),z)^3 = 1$ *and* (ix) $(((w, x), y), z) = ((w, x), (y, z)) = 1$. *Hence if the mapping $x \to x^2$ is a semi-endomorphism of G, then G is nilpotent of class at most 3.* (Levi [B 8].)

Proof. From (vii) with $z = y$ we get (iii) of Lemma 5.1. From (5.9) we get the identity

$$(xy, z) = (x, z)((x, z), y)(y, z)\,, \tag{5.11}$$

valid in any group. Now we use (i)—(vi) of Lemma 5.1. By (iv) and (5.11) we get $(xy, z) = (y^{-1}x^{-1}, z)^{-1} = (x^{-1}, z)^{-1}((y^{-1}, z), x^{-1})^{-1}(y^{-1}, z)^{-1}$ $= (x, z)((y, z), x)^{-1}(y, z)$. Comparison with (5.11) gives $((x, z), y) = ((y, z), x)^{-1}$. Interchange of y, z gives (vii). By (iv) and (vii), $((x, y), z)$ is skew-symmetric. Hence $((w, x), (y, z)) = (((y, z), x), w)^{-1}$ is skew-symmetric in w, x and in y, z, x and thus in w, x, y, z. The even permutation $(wy)(xz)$ must leave $((w, x), (y, z))$ fixed, yet replaces it by $((y, z), (w, x)) = ((w, x), (y, z))^{-1}$. Therefore

$$(((w, x), y), z)^2 = ((w, x), (y, z))^2 = 1 . \tag{5.12}$$

By (iii), (x, y) lies in the centre of the subgroup generated by x, y. By this and (vii), $((x, y), z)$ lies in the centre of the subgroup generated by x, y, z. Moreover, $((x, y), (x, z)) = ((x, (x, z)), y)^{-1} = 1$. Hence, from (5.11), $(xy, z)^2 = (x, z)^2(y, z)^2((x, y), z)^{-2}$. On the other hand, by (v), $(xy)^2 = x^2y^2(y, x)$ and hence, by (iv), (5.11), (vii), $(xy, z)^2 = (x^2y^2(y, x), z)$ $= (x^2y^2, z)((x^2y^2, (y, x)), z)^{-1}((y, x), z) = (x^2, z)((x^2, z), y^2)(y^2, z)((y, x), z)$ $= (x, z)^2(y, z)^2((x, y), z)^{-5}$. Comparison gives (viii). And (ix) follows from (viii), (5.12). The concluding statement of Lemma 5.3 is a direct consequence of (ix). This completes the proof.

Example 1. Let G be the free Burnside group of prime exponent p, $p > 3$, and with two or more generators. Clearly $x \to x^2$ is a permutation of G. It is known that G is not nilpotent of class 3 or less; hence, by Lemma 5.3, $x \to x^2$ is not a semi-automorphism of G. Therefore, by Theorem 5.2, 5.3, the loop $G(^1/_2)$ exists, has the left inverse property — satisfies, indeed, the almost-Moufang identity (5.7), — but is not a Moufang loop.

In the rest of this section we determine the structure of Moufang loops G for which the mapping $x \to x^2$ is a semi-automorphism. The case that G is associative is covered by Lemma 5.3. For the non-associative case we shall need some additional machinery.

Lemma 5.4. *Every Moufang loop satisfies the following identities:*

$$R(x^{-1}, y^{-1}) = L(x, y) = L(y, x)^{-1} ; \tag{5.13}$$

$$L(x, y) = L(xy, y) = L(x, yx) ; \tag{5.14}$$

$$L(x^{-1}, y^{-1})L(x^{-1}, y) = L((x, y), y) ; \tag{5.15}$$

$$xL(z, y) = x(x, y, z)^{-1} ; \tag{5.16}$$

$$(x, y, z) = (x, yz, z) = (x, y, zy) ; \tag{5.17}$$

$$(x, y, z) = (xy, z, y)^{-1} ; \tag{5.18}$$

$$(x, y, z) = (x, y, zx) ; \tag{5.19}$$

$$y[x(x, y, z)^{-1}] = (yx)(y, x, z) . \tag{5.20}$$

Proof. $R(x^{-1}, y^{-1}) = L(x, y)$ by Lemma 2.2. Set $w = zL(x, y)$. Then $w = (x^{-1}y^{-1})[y(xz)]$, $yw = [y(x^{-1}y^{-1})y](xz) = (yx^{-1})(xz)$. Hence

$w = zL(x, yx^{-1})$, so $L(x, y) = L(x, yx^{-1})$. Again, $x(yw) = [x(yx^{-1})x]z = (xy)z$. Thus $z = wL(y, x)$, so $L(x, y) = L(y, x)^{-1}$. This is enough to prove (5.13), (5.14). By definition, $(xy)z = [x(yz)](x, y, z)$. Hence $(x,y,z)^{-1} = [z^{-1}(y^{-1}x^{-1})][x(yz)]$ and $x(x,y,z)^{-1} = \{x[z^{-1}(y^{-1}x^{-1})]x\}(yz) = [(xz^{-1})y^{-1}](yz) = xR(z^{-1}, y^{-1}) = xL(z, y)$. This proves (5.16), which, with (5.14), implies (5.17). Next compute $p = [(xy)z]y$ in two ways. On the one hand, $p = [(xy)(zy)](xy, z, y)$. On the other hand, $p = x(yzy) = [(xy)(zy)](x,y,zy)^{-1}$. Since, by (5.17), $(x,y,zy) = (x,y,z)$, comparison gives (5.18). If $q = (x, y, z)^{-1}$, then, by (5.16), $xq = xL(z,y) = (z^{-1}y^{-1})[y(zx)]$ and hence $y(xq) = [y(z^{-1}y^{-1})y](zx) = (yz^{-1})(zx) = (yx)(y, z^{-1}, zx)$. By (5.17), $(y, z^{-1}, zx) = (y, x, zx) = (y, x, z)$. Hence we have (5.20). Replacing z by zx in (5.20) and using (5.17), we get (5.19). By (5.18), (5.19), $(y, x, z) = (yx, z, x)^{-1} = (yx, z, y^{-1})^{-1}$. By this and (5.16), we see that (5.20) can be rewritten as $y[xL(z, y)] = (yx)L(y^{-1}, z)$. By Lemma 2.2, $\theta = L(y^{-1}, z)$ is a pseudo-automorphism with companion $c = (y, z^{-1})$. Hence $\{y[xL(z, y)]\}c = [(yx)\theta]c = (y\theta)[(x\theta)c]$. However, $y\theta = y$. Therefore, if we replace x by $x\theta^{-1} = xL(z, y^{-1})$, we get $xL(z, y^{-1})L(z, y) = y^{-1}\{[y(xc)]c^{-1}\} = xs^{-1}$, where, by (5.18), (5.19), $s = (y^{-1}, y(xc), c^{-1}) = (xc, c^{-1}, y(xc))^{-1} = (xc, c^{-1}, y)^{-1} = (x, y, c^{-1})$. Therefore, by (5.16), $xs^{-1} = x(x, y, c^{-1})^{-1} = xL(c^{-1}, y)$. Hence $L(z, y^{-1})L(z, y) = L((z^{-1}, y), y)$. This proves (5.15) and completes the proof of Lemma 5.4.

Lemma 5.5. *Let G be a Moufang loop. Then G satisfies all or none of the following identities:* (i) $((x, y, z), x) = 1$; (ii) $(x, y, (y, z)) = 1$; (iii) $(x, y, z)^{-1} = (x^{-1}, y, z)$; (iv) $(x, y, z)^{-1} = (x^{-1}, y^{-1}, z^{-1})$; (v) $(x, y, z) = (x, zy, z)$; (vi) $(x, y, z) = (x, z, y^{-1})$; (vii) $(x, y, z) = (x, xy, z)$. *When these identities hold, the associator* (x, y, z) *lies in the centre of the subloop generated by* x, y, z; *and the following identities hold for all integers* n:

$$(x, y, z) = (y, z, x) = (y, x, z)^{-1} \tag{5.21}$$

$$(x^n, y, z) = (x, y, z)^n \tag{5.22}$$

$$(xy, z) = (x, z)((x, z), y)(y, z)(x, y, z)^3. \tag{5.23}$$

Proof. Since $L(z, y)$ is a pseudo-automorphism of G, (5.16) and (4.1) yield $x^n(x^n, y, z)^{-1} = [x(x, y, z)^{-1}]^n$ for any integers n. Taking $n = -1$, we see that (i), (iii) are equivalent; moreover, (i) implies (5.22). By (5.16), (5.15), (5.14), we see that (ii) is equivalent to the identity $L(y^{-1}, z^{-1}) = L(z, y^{-1})$ or $L(z, y) = L(y^{-1}, z)$. Hence, by (5.16), (ii) is equivalent to (vi). For the remaining equivalence proofs we use (5.17), (5.18), (5.19) without mention. By (iii), $(x, y, z) = (xy, z, y)^{-1} = (y^{-1}x^{-1}, z, y) = (y^{-1}x^{-1}, z, x^{-1}) = (y^{-1}, x^{-1}, z)^{-1}$. Consequently, $(x,y,z) = (xy, z, y)^{-1} = (z^{-1}, y^{-1}x^{-1}, y) = (z^{-1}, y^{-1}x^{-1}, x^{-1}) = (z^{-1}, y^{-1}, x^{-1}) = (y, z, x^{-1})^{-1}$. From this, by repetition, $(x, y, z) = (y, z, x^{-1})^{-1} = (z, x^{-1}, y^{-1}) = (x^{-1}, y^{-1}, z^{-1})^{-1}$, which gives (iv). By (iv), $(x, zy, z) = (x^{-1}, y^{-1}z^{-1}, z^{-1})^{-1} = (x^{-1}, y^{-1}, z^{-1})^{-1} = (x, y, z)$, whence we have (v).

By (v), $(x, z, y^{-1}) = (x, yz, y^{-1}) = (x, yz, z) = (x, y, z)$, which is (vi). By (vi), $(x, xy, z) = (x, z, y^{-1}x^{-1}) = (x, z, y^{-1}) = (x, y, z)$, giving (vii). By (vii), $(x, y, z) = (x, x^{-1}y, z) = (y, z, x^{-1}y)^{-1} = (y, z, x^{-1})^{-1}$, whence, by repetition, we get (iv) and therefore (vi). By (iv), (vi) $(x, y, z)^{-1} = (x^{-1}, y^{-1}, z^{-1}) = (x^{-1}, z^{-1}, y) = (x^{-1}, y, z)$, which is (iii). Therefore the seven identities are equivalent and imply (5.22).

For the rest of the proof we assume (i)—(vii). Using (iii), we get $(x, y, z) = (x^{-1}, y, z)^{-1} = (x^{-1}y, z, y) = (y^{-1}x, z, y)^{-1} = (y^{-1}x, z, x)^{-1} = (y^{-1}, x, z) = (y, x, z)^{-1}$; this, applied to (vi), gives $(x, y, z) = (x, z, y^{-1}) = (z, x, y^{-1})^{-1} = (z, y, x)^{-1} = (y, z, x)$. Therefore we have (5.21). Now (5.20) can be rewritten as $y[x(x, y, z)^{-1}] = (yx)(x, y, z)^{-1}$, which, by Moufang's Theorem, is equivalent to

$$(x, y, (x, y, z)) = 1\,. \tag{5.24}$$

Set $a = (x, y, z)$. Let H be the subloop generated by x, y, z. By (i), $(a, x) = 1$. Hence, by Lemma 2.2, $R(a, x)$ is an automorphism of G and the set S, of all s in H such that $(s, a, x) = 1$, is a subloop of H. By (5.24) and (5.21), S contains x, y, z, so $S = H$. By this and symmetry, $(h, a, x) = (h, a, y) = (h, a, z) = 1$ for every h in H. Next let P be the set of all p in H such that $(p, a) = 1$ and let F be the set of all f in H such that $fP \subset P$. By Lemma 3.4, $F \subset P$ and F is a subloop of H. By Lemma 2.2, $T(a)$ is a pseudo-automorphism of G with companion a^{-3}. For any p in P, since $x^{-1} \cdot xT(a) = 1 = (x, p, a^{-3})$, we have $[(xp)T(a)]a^{-3} = (xT(a))[(pT(a))a^{-3}] = (xp)a^{-3}$. Hence xp is in P, x is in F. By (i) and (5.21), y, z are also in F, so $F = H$. Hence $(h, a) = 1$ for every h in H. Now, since $(x, a, h) = (y, a, h) = (z, a, h) = (a, h) = 1$, we conclude as before that $(h', a, h) = 1$ for all h, h' in H. Therefore a is in the centre of H.

Since $\theta = T(z)$ is a pseudo-automorphism with companion z^{-3}, $(xy)(xy, z) = (xy)\theta = \{(x\theta)[(y\theta)z^{-3}]\}z^3 = [(x\theta)(y\theta)]w$ where $w = (x\theta, (y\theta)z^{-3}, z^3) = (z^{-1}xz, z^{-1}yz^{-2}, z)^3 = (x, y, z)^3$. Therefore, since $a = (x, y, z)$ is in the centre of H,

$$(xy)[(xy, z)(x, y, z)^{-3}] = (x\theta)(y\theta) = [x(x, z)][y(y, z)]\,. \tag{5.25}$$

By several uses of (ii), the right hand side of (5.25) becomes $x\{(x, z)[y(y, z)]\} = x\{[(x, z)y](y, z)\} = x\{y[(x, z)T(y) \cdot (y, z)]\}$. By this and (5.24), (5.25) may be written

$$(xy, z)(x, y, z)^{-3} = [(x, z)T(y) \cdot (y, z)]p^{-1} \tag{5.26}$$

where, by several uses of (ii), (vii) and (5.21), $p = (x, y, (x, z)T(y) \cdot (y, z)) = (x, y, (y \cdot (x, z)T(y))(y, z)) = (x, y, (x, z)y(y, z)) = (x, y, [x(x, z)][y(y, z)]) = (x, y, (x\theta)(y\theta))$. Since (x, y, z) is in the centre of H, we use (5.25) to get $p = (x, y, (xy)(xy, z)) = (x, xy, (xy)(xy, z)) = (x, xy, (xy, z)) = 1$, by (ii). Setting $p = 1$ and $(x, z)T(y) = (x, z)((x, z), y)$ in (5.26), we get (5.23). This completes the proof of Lemma 5.5.

Lemma 5.6. *Let the mapping $x \to x^2$ be a permutation of the Moufang loop G. Then the following conditions are necessary and sufficient in order that $G(^1/_2)$ be an abelian group:* (a) *every commutator (x, y) lies in the nucleus N of G;* (b) *every associator (x, y, z) lies in the centre Z of G;* (c) $((x, y), z) = (x, y, z)^2$ *for all x, y, z of G.*

Proof. The loop $G(^1/_2)$, defined by (5.5), has operation $(*)$ where $x*y = x^{1/2} y x^{1/2} = y P(x^{1/2})$. It is easily verified that $G(^1/_2)$ is an abelian group if and only if $y*(x*z) = x*(y*z)$ for all x, y, z of G. Since $x \to x^2$ is a permutation of G, an equivalent condition is

$$P(x)\, P(y) = P(y)\, P(x) \tag{5.27}$$

for all x, y of G.

Sufficiency. Assume (a), (b), (c). Since (a) implies (ii) of Lemma 5.5, all the identities of Lemma 5.5 are available. From (c) with $z = y$, $((x, y), y) = 1$ and hence $x y^2 x = y x^2 y$ for all x, y of G. From (c) with z in N, and from (a), each commutator is in the centre of N. Thus

$$\begin{aligned} z P(x)\, P(y) &= y(x z x) y = y[(x^2 z)(z, x)] y = [y(x^2 z)][(z, x) y] \\ &= [y(x^2 z) y](z, x)((z, x), y) = \{y[(x^2 z) y]\}(z, x)(x, y, z)^2 \\ &= \{y[x^2(z y)]\}(x^2, z, y)(z, x)(x, y, z)^2 \\ &= \{y[x^2(y z)]\}(z, y)(z, x) \\ &= [(y x^2 y) z](z, y)(z, x)\,. \end{aligned}$$

The last expression is symmetric in x, y, proving (5.27).

Necessity. Assume (5.27). Since $(L(x), R(x), P(x))$ is an autotopism of G for all x, so, by Lemma 2.1, is $A(x) = (P(x), L(x)^{-1}, L(x))$. Hence the commutator-triple $(A(y), A(x)) = (I, (L(y^{-1}), L(x^{-1})), (L(y), L(x)))$ is also an autotopism of G. We deduce that $(L(y^{-1}), L(x^{-1})) = (L(y), L(x)) = R(c)$ where $c = 1(L(y^{-1}), L(x^{-1})) = (x, y)$ is in the nucleus N of G. This proves (a). In addition,

$$x^{-1}(y^{-1}(x(y z))) = z(x, y) \tag{5.28}$$

for all x, y, z. When z is in N, (5.28) reduces to $(x, y) z = z(x, y)$. Thus every commutator is in the centre C of N. Since (x, y) is in N, (5.28) can be rewritten as $x(y z) = [y(x z)](x, y)$; and this, when z is replaced by $x z$, becomes

$$(x y x) z = [y(x^2 z)](x, y)\,. \tag{5.29}$$

Now compute $(y x^2) z$ in two ways. On the one hand, since (a) implies (5.21), (5.22), we get $(y x^2) z = [y(x^2 z)](y, x^2, z) = [y(x^2 z)](y, x, z)^2$. On the other hand, since, by Theorem 5.3, $((x, y), x) = 1$, we find from (a) and (5.29) that $(y x^2) z = [x y x(y, x)] z = (x y x)[(y, x) z] = [(x y x) z](y, x)((y, x), z) = [y(x^2 z)]((y, x), z)$. Comparison gives $(y, x, z)^2 = ((y, x), z)$, or (c). Since $(p^2, q) = (p, q)^2 = 1$ implies $(p, q) = 1$ and, similarly, $(p^2, q, r) = 1$ implies $(p, q, r) = 1$, we see as well that all associators lie in C. Since (5.22) holds, we can rewrite (5.16) as $x\theta = x(x, y, z)$ where $\theta = L(z^{-1}, y)$. By Lemma 2.2 and (a), θ is an automorphism of G. Thus $(wx)(wx, y, z) = (wx)\theta = (w\theta)(x\theta) = [w(w, y, z)][x(x, y, z)] = (wx)[(w, y, z)((w, y, z), x)(x, y, z)]$ or

$$(wx, y, z) = ((w, y, z), x)(w, y, z)(x, y, z) \tag{5.30}$$

for all w, x, y, z of G. In particular, if a is in N, $(ax, y, z) = (xa, y, z) = (x, y, z)$ for all x, y, z. Thus, since $wx = xw(w, x)$ and (w, x) is in N, $(wx, y, z) = (xw, y, z)$. Consequently,

$$((w, y, z), x) = ((x, y, z), w) \tag{5.31}$$

for all w, x, y, z. By (5.31), $((w, y, z), x)$ is symmetric in w, x. By (5.31) and (5.22), $((w, y, z), x)$ is skew-symmetric in w, y, z and in x, y, z and hence in w, x. Therefore, since $p^2 = 1$ implies $p = 1$, $((w,y,z),x) = ((w, y, z), x)^{-1} = 1$. This completes the proof of (b) and of Lemma 5.6.

Example 2. Let G be a nilpotent group of class 2 all of whose elements have finite odd order. Then $x \to x^2$ is a permutation of G and G satisfies (a), (b), (c), of Lemma 5.6, so $G(^1/_2)$ is an abelian group. Since G is a non-commutative group, G and $G(^1/_2)$ are not isomorphic and hence not isotopic, even though both have the same core.

Example 3. We may construct as follows a Moufang loop G such that $G(^1/_2)$ is an abelian group but G is not associative. Let F be a field of characteristic not two and let R be an associative algebra over F containing a vector space A over F such that (i) $ab = -ba$ for all a, b in A; (ii) $abc \neq 0$ for some a, b, c in A. For each positive integer n, let R_n denote the vector space over F spanned by all products of n or more factors from A. Let G be the set of all ordered triples (a, p, x), a in A, p in R_2, x in R_3, with equality componentwise and with multiplication defined by $(a, p, x)(b, q, y) = (a + b, p + q + ab, x + y + pb)$. Then, as tedious calculation would show, G is a Moufang loop satisfying the requirements that $G(^1/_2)$ exist and be an abelian group. The commutators are the elements $(0, 2ab, pb - qa)$ and the associators are the elements $(0, 0, abc)$, with a, b, c in A, p, q in R_2. Note that, if F has characteristic zero, no element of G has finite order except the identity element.

Lemma 5.7. *If G is a commutative Moufang loop, then:* (i) *the mapping $x \to x^3$ is a centralizing endomorphism of G;* (ii) *$x^3 = 1$ for every x in the (commutator-) associator subloop G.*

Proof. By Moufang's Theorem and commutativity, $(xy)^3 = x^3y^3$ for all x, y of G. For each x in G, by commutativity, $T(x) = R(x)L(x)^{-1} = I$ and, by Lemma 2.2, $T(x)$ is a pseudo-automorphism with companion x^{-3}. Hence x^3 lies in the nucleus, which coincides with the centre, of G. This proves (i). The kernel of the endomorphism $x \to x^3$ must contain G', which proves (ii).

Theorem 5.4. *Let the mapping $x \to x^2$ be a semi-automorphism of the Moufang loop G. Then:* (i) *The set T, consisting of all t in G such that $t^3 = 1$, is a normal subloop of G and $(G/T)\,(^1/_2)$ is an abelian group.* (ii) *If N is the nucleus of G, the mapping $xN \to x^3N$ is a centralizing endomorphism of G/N.*

Proof. By Theorem 2.1, N is normal in G. Since $((x, y), x) = 1$, we have $(xy)^3 = x^3y^3(x, y)^{-3}$ and $(x, y)^3 = (x^3, y) = (x, y^3)$ for all x, y of G. This is enough to show that T is a subloop of G. If θ is an inner mapping of G, $(x\theta)^3 = x^3\theta$ for all x in G, whence T is normal in G. If $x^2 = y^2t$ for t in T, then $x^6 = (y^2t)^3 = y^6t^3(x^2, t)^{-3} = y^6$, so $x^3 = y^3$ and $(xy^{-1})^3 = x^3y^{-3}(x, y^{-1})^{-3} = (x, y^3) = 1$. Thus $x^2 \equiv y^2 \mod T$ implies $x \equiv y \mod T$. Consequently, the mapping $xT \to x^2T$ is a semi-automorphism of G/T. Thus $(G/T)\,(^1/_2)$ exists and is a homomorphic image of $G(^1/_2)$. By Theorem 5.3 and Lemma 5.7 (ii), T contains the associator subloop of $G(^1/_2)$, so $(G/T)\,(^1/_2)$ is an abelian group. This proves (i). We now apply Lemma 5.6 to G/T. As a first consequence, if c is any commutator and a any associator of G,

$$(c, x, y) \equiv (a, x, y) \equiv (a, x) \equiv 1 \mod T \tag{5.32}$$

for all x, y of G. If $\theta = L(z, y)$, $a = (x, y, z)$, then (5.15) gives $x^3(x^3,y,z)^{-1} = x^3\theta = (x\theta)^3 = (xa^{-1})^3 = x^3a^{-3}(x, a)^3$. By (5.32), $(x, a)^3 = 1$, so

$$(x^3, y, z) = (x, y, z)^3 \tag{5.33}$$

for all x, y, z. By (5.33), (5.32), if a is any associator, $(a^3, x, y) = (a, x, y)^3 = 1$ for all x, y; showing that a^3 is in the nucleus N of G. Thence, by (5.33), $(x^3, y, z) \equiv 1 \mod N$ for all x, y, z, so x^3N is in the nucleus of G/N for all x. If c is any commutator of G, (5.33), (5.32) yield $(c^3, x, y) = (c, x, y)^3 = 1$ for all x, y; thus c^3 is in N. In particular, $(x^3, y) = (x, y)^3 \equiv 1 \mod N$, so x^3N is in the centre of G/N for all x. And, finally, $(xy)^3 = x^3y^3(x, y)^{-3} \equiv x^3y^3 \mod N$, so the mapping $xN \to x^3N$ is a centralizing endomorphism of G/N. This completes the proof of Theorem 5.4.

Theorem 5.4 will now be interpreted in another way in terms of commutative Moufang loops.

Lemma 5.8. *Every loop isotopic to a Moufang loop G is isomorphic to a principal isotope (G, o) with operation (o) given by $x o y = (xf)\,(f^{-1}y)$ for some fixed element f of G.*

Proof. We need only consider a principal isotope H with operation $(\times)$ given by $x \times y = xR(v)^{-1} \cdot yL(u)^{-1}$. The identity element of H is $e = uv$. Hence, if we define (G, o) by $(x o y)e = (xe) \times (ye)$, (G, o) is isomorphic to H and has the same identity element, 1, as G. Since $(pq)e^{-1} = (pe)\,(e^{-1}qe^{-1})$ for all p, q of G, we have $x o y = [(xe)R(v)^{-1} \cdot (ye)L(u)^{-1}]e^{-1} = (x\theta)\,(y\varphi)$ for all x, y, where θ, φ are suitable permutations of G. If $1\theta = f$, then $y = 1 o y = f(y\varphi)$, so $y\varphi = f^{-1}y$ for all y. In particular,

$1\varphi = f^{-1}$ and hence $x = x o 1 = (x\theta) f^{-1}$, $x\theta = xf$ for all x. Thus $x o y = (xf)(f^{-1}y)$ for all x, y. This completes the proof of Lemma 5.8.

Theorem 5.5. *A necessary and sufficient condition that the Moufang loop G have the property that $x \to x^3$ is a centralizing endomorphism of G is that G be either an isotope of a commutative Moufang loop or a normal subloop of index 3 in such a loop-isotope.*

Proof. Let P denote the property described in the theorem. If G has P, consider the isotope (G, o) defined by $x o y = (xf)(f^{-1}y)$. For all integers m, n, and all x in G, $x^m o x^n = x^m f f^{-1} x^n = x^{m+n}$; hence powers are the same in G and (G, o). Again, $x^3 o y = x^3 f f^{-1} y = x^3 y$ and $y o x^3 = y f f^{-1} x^3 = y x^3$. Also $(x^3 o y) o z = (x^3 y) o z = [(x^3 y) f](f^{-1}z) = x^3[(yf)(f^{-1}z)] = x^3(y o z) = x^3 o (y o z)$. Hence x^3 is in the centre of (G, o) for all x. Since $(x o y)^3 = (xf)^3 (f^{-1}y)^3 = x^3 f^3 f^{-3} y^3 = x^3 y^3$, we see that (G, o) has property P. Therefore, by Lemma 5.8, P is an isotopic invariant. In particular, by Lemma 5.7, every loop-isotope of a commutative Moufang loop has P. Consequently, every subloop of such an isotope also has P. Note, however, that the non-commutative groups with P are not isotopic to any commutative loop. The rest of the proof will merely be sketched; for details see BRUCK [70], Chapter II, § 8. Let G be any Moufang loop with property P and let A be the additive group of integers mod 3. Let E be the set of all couples (x, p), x in G, p in A, with equality componentwise and with multiplication defined by

$$(x, p)(y, q) = (\varphi_{q-p}(x, y), p + q), \tag{5.34}$$

where, for each p in A, φ_p is the function from $G \times G$ to G defined by

$$\varphi_0(x, y) = x^{-1} y x^2, \quad \varphi_1(x, y) = xy, \quad \varphi_2(x, y) = yx. \tag{5.35}$$

Tedious calculations show that E is a commutative Moufang loop. Accepting this, consider the isotope (E, o) defined by

$$(x, p) o (y, q) = [(x, p)(1, 1)][(1, 2)(y, q)] = (\varphi_{q-p+1}(x, y), p + q). \tag{5.36}$$

Since $(x, o) o (y, o) = (xy, o)$ for all x, y, the homomorphism $(x, p) \to p$ of (E, o) upon A has kernal isomorphic to G. Thus G has been embedded as a normal subloop of index 3 in an isotope of a commutative Moufang loop. This completes the proof of Theorem 5.5.

VIII. Commutative Moufang Loops

1. Examples

BOL [66] was the first to construct a commutative Moufang loop which is not an abelian group. Each of Bol's examples is centrally nilpotent of class 2. BRUCK [70] showed how to construct examples which are centrally nilpotent of class 3. In his University of Wisconsin thesis (1953), T. SLABY formulated the following theorem: *Every commutative*

Moufang loop which can be generated by n elements ($n > 1$) is centrally nilpotent of class at most $n - 1$. In collaboration with the author, SLABY proved this theorem for $n = 4, 5$ as well as for the previously known cases $n = 2, 3$.

The present chapter will reobtain Slaby's results and go on to prove the theorem for every positive integer n. Moreover, it will appear that the class, $k(n)$, of the free commutative Moufang loop on n generators satisfies the inequalities

$$c(n) \leqq k(n) \leqq n - 1 \qquad \text{for } n \geqq 3 \quad (1.1)$$

where

$$c(n) = 1 + [n/2]. \quad (1.2)$$

It seems worth remarking here on an interesting situation for $n = 5$ which will not be mentioned later. From (1.1) with $n = 5$ we get $3 \leqq k(5) \leqq 4$. The exact value of $k(5)$ is still uncertain. It can be shown, however, that if $k(5) = 3$ then $k(n) = c(n)$ for every $n \geqq 3$. Indeed, a simple proof that $k(5) = 3$ would greatly shorten the present chapter, but in the absence of such proof I have come to believe that $k(5) = 4$.

Once the existence of $k(n)$ is known, the inequality $k(n) \geqq c(n)$ may be deduced from the following construction, which shows the existence of an (infinitely generated) commutative Moufang loop H containing, for every positive integer n, a subloop generated by n elements which is centrally nilpotent of class $c(n)$. It also turns out that H is transfinitely centrally nilpotent of class ω and that the centre of H has order 1.

Let K be the field of three elements. Let Σ be any infinite set of positive ordinals. For the sake of simplicity we may take Σ to be the set of all positive integers, but this is not essential. Let F be the set of all finite subsets of Σ, including the null set. Denote the null set by 0 and, for each element n of Σ, denote the one-element subset (n) by n. Construct a symbol $e(S)$ for each member S of F and let R be the vector space over K consisting of all finite linear combinations of the $e(S)$, S in F, with coefficients in K, equality being componentwise. Define multiplication in R by the distributive laws and the following rules:

(i) If $S, T \in F$ and $S \cap T \neq 0$, then $e(S)\,e(T) = 0$, the zero vector.

(ii) If $S, T \in F$ and $S \cap T = 0$, then $e(S)\,e(T) = (-1)^p e(S \cup T)$ where p is the number of pairs (s, t), s in S, t in T, with $t < s$.

It is easily verified that R is an associative algebra over K — a type of "exterior" algebra of characteristic 3 — with $e(0)$ as the multiplicative identity element. If the member S of F consists of the ordinals $n_1, \ldots, n_t$, with $n_1 < n_2 < \cdots < n_t$, then $e(S) = e(n_1)\,e(n_2) \ldots e(n_t)$. We denote by A the vector subspace with the elements $e(n)$, n in Σ, as basis. And, for each positive integer k, we denote by RA^k the vector subspace consisting of all finite linear combinations of products of form $x a_1 a_2 \ldots a_k$ where x is in R and $a_1, \ldots, a_k$ are in A. Furthermore we define RA^0 to be R. The following significant facts are readily verified:

(I) $$ab = -ba\,, \quad \text{all } a, b \text{ in } A\,.$$

(II) To each nonzero x in R there corresponds a unique "degree" n, a non-negative integer such that x is in RA^n but not in RA^{n+1}.

Let H be the set of all ordered couples (a, x), a in A, x in R, with equality componentwise and with multiplication defined by

$$(a, x)\,(b, y) = (a + b,\, x + y + (x - y)\,ab)\,. \tag{1.3}$$

In view of (I), multiplication in H is commutative. A more extended calculation shows that H is a Moufang loop. In particular, the identity element is $(0, 0)$, the inverse of (a, x) is $(a, x)^{-1} = (-a, -x)$ and the associator of three elements (a, x), (b, y), (c, z) is

$$((a, x), (b, y), (c, z)) = (0,\, xbc + yca + zab)\,. \tag{1.4}$$

Let $\{H_\alpha\}$ denote the lower central series of H, so that $H_0 = H$ and $H_1 = H'$ is the associator subloop of H. From (1.4), (1.3) we see that H' is an abelian group. Indeed, from (1.4) with $y = z = 0$, H' contains $(0, xbc)$ for all x in R, b, c in A. By taking products of such elements we deduce that $H_1 = H' = (0, RA^2)$. More generally, $H_n = (0, RA^{2n})$ for every positive integer n. Since Σ is infinite, $H_n \neq 1$ for every positive integer n. On the other hand, by (II), $H_\omega = \bigcap_{n=0}^{\infty} H_n = 1$. Therefore H is transfinitely centrally nilpotent of class ω.

Using (1.4) we may verify that the element (a, x) lies in the centre $Z(H)$ of H if and only if both $xbc = 0$ for all b, c in A and $yca = 0$ for all y in R, c in A. Since Σ is infinite, these requirements force $x = a = 0$. Therefore $Z(H) = 1$.

Finally, for any preassigned integer $n \geqq 3$, we let G be a subloop of H generated by n elements of form $(e(i), e(0))$ where i ranges over n distinct elements of Σ. A straightforward calculation, which we omit, shows that G is centrally nilpotent of class $c(n)$. This completes our discussion of the construction.

In what follows we develop identities in stages, proving Slaby's theorem successively for $n = 2, 3, 4, 5$ and then going on to the general case.

2. Two and three generators

Henceforth let G be a commutative Moufang loop. We begin by adapting some material from Chapter VII. The identity VII, (3.1) becomes

$$x^2(yz) = (xy)\,(xz)\,. \tag{2.1}$$

By VII, Lemma 3.3, every inner mapping or pseudo-automorphism of G is an automorphism of G. By VII, Lemma 5.6, since $R(x) = L(x)$ for all x, we have

$$xL(y, z) = xR(y, z) = x(x, y, z)\,, \qquad (2.2)$$

$$(x, y, z) = (x, xy, z) = (y, x, z)^{-1} = (y, z, x)\,, \qquad (2.3)$$

$$(x^m, y^n, z^p) = (x, y, z)^{mnp}\,, \qquad (x, y, z)^3 = 1\,, \quad (2.4)$$

for all x, y, z in G and all integers m, n, p. Moreover, Moufang's Theorem and VII, Lemma 5.5 imply the following:

Theorem 2.1. *If $n = 2$ or 3, every commutative Moufang loop G which can be generated by n elements is centrally nilpotent of class at most $n - 1$.*

As a consequence of (2.1)—(2.4) and the theorem just stated,

$$R(x, y) = R(y, x)^{-1} = R(x, xy)\,, \quad R(x^n, y) = R(x, y^n) = R(x, y)^n \qquad (2.5)$$

for all x, y in G and all integers n. Another consequence is the validity of the equations

$$\begin{aligned}(xy)z &= [x(yz)]\,(x, y, z) = [x(x, y, z)]\,(yz) \qquad (2.6)\\ &= x\,\{[y(x, y, z)]z\} = x\,\{y[(x, y, z)z]\}\,.\end{aligned}$$

3. Four generators

We begin with an awkward form of an important lemma. A more convenient form will be obtained in Lemma 4.6.

Lemma 3.1. *If w, x, a, b are elements of a commutative Moufang loop G, then $(wa)(xb) = (wx)c$ where $c = pq^{-1}$ and*

$$p = [aR(w, x)]\,[bR(x, w)] = [a(a, w, x)]\,[b(b, x, w)]\,, \qquad (3.1)$$

$$q = (w^{-1}x, aR(w, x), bR(x, w))\,. \qquad (3.2)$$

Proof. We multiply the equation $(wx)c = (wa)(xb)$ by w^{-2} and use (2.1) to get $x(w^{-1}c) = a[w^{-1}(xb)]$. If $\theta = R(w, x)$ then, by (2.5), $R(w^{-1}, x) = \theta^{-1}$, $R(x, w^{-1}) = \theta$. Hence $x(w^{-1}c) = (xw^{-1})\,(c\theta^{-1})$ and $w^{-1}(xb) = (w^{-1}x)\,(b\theta)$, so our equation becomes $(w^{-1}x)\,(c\theta^{-1}) = a[(w^{-1}x)\,(b\theta)]$. By (2.2), (2.4), $\theta^3 = I$ and $\theta^2 = \theta^{-1}$. Moreover, by (2.2), θ leaves w and x fixed. Hence $(w^{-1}x)c = [(w^{-1}x)\,(c\theta^{-1})]\theta = \{a[(w^{-1}x)\,(b\theta)]\}\theta = (a\theta)\,[(w^{-1}x)\,(b\theta^{-1})] = [(w^{-1}x)\,(b\theta^{-1})]\,(a\theta)$. Now applying (2.6), we get $(w^{-1}x)c = (w^{-1}x)\,\{[(b\theta^{-1})\,(a\theta)]\,(w^{-1}x, b\theta^{-1}, a\theta)\}$, whence $c = [(a\theta)\,(b\theta^{-1})] \times \times (w^{-1}x, a\theta, b\theta^{-1})^{-1}$. Since $\theta = R(w, x)$, $\theta^{-1} = R(x, w)$, the proof is complete.

Our first application of Lemma 3.1 is to an expansion formula for (wx, y, z). By (2.2), since $\varphi = R(y, z)$ is an automorphism, $(wx)(wx, y, z) = (wx)\varphi = (w\varphi)\,(x\varphi) = [w(w, y, z)]\,[x(x, y, z)]$. We apply Lemma 3.1 with $a = (w, y, z)$, $b = (x, y, z)$, $c = (wx, y, z)$. In this case, replacement of v by y^{-1} replaces a, b, c, p by their inverses but leaves q fixed. Hence we have both $c = pq^{-1}$ and $c^{-1} = p^{-1}q^{-1}$. By multiplication, $1 = q^{-2}$. However, $q^3 = 1$ by (2.4), so $q = 1$. Therefore $c = p$, $1 = q$; that is,

$$(wx, y, z) = [(w, y, z)\, R(w, x)]\, [(x, y, z)\, R(x, w)] \tag{3.3}$$

$$= [(w, y, z)\, ((w, y, z), w, x)]\, [(x, y, z)\, ((x, y, z), x, w)]\,,$$

$$1 = (w^{-1}x, (w, y, z)\, R(w, x), (x, y, z)\, R(x, w))\,. \tag{3.4}$$

We consider (3.4), taking $a = (w, y, z)$, $b = (x, y, z)$, $\theta = R(w, x)$. First we replace x by wx in (3.4). Since $R(w, wx) = \theta$ and $R(wx, w) = \theta^{-1}$, (3.4) yields $1 = (x, a\theta, (wx, y, z)\theta^{-1})$, whence $1 = 1\theta^{-1} = (x, a, (wx, y, z)\theta)$. By (3.3), $(wx, y, z)\theta = [(a\theta)(b\theta^{-1})]\theta = (a\theta^{-1})b$. Therefore $1 = (x, a, (a\theta^{-1})b)$ $= ((a\theta^{-1})b, x, a)$. Setting $\psi = R(a\theta^{-1}, b)$ and applying (3.3), we get $1 = [(a\theta^{-1}, x, a)\psi]\, [(b, x, a)\psi^{-1}]$. However, by Theorem 2.1, (a, x, w) lies in the centre of the subloop generated by a, x, w; consequently, $(a\theta^{-1}, x, a) = (a(a, x, w), x, a) = (a, x, a) = 1$. This leaves $1 = (b, x, a)\psi^{-1}$, whence $1 = 1\psi = (b, x, a) = (a, b, x)$. In detail, we have the identity

$$((w, y, z), (x, y, z), x) = 1\,. \tag{3.5}$$

Since $(xz, y, z) = (x, y, z)$, we may replace x by xz in (3.5) without altering a, b. Thus $1 = (a, b, xz) = (xz, a, b)$, whence by (3.3), (3.5), $1 = [(x, a, b)R(x, z)]\, [(z, a, b)R(z, x)] = (z, a, b)R(z, x)$. Therefore $1 = (z, a, b) = (a, b, z)$; that is,

$$((w, y, z), (x, y, z), z) = 1\,. \tag{3.6}$$

Lemma 3.2. *If the commutative Moufang loop G is generated by a set S consisting of four elements, then*

$$((S, S, S), (S, S, S), G) = 1\,. \tag{3.7}$$

Proof. Let s_i $(i = 1, 2, \ldots, 7)$ denote elements of S. Set $u = (s_1, s_2, s_3)$, $v = (s_4, s_5, s_6)$ and consider $w = (u, v, s_7)$. If two of s_1, s_2, s_3 are equal, $u = 1$ and $w = 1$. Similarly, if two of s_4, s_5, s_6 are equal, $v = w = 1$. If s_1, s_2, s_3 are distinct and s_4, s_5, s_6 are s_1, s_2, s_3 in some order, then $v = u$ or u^{-1} and $w = 1$. In the remaining case, by skew-symmetry of the associator, we may assume that s_1, s_2, s_3, s_4 are the four distinct elements of S and that $s_5 = s_2$, $s_6 = s_3$. Then, if $s_7 = s_1$, $w = 1$ by (3.5) and, if $s_7 = s_2$ or s_3, $w = 1$ by (3.6). Hence $w = 1$ in all cases. Since $R(u, v)$ is an automorphism, the set H of all h in G such that $(u, v, h) = 1$ is a subloop of G. Since H contains S, then $H = G$. This completes the proof of Lemma 3.2.

Now set $a = (w, y, z)$, $b = (x, y, z)$, $c = (a, w, x)$, $d = (b, x, w)$. By (3.3) and Lemma 3.1, $(wx, y, z) = (ac)(bd) = (ab)(pq^{-1})$ where $p = [cR(a, b)]\, [dR(b, a)]$ and $q = (a^{-1}b, cR(a, b), dR(b, a))$. By Lemma 3.2, $(a, b, h) = 1$ for all h in the subloop generated by w, x, y, z. In particular, $R(a, b) = R(b, a)^{-1}$ leaves c, d fixed, so $p = cd$, $q = (a^{-1}b, c, d)$.

By Lemma 3.2 again, $(c, d, k) = 1$ for all k in the subloop generated by a, b, w, x; in particular, $q = (a^{-1}b, c, d) = 1$. Thus (3.3) can be replaced by the identity

$$(wx, y, z) = [(w, y, z)(x, y, z)][((w, y, z), w, x)((x, y, z), x, w)]. \tag{3.8}$$

We define the function h on the loop G by

$$(wx)(yz) = [(wy)(xz)]h(w, x, y, z). \tag{3.9}$$

Lemma 3.3. *For any commutative Moufang loop G, the function h is skew-symmetric in its four arguments. In particular, if a, b, c, d are elements of G such that $(ab)(cd) = (ac)(bd)$, then $(ab)(cd) = (ad)(bc)$.*

Proof. We note from (3.9) that interchange of w and z replaces h by its inverse. Now we consider the result of replacing x, y, z in (3.9) by wx wy, wz respectively. By (2.1), $[w(wx)][(wy)(wz)] = (w^2x)[w^2(yz)] = w^4[x(yz)]$ and similarly for the coefficient of h. Hence $h(w, wx, wy, wz) = \{[x(yz)]w^4\}\{w^{-4}[y(zx)]^{-1}\} = pq$ where $p = [x(yz)][y(zx)]^{-1}$ and $q = (x(yz), w^4, w^{-4}[y(zx)]^{-1})$. Since $x(yz) = (yz)x = [y(zx)](y, z, x), p = (x, y, z)$. Also $q = (x(yz), w, y(zx))^{-4} = (x(yz), w, [x(zy)](x, z, y))^{-1} = (x(yz), w, (x, y, z))$. Both p and q are now seen to be skew-symmetric in x, y, z. Hence $h(w, wx, wy, wz)$ and, therefore, $h = h(w, x, y, z)$, is skew-symmetric in x, y, z. Since h is skew-symmetric in w, z and in x, y, z, then h is skew-symmetric in w, x, y, z. If $(ab)(cd) = (ac)(bd)$ then $h(a, b, c, d) = 1$; hence $h(a, b, d, c) = 1$, so $(ab)(cd) = (ab)(dc) = (ad)(bc)$. This completes the proof of Lemma 3.3.

A comparison of (3.3), (3.8) yields our simplest application of Lemma 3.3:

Lemma 3.4. *If G is a commutative Moufang loop, the identity* (3.8) *remains true when the four associators on the right hand side are permuted arbitrarily.*

The next two lemmas give important methods of generating new identities.

Lemma 3.5. *Let G be a commutative Moufang loop containing a subloop K and elements a, b such that (x, a, b) is in K for each element x in G. Then the elements $((a, b, x_1), x_1, x_2)$, $(((a, b, x_1), x_1, x_2), x_2, x_3), \ldots$, are in K for all elements $x_1, x_2, x_3, \ldots$ in G.*

Proof. Set $p = ((w, a, b), w, x)$, $q = ((x, a, b), x, w)$. Then, by (3.8), K contains $(wx, a, b) = [(w, a, b)(x, a, b)](pq)$, and hence pq, for all w, x in G. Replacing x by x^{-1}, we see that K contains $p^{-1}q$. Hence K contains $(pq)^{-1}(p^{-1}q) = p^{-2} = p = ((a, b, w), w, x)$ for all w, x in K. Thus K contains $((a, b, x_1), x_1, x_2)$ for all x_1, x_2 in K. Now we can repeat the process, for any fixed x_1, with a, b replaced by (a, b, x_1) and x_1 respectively. Hence, by iteration, we get the full conclusion of Lemma 3.5.

Lemma 3.6. *If a, b, c, d are elements of the commutative Moufang loop such that $((x, a, b), c, d) = 1$ for every element x of G, then*

$1 = (((a, b, x_1), x_1, x_2), c, d) = ((((a, b, x_1), x_1, x_2), x_2, x_3), c, d) = \ldots$, *for all elements* $x_1, x_2, x_3, \ldots$, *of* G.

Proof. The set K of all elements k in G such that $(k, c, d) = 1$ is a subloop of G which satisfies the hypotheses of Lemma 3.5.

Let $\{G_i\}$ and $\{Z_i\}$ denote the lower and upper central series, respectively, of G. Moreover, let $G'(= G_1)$ and $Z(= Z_1)$ denote the associator subloop and centre, respectively, of G.

Lemma 3.7. *Let* G *be a commutative Moufang loop generated by a subset* S *and let* n *be a positive integer. A necessary and sufficient condition that the element* a *of* G *should be in* Z_n *is that*

$$((\ldots((a, s_1, s_2), s_3, s_4), \ldots), s_{2n-1}, s_{2n}) = 1 \tag{3.10}$$

for all $s_1, s_2, \ldots, s_{2n}$ *in* S.

Proof. The necessity of the condition follows from the definition of the upper central series. The case $n = 1$ of (3.10) may be written $(a, s, s') = 1$ for all s, s' in S; we now assume this special case of (3.10). The set H of all h in G such that $(a, S, h) = 1$ is a subloop of G containing S, so $H = G$. Therefore the set K of all k in G such that $(a, k, G) = 1$ is a subloop containing S, so $K = G$. Hence $(a, G, G) = 1$, $a \in Z = Z_1$. Therefore Lemma 3.7 is true for $n = 1$. Now assume inductively that Lemma 3.7 is true for n and let a be an element which satisfies (3.10) when n is replaced by $n + 1$. Equivalently, if w denote the left hand side of (3.10), $(w, s, s') = 1$ for all s, s' in S. Hence, by the case $n = 1$, w is in Z. That is, (3.10) holds modulo the centre Z for all s_i in S. Consequently, aZ is in $Z_n(G/Z) = Z_{n+1}/Z$, so a is in Z_{n+1}. This completes the proof of Lemma 3.7.

Lemma 3.8. *Let* G *be a commutative Moufang loop generated by a subset* S *and let* n *be a positive integer. A necessary and sufficient condition that* $G_n = 1$ *is that*

$$((\ldots((s_0, s_1, s_2), s_3, s_4), \ldots), s_{2n-1}, s_{2n}) = 1$$

for all $s_0, s_1, s_2, \ldots, s_{2n}$ *in* S.

Proof. By Lemma 3.7, the condition is necessary and sufficient in order that $S \subset Z_n$. By Lemma 1.2 of Chapter IV, $G_n = 1$ if and only if $G = Z_n$. However, $G = Z_n$ if and only if $S \subset Z_n$. This completes the proof of Lemma 3.8.

Theorem 3.1. *Every commutative Moufang loop generated by four elements is centrally nilpotent of class at most* 3.

Proof. In view of Lemma 3.8, an analysis along the lines of the proof of Lemma 3.2 shows that we need only prove that the following identities hold in every commutative Moufang loop G:

(a) $(((w, x, y), w, z), x, y) = 1$.

(b) $(((w, x, y), w, z), w, x) = 1$.

(c) $(((w, x, y), w, z), x, z) = 1$.

Proof of (a). By Theorem 2.1, $((w, x, y), x, y) = 1$ for all w in G. Hence, by Lemma 3.6, $(((w, x, y), w, z), x, y) = 1$ for all w, z in G.

Proof of (b). Set $a = (w, x, y)$, $b = (z, w, x)$, $p = (a, w, z)$, $\theta = R(w,x)$. Certainly $w\theta = w$ and, by Theorem 2.1, $a\theta = a$. Therefore $p(p, w, x) = p\theta = (a, w, z)\theta = (a\theta, w\theta, z\theta) = (a, w, zb) = (zb, a, w)$. Expanding the last associator by (3.3), we have $p(p,w,x) = [(z,a,w)R(z,b)][(b,a,w)R(b,z)]$. However, by Lemma 3.2, $(b, a, w) = ((z, w, x), (w, x, y), w) = 1$. Therefore $p(p, w, x) = (z, a, w)R(z, b) = p(p, z, b)$ and $(p, w, x) = (p, z, b)$. By Lemma 3.2 again, $(p, z, b) = ((a, w, z), z, (z, w, x)) = 1$. That is, $(p, w, x) = 1$, which proves (b).

Proof of (c). With a, b, p as before, define $q = (a, x, z)$, $\varphi = R(x, z)$. Then $a\varphi = aq$, $w\varphi = wb$, $z\varphi = z$, so $p(p, x, z) = p\varphi = (a, w, z)\varphi = (a\varphi, w\varphi, z\varphi) = (aq, wb, z)$ Expanding the latter associator by three applications of (3.3), we get

$$\begin{aligned} p(p, x, z) &= [(a, wb, z)\lambda]\,[(q, wb, z)\lambda^{-1}] \\ &= \{[(a, w, z)\mu]\,[(a, b, z)\mu^{-1}]\}\lambda \cdot \{[(q, w, z)\mu]\,[(q, b, z)\mu^{-1}]\}\lambda^{-1} \\ &= \{(p\mu\lambda)\,[(a, b, z)\mu^{-1}\lambda]\}\,\{[(q, w, z)\mu\lambda^{-1}]\,[(q, b, z)\mu^{-1}\lambda^{-1}]\} \end{aligned}$$

where $\lambda = R(a, q)$, $\mu = R(w, b)$. By Lemma 3.2, $(a, b, z) = ((w, x, y), (z, w, x), z) = 1$ and $(q, b, z) = ((a, x, z), (w, x, z), z) = 1$. By the same authority, $p\mu = p(p, w, b) = p((a, w, z), w, (z, w, x)) = p$ and hence $p\mu\lambda = p\lambda = p(p, a, q) = p((a, w, z), a, (a, x, z)) = p$. Similarly, $(q, w, z)\mu = (q, w, z)((q, w, z), w, (z, w, x)) = (q, w, z)$ and hence $(q, w, z)\mu\lambda^{-1} = (q, w, z)\lambda^{-1} = (q, w, z)((q, w, z), q, a)$. Therefore $p(p, x, z) = p[(q,w,z)((q, w, z), q, a)]$ or

$$(p, x, z) = (q, w, z)\,((q, w, z), q, a)\,. \tag{3.11}$$

Replacement of x by x^{-1} replaces a, p by their inverses and leaves q fixed. Hence from (3.11) we get

$$(p, x, z) = (q, w, z)\,((q, w, z), q, a)^{-1}.$$

And now the device of multiplying corresponding members of the two equations gives

$$(p, x, z) = (q, w, z) = k, \text{ say.} \tag{3.12}$$

Next we compute $E = ((a, wx, z), wx, z)$ in two ways. On the one hand, by Theorem 2.1, $E = 1$. Since, by (3.3), $(a,wx,z) = [pR(w,x)][qR(x,w)]$ and, by (b), $(p,w,x) = (q,x,w) = 1$, we have $(a,wx,z) = pq$, $E = (pq,wx,z)$. Expanding the last associator by three applications of (3.3), we get $E = \{[(p, w, z)\beta]\,[(q, w, z)\beta^{-1}]\}\alpha \cdot \{[(p, x, z)\beta]\,[(q, x, z)\beta^{-1}]\}\alpha^{-1}$ where $\alpha = R(w, x)$, $\beta = R(p, q)$. By Theorem 2.1, $(p, w, z) = ((a, w, z), w, z) = 1$ and, similarly, $(q, x, z) = 1$. By (3.12), $(p, x, z) = (q, w, z) = k$; moreover, by Lemma 3.2, $k\beta = k(k, p, q) = k((q, w, z), (a, w, z), q) = k$. Therefore $1 = E = (k\alpha)(k\alpha^{-1}) = (k(k, w, x))(k(k, w, x)^{-1}) = k^2$ and hence $k = k^{-2} = 1$.

By this and (3.12), $(p, x, z) = 1$, proving (c). And now the proof of Theorem 3.1 is complete.

Theorem 3.1 allows us to derive two very useful lemmas.

Lemma 3.9. *In any commutative Moufang loop the associator* $((x, y, w), w, z)$ *is skew-symmetric in* x, y, z. *Equivalently, G satisfies the identity*

$$((x, y, w), w, z) = ((y, z, w), w, x) . \tag{3.13}$$

Proof. We work entirely within the subloop H generated by the four elements w, x, y, z. By Theorem 3.1, $H_3 = 1$. If h, h' are in H and k, k' are in H_1, consider the identity (3.8): $(kk', h, h') = [(k, h, h')(k', h, h')][((k, h, h'), k, k')((k', h, h'), k', k)]$. The last two factors are in $H_3 = 1$, so $(kk', h, h') = (k, h, h')(k', h, h')$. If, further, k' is in H_2, then (k', h, h') is in $H_3 = 1$ and we can drop still another factor. We apply these remarks as follows: By Theorem 2.1, $((w, h, y), w, h) = 1$ for every h in H; we use this fact with $h = xz$, x and z. By (3.8), $(w, xz, y) = (ab)c$ where $a = (w, x, y)$, $b = (w, z, y)$ are in H_1 and $c = (a, x, z)(b, z, x)$ is in H_2. Hence $1 = ((w, xz, y), w, xz) = ((ab)c, w, xz) = (ab, w, xz) = (a, w, xz)(b, w, xz)$. Since $(a, w, x) = ((w, x, y), w, x) = 1$ and $((a, w, z), z, x)$ is in $H_3 = 1$, (3.8) gives $(a, w, xz) = (a, w, z) = ((w, x, y), w, z) = ((x, y, w), w, z)$. Similarly, $(b, w, xz) = (b, w, x) = ((w, z, y), w, x) = ((y, z, w), w, x)^{-1}$. Hence $1 = ((x, y, w), w, z)((y, z, w), w, x)^{-1}$, proving (3.13).

Lemma 3.10. *Every commutative Moufang loop G satisfies the identity*

$$(p, (p, w, x), (p, y, z)) = 1 . \tag{3.14}$$

Proof. For any fixed element p, denote the left hand side of (3.14) by $F(w, x, y, z)$. Clearly F is skew-symmetric in y, z. By Lemma 3.9, $F(w, x, y, z) = (p, z, (p, y, (p, w, x))^{-1}$; hence, by Lemma 3.9, F is skew-symmetric in w, x, y. Consequently F is skew-symmetric in w, x, y, z and, since the permutation $(wy)(xz)$ is even, $F(w, x, y, z) = F(y, z, w, x)$. On the other hand, $F(y, z, w, x) = (p, (p, y, z), (p, w, x)) = F(w, x, y, z)^{-1}$. Hence $F^2 = 1$, $F = 1$, proving Lemma 3.10.

4. A calculus of associator subloops

At this point we are able to introduce a calculus of associator subloops of a commutative Moufang loop which bears striking resemblance to P. Hall's calculus of commutator subgroups of a group.

If A, B, C are normal subloops of a commutative Moufang loop we define the *associator subloop of* A, B, C, which we denote by (A, B, C), to be the subloop generated by all associators (a, b, c) with a, b, c in A, B, C respectively. By (2.3),

$$(A, B, C) = (B, A, C) = (C, B, A) . \tag{4.1}$$

Lemma 4.1. *If the normal subloops A, B, C of the commutative Moufang loop G are generated by the self-conjugate subsets U, V, W respectively, then their associator subloop (A, B, C) is generated by the set of all (u, v, w) with u, v, w in U, V, W respectively. In particular, (A, B, C) is normal in G.*

Proof. Let H be the subloop generated by the set P consisting of all associators (u, v, w), u, v, w in U, V, W respectively. Since U, V, W are self-conjugate and since every inner mapping of G is an automorphism of G, P is self-conjugate also. Therefore, by the Corollary to Lemma 1.4 of Chapter IV, H is normal in G. Since G/H is a commutative Moufang loop, the set X of all x in G such that $(x, V, W) \equiv 1 \bmod H$ is a subloop of G. Since X contains U, then X contains A. Therefore $(A, V, W) \equiv 1 \bmod H$. In similar fashion we deduce that $(A, B, W) \equiv 1 \bmod H$ and then that $(A, B, C) \equiv 1 \bmod H$. Therefore $(A, B, C) \subset H \subset (A, B, C)$, so $(A, B, C) = H$. This completes the proof of Lemma 4.1.

Lemma 4.2. *If A, B, C, X, Y are normal subloops of the commutative Moufang loop G, then*

$$((A,B,C),X,Y) \subset ((A,X,Y),B,C)\ ((B,X,Y),C,A)\ ((C,X,Y),A,B). \tag{4.2}$$

Proof. By Lemma 4.1, $((A, B, C), X, Y)$ is generated by the set of all elements $((a, b, c), x, y)$ with a, b, c, x, y in A, B, C, X, Y respectively. If a is in A and p, q are in G, then (a, p, q) is in A. By (3.8) and (3.14), $(a(a, x, y), b, c) = [(a, b, c)\ ((a, x, y), b, c)]p$ where $p = (((a, x, y), b, c), (a, x, y), a)$. Therefore $(a(a, x, y), b, c) \equiv (a, b, c) \bmod ((A, X, Y), B, C)$. If P denotes the right hand side of (4.2), we thus have

$$(a(a, x, y), b, c) \equiv (a, b, c) \bmod P \tag{4.3}$$

and similarly for any permutation of a, b, c. If $\theta = R(x, y)$, then $(a,b,c)\ ((a,b,c),x,y) = (a,b,c)\theta = (a\theta, b\theta, c\theta) = (a(a,x,y), b(b,x,y), c(c,x,y))$. Therefore, by three uses of (4.3), $(a, b, c)\ ((a, b, c), x, y) \equiv (a, b(b, x, y), c(c, x, y)) \equiv (a, b, c(c, x, y)) \equiv (a, b, c) \bmod \mathrm{P}$. THUS $((a, b, c), x, y) \equiv 1 \bmod \mathrm{P}$. This completes the proof of Lemma 4.2.

In terms of associator subloops, the lower central series $\{G_\alpha\}$ of a commutative Moufang loop G may be defined by

$$G_0 = G, \quad G_{i+1} = (G_i, G, G), \qquad i = 0, 1, 2, \ldots; \tag{4.4}$$

$$G_\omega = \bigcap_{i=0}^{\infty} G_i \tag{4.5}$$

where ω is the first limit ordinal, and so on. If $\{Z_\alpha\}$ is the upper central series of G we shall need to define $Z_i = Z_0 = 1$ for all negative integers i.

The *derived series* $\{G^{(n)}\}$ of G may be defined as follows:

$$G^{(0)} = G,\ G^{(i+1)} = (G^{(i)})' = (G^{(i)}, G^{(i)}, G^{(i)}), \quad i = 0, 1, 2, \ldots, \tag{4.6}$$

$$G^{(\omega)} = \bigcap_{i=0}^{\infty} G^{(i)}, \tag{4.7}$$

and so on.

Lemma 4.3. *If G is a commutative Moufang loop,*

$$(G_i, G_j, G_k) \subset G_{i+j+k+1}, \tag{4.8}$$

$$(G_i, G_j, Z_k) \subset Z_{k-i-j-1}, \tag{4.9}$$

$$G^{(i)} \subset G_{(3^i-1)/2} \tag{4.10}$$

for all non-negative integers i, j, k.

Proof. The identity

$$(G_i, G_j, G) \subset G_{i+j+1} \tag{4.11}$$

holds for all non-negative integers i when $j = 0$. If (4.11) holds for some j and all i, then, by (4.2), $(G_i, G_{j+1}, G) = ((G_j, G, G), G_i, G) \subset \subset ((G_j, G_i, G), G, G)\ ((G, G_i, G), G_j, G) \subset (G_{i+j+1}, G, G)\ (G_{i+1}, G_j, G) \subset G_{i+j+2}$. Therefore (4.11) holds for all non-negative integers i, j. Consequently, (4.8) holds for all i, j when $k = 0$. Now we establish (4.8) for all i, j, k by a similar use of (4.2). Next we observe that (4.10) holds for $i = 0$ and establish it for all $i \geqq 0$ by induction and (4.8). Finally, we note, from the definition of the upper central series, that (4.9) holds for all k if $i = j = 0$; then we establish (4.9) for all non-negative i, j by (4.2) and induction.

We are now in a position to strengthen some of the results of § 3. Instead of Lemma 3.2 we have:

Lemma 4.4. *If the commutative Moufang loop G can be generated by 4 elements, then $(G, G', G') = 1$. In particular, the associator subloop G' is an abelian group.*

Proof. $(G, G', G') = (G_0, G_1, G_1) \subset G_3 = 1$ by (4.8) and Theorem 3.1.

By using (3.8), Theorem 3.1 and Lemmas 3.9, 4.4 we may obtain the following:

Lemma 4.5. *The function h of Lemma 3.3 satisfies*

$$h(w, x, y, z) = (w, x, y)^{-1} (x, y, z)\ (y, z, w)^{-1} (z, w, x)\,. \tag{4.12}$$

Lemma 4.6. *For all w, x, a, b of a commutative Moufang loop G we have*

$$(wa)\ (xb) = (wx) \left\{ (ab) \left[\begin{matrix} (w,a,b)\ (x,b,a)\ ((w,x,a),a,b)\ ((x,w,b),b,a) \\ (a,w,x)\ (b,x,w)\ ((a,b,w),w,x)^{-1}((b,a,x),x,w)^{-1} \end{matrix} \right] \right\}.$$

Note that for (4.13), we need merely evaluate the p, q of Lemma 3.1.

5. The distributor

We shall need the following:

Lemma 5.1. *Every commutative Moufang loop G satisfies the identity*

$$((u, x, y), (v, x, y), (w, x, y)) = 1 . \tag{5.1}$$

Proof. Define the mapping $\theta = \theta(x, y)$ of G by $u\theta = (u, x, y)$, all u in G. Then $uR(x,y) = u(u\theta)$, $u\theta R(x,y) = u\theta$, $u\theta^2 = 1$. Set $a = (u\theta, v, w)$, $b = (u\theta, v\theta, w)$, $c = (u\theta, v, w\theta)$, $d = (u\theta, v\theta, w\theta)$. If H is the subloop generated by u, v, x, y, then, by Lemma 4.3, $(u\theta, v, v\theta)$ is in $(H_1, H, H_1) \subset H_3$ and, by Theorem 3.1, $H_3 = 1$. Consequently, $(u\theta, v, v\theta) = 1$. (Henceforth, we shall refer to this type of argument as a *class argument.*) Since $(u\theta, v, v\theta) = 1$, then, by Lemma 3.9, $(a, v, v\theta) = ((u\theta, v, w), v, v\theta) = ((u\theta, v, v\theta), v, w)^{-1} = 1$ and $(b, v\theta, v) = ((u\theta, v\theta, w), v\theta, v) = ((u\theta, v\theta, v), v\theta, w)^{-1} = 1$. Therefore, by (3.8), $(u\theta, v(v\theta), w) = (ab)[(a, v, v\theta)(b, v\theta, v)] = ab$. Similarly, $(u\theta, v, w(w\theta)) = ac$, $(u\theta, v(v\theta), w\theta) = cd$, $(u\theta, v\theta, w(w\theta)) = bd$. Since $a = (u\theta, v, w)$, then $a(a\theta) = aR(x, y) = (u\theta, v(v\theta), w(w\theta))$. Expanding the latter associator in two ways we get, on the one hand, $a(a\theta) = (u\theta, v, w(w\theta))(u\theta, v\theta, w(w\theta)) = (ac)(bd)$ and, on the other hand, $a(a\theta) = (u\theta, v(v\theta), w)(u\theta, v(v\theta), w\theta) = (ab)(cd)$. Thus, by Lemma 3.3,

$$a(a\theta) = (ab)(cd) = (ac)(bd) = (ad)(bc) . \tag{5.2}$$

Replacement of x by x^{-1} replaces a, d by their inverses but leaves $a\theta$, b, c fixed. In particular, we have both $a(a\theta) = (ad)(bc)$ and $a^{-1}(a\theta) = (ad)^{-1}(bc)$, whence $a^2 = [a(a\theta)][a^{-1}(a\theta)]^{-1} = [(ad)(bc)][(ad)^{-1}(bc)]^{-1} = (ad)^2 = a^2d^2$. This gives $d^2 = 1$, $d = d^{-2} = 1$. However, $d = 1$ is equivalent to (5.1). The proof of Lemma 5.1 is now complete. We note for future reference the identity $a(a\theta) = a(bc)$ or $a\theta = bc$ or

$$\begin{aligned} &(((u, x, y), v, w), x, y) \\ &= ((u, x, y), (v, x, y), w)((u, x, y), v, (w, x, y)) . \end{aligned} \tag{5.3}$$

The *distributor*, $D = D(G)$, of a commutative Moufang loop G, is the set of all elements d of G which satisfy the equivalent equations (i)—(vi) of the next lemma. The name comes from (i).

Lemma 5.2. *If the element d of the commutative Moufang loop G satisfies one of the following equations for all w, x, y, z in G, then d satisfies all of them:* (i) $(xy, z, d) = (x, z, d)(y, z, d)$; (ii) $(xd, y, z) = (x, y, z)(d, y, z)$; (iii) $((w, x, y), z, d) = 1$; (iv) $((d, w, x), y, z) = ((d, y, x), w, z)^{-1}$; (v) $((x, y, z), x, d) = 1$; (vi) $((d, x, y), x, z) = 1$. *The set $D = D(G)$ of all d in G which satisfy* (iii) *is a characteristic normal subloop of G, called the distributor of G.*

Proof. When we say that d satisfies (i) we mean that (i) holds for the given element d of G and for all x, y, z in G. Similarly for (ii)—(vi).

We note that (v), (vi) are equivalent by Lemma 3.9. If d satisfies (i), then, for any fixed z, the mapping $x \to (x, z, d)$ is an endomorphism of G and, in particular, $((w, x, y), z, d) = ((w, z, d), (x, z, d), (y, z, d))$. Therefore, by Lemma 5.1, $((w, x, y), z, d) = 1$. That is, (i) implies (iii). If d satisfies (iii), then d satisfies (v) and hence (vi). If d satisfies (v), (vi), then, by (3.8), d satisfies both (i) and (ii). If d satisfies (ii), set $p = ((x,y,z),x,d)$, $q = ((d, y, z), d, x)$. Then, by (3.8), $pq = 1$ for all x, y, z in G. Replacing x by x^{-1}, we get $pq^{-1} = 1$. Thus $p = p^{-2} = (pq)^{-1}(pq^{-1})^{-1} = 1$, so d satisfies (v). Hence (i), (ii), (iii), (v), (vi) are equivalent. The set D of all d in G which satisfy (iii) is clearly a characteristic (hence normal) subloop of G. If d is in D, so are $d' = (d, w, x)$, $d'' = (d, y, x)$ and the product $d'd''$. Then, by (v) followed by two uses of (i) and two of (ii), $1 = ((d, wy, x), wy, z) = (d'd'', wy, z) = (d'd'', w, z)\ (d'd'', y, z) = [(d',w,z)\ (d'',w,z)]\ [(d',y,z)\ (d'',y,z)]$. By (vi), $(d',w,z) = (d'',y,z) = 1$. Therefore $1 = (d'', w, z)\ (d', y, z)$, whence d satisfies (iv). Finally, if d satisfies (iv) then $((d, x, y), x, z) = ((d, x, x), y, z)^{-1} = 1$, so d satisfies (vi). This completes the proof of Lemma 5.2.

If $\mathfrak{L}$ is the class of all commutative Moufang loops and if $f(G) = D(G)$ for each G in $\mathfrak{L}$, then f is a nilpotency function for $\mathfrak{L}$ in the sense of Chapter IV, § 1. Hence we may define the *lower distributor series* $\{L_\alpha\}$ and the *upper distributor series* $\{D_\alpha\}$ for any commutative Moufang loop G. By (iii) of Lemma 5.2,

$$L_0 = G\,, \quad L_{i+1} = (L_i, G, G')\,, \qquad i = 0, 1, 2, \ldots, \tag{5.4}$$

where $G' = G_1$ is the associator subloop of G. In particular, $L_1 = (G, G, G_1) = G_2$. If $L_i \subset G_{2i}$ for some non-negative integer i, then $L_{i+1} \subset (G_{2i}, G, G_1) \subset \subset G_{2i+2}$ by Lemma 4.3. Therefore

$$L_1 = G_2\,, \quad L_i \subset G_{2i}\,, \qquad i = 0, 1, 2, \ldots, \tag{5.5}$$

and

$$L_\omega \subset G_\omega\,. \tag{5.6}$$

For any positive integer n let $c(n)$ denote the least positive integer k such that $2k$ exceeds n. Equivalentlv,

$$c(n) = 1 + [n/2]\,. \tag{5.7}$$

Lemma 5.3. *Let G be a commutative Moufang loop and n be a positive integer. Then:* (a) $Z_2 \subset D$. (b) *If G has n generators, $D \subset Z_{c(n)}$ and $G_{2+ic(n)} \subset \subset L_{i+1}$ for $i = 0, 1, 2, \ldots$.* (c) *If G is finitely generated, $G_\omega = L_\omega$ where ω is the first limit ordinal.*

Proof. (a) If d is in Z_2 then (d, x, y) is in the centre $Z = Z_1$ of G and hence d satisfies (vi) of Lemma 5.2. Therefore $Z_2 \subset D$.

(b) Let S be a set of n elements which generate G and set $t = c(n)$. If $s_1, s_2, \ldots, s_{2t}$ are elements of S, some two must be equal. Since D

is normal in G we see from (iv) of Lemma 5.2 that, for d in D, the associator

$$((\ldots((d, s_1, s_2), s_3, s_4), \ldots), s_{2t-1}, s_{2t})$$

is skew-symmetric in the s_i and hence equal to 1. Therefore, by Lemma 3.7, $D \subset Z_{c(n)}$. By (5.5), $G_2 \subset L_1$. Now assume inductively that $G_{2+ic(n)} \subset \subset L_{i+1}$ for some non-negative integer i. If $L_{i+2} = 1$ then $G_{2+ic(n)} \subset L_{i+1} \subset \subset D \subset Z_{c(n)}$ and hence $G_{2+ic(n)+c(n)} = 1$. Therefore, in general, $G_{2+(i+1)c(n)} \subset \subset L_{i+2}$. This proves (b).

(c) By (b), $G_\omega \subset L_\omega$. Hence, by (5.6), $G_\omega = L_\omega$. This proves (c) and completes the proof of Lemma 5.3.

At this point it is interesting to consider the construction in § 1. We readily check that the loop H satisfies $(H, H', H') = 1$. Moreover $H_2 = L_1(H) \neq 1$ but $L_2(H) = 1$. Hence H is distributor-nilpotent of class 2 but (as previously shown) is transfinitely centrally nilpotent of class ω. We also observe that $D_2(H) = H$ whereas $1 = Z_1(H) = Z_2(H) = \cdots$. This is enough to show that central nilpotence and distributor nilpotence are not too closely linked. We note further that, for each integer $n \geqq 3$, H has an n-generator subloop G with central class $c(n)$, distributor class 2. As a matter of fact the construction in § 1 arose from the concept of the distributor and, in particular, from the following lemma:

Lemma 5.4. *Let G be a commutative Moufang loop with $n \geqq 3$ generators such that $(G, G', G') = 1$. Then $G_{c(n)} = 1$. Moreover, for a suitable choice of G, $G_k \neq 1$ if $k < c(n)$.*

Proof. Let G be generated by a set S of n elements. Set $t = c(n)$ and consider the associator

$$a = ((\ldots((s_1, s_2, s_3), s_4, s_5), \ldots), s_{2t}, s_{2t+1})$$

where the s_i are in S. Since $(G, G', G') = 1$ we see by (iii) of Lemma 5.2 that (s_1, s_2, s_3) is in D. Hence a is skew-symmetric in $s_4, \ldots, s_{2t+1}$ and therefore $a = 1$ unless $s_4, \ldots, s_{2t+1}$ are distinct. Also $a = 1$ unless s_1, s_2, s_3 are distinct. Since $2t + 1 \geqq n + 2$ we may assume that $s_4 = s_2$, $s_5 = s_3$; but then $a = 1$ by Theorem 2.1. Hence $a = 1$ in all cases. Therefore, by Lemma 3.8, $G_t = 1$. That is, $G_{c(n)} = 1$. The concluding statement of Lemma 5.4 can be verified in terms of the loop G defined in § 1.

It is natural to inquire whether there exists an analogue of the distributor in the theory of groups. One might define the distributor D of a group G as the set of all elements d in G such that $(xy, d) = (x,d)(y,d)$ for all x, y in G. But then one sees easily that $D = Z_2$. In view of this fact, distributor series for a group offer little in the way of novelty.

6. Five generators

Our only 5-element identities so far are (3.14), (5.1), (5.3) and those which can be obtained from Lemma 3.6 with the help of Theorems 2.1, 3.1.

Lemma 6.1. *Let G be a commutative Moufang loop generated by five elements u, v, w, x, y. Then the element $p = ((u, v, w), u, x)$ is in Z_2.*

Proof. If S is the subset consisting of u, v, w, x, y, we must show that $(p, s, s') \in Z$ for all $s, s' \in S$. If s, s' are chosen from u, v, w, x then $(p, s, s') = 1$ by Theorem 3.1. Hence we need only prove that $(p, s, y) \in Z$ for each s in S. The cases $s = v$, $s = w$ are equivalent, by the form of p, and the case $s = y$ is immediate by Theorem 2.1. Hence it will be enough to prove the following:

(a) $(p, u, y) \in Z$.

(b) $(p, v, y) \in Z$.

(c) $(p, x, y) \in Z$.

Proof of (a). By Lemmas 3.9, 3.10, $(p, u, y) = (((u, v, w), u, x), u, y) = ((y, u, x), u, (u, v, w))^{-1} = 1$. Hence $(p, u, y) \in Z$.

Proof of (b). If k, k' are chosen from u, w, x, Theorem 3.1 implies that $(((u, x, w), u, v), k, k') = 1$. Taking $a = (u, x, w)$, $b = u$, $c = k$, $d = k'$, we have $((a, b, v), c, d) = 1$ for all v in any commutative Moufang loop. Hence, by Lemma 3.6, $(((a, b, v), v, t), c, d) = 1$ for all v, t; and we take $t = y$ so that $((a, b, v), v, t) = (((u, x, w), u, v), v, y) = (((u, v, w), u, x), v, y)^{-1} = (p, v, y)^{-1}$. Therefore

$$((p, v, y), k, k') = 1 , \quad k, k' = u, w, x . \tag{6.1}$$

In particular, $((p, v, y), u, s) = 1$ for $s = u, w, x$. Again, $((p, v, y), u, v) = ((p, v, u), y, v)^{-1}$ and $((p, v, y), u, y) = ((p, u, y), v, y)^{-1}$ by Lemma 3.9. Since $(p, v, u) = 1$ by Theorem 3.1 and $(p, u, y) \in Z$ by (a), we have

$$((p, v, y), u, S) = 1 . \tag{6.2}$$

Again, $((p, v, y), v, y) = 1$ trivially; and, for $s = u, v, w, x$, $((p, v, y), v, s) = ((p, v, s), v, y)^{-1} = 1$ by Lemma 6.9 and Theorem 3.1. Hence

$$((p, v, y), v, S) = 1 . \tag{6.2'}$$

By Lemma 4.4, every commutative Moufang loop satisfies the identity $((u,v,w),(w,v,y),y) = 1$. Hence, by Lemma 3.6, $1 = (((u,v,w),u,x),(w,v,y),y) = (p,(w,v,y),y)$ and thence, by Lemma 3.9, $1 = ((p, v, y), w, y)$. Moreover, $((p, v, y), w, t) = 1$ for $t = u, w, x$ by (6.1) and for $t = v$ by (6.2'). Hence

$$((p, v, y), w, S) = 1 . \tag{6.3}$$

In view of Lemma 3.9, interchange of w and x replaces p by p^{-1}. Hence (6.3) implies $((p, v, y), x, S) = 1$. By this, together with (6.1), (6.2), (6.2'), (6.3), we have $((p, v, y), S, S) = 1$. The proof of (b) is now complete.

Proof of (c). For $t,t' = u,v,w$, Theorem 3.1 implies $(((u,v,w),u,x),t,t') = 1$ for all x and thence Lemma 3.6 implies $((((u, v, w), u, x), x, y), t, t') = 1$ for all x, y. Thus

$$((p, x, y), t, t') = 1, \quad t, t' = u, v, w. \tag{6.4}$$

If $t = u, v, w$ we note the following: $((p, x, y), t, x) = ((p, x, t), y, x)^{-1} = 1$ by Lemma 3.9, Theorem 3.1; $((p, x, y), t, y) = ((p, t, y), x, y)^{-1} = 1$ by Lemma 3.9 followed by (a) (if $t = u$) or by (b) (if $t = v$) or by (b) with v, w interchanged (if $t = w$). By this and (6.4), $((p, x, y), t, S) = 1$ for $t = u, v, w$. Since $((p, x, y), x, y) = 1$ trivially, we have $((p, x, y), S, S) = 1$. This proves (c) and completes the proof of Lemma 6.1.

Lemma 6.2. *Every commutative Moufang loop G satisfies the identities*

$$((u, x, y), (v, x, y), w) = ((v, x, y), (w, x, y), u), \tag{6.5}$$

$$(((u, x, y), v, w), x, y) = ((u, x, y), (v, x, y), w)^{-1}, \tag{6.6}$$

$$((((u, x, y), v, w), x, y), v, w) = 1. \tag{6.7}$$

Proof. By Lemma 4.4, G satisfies the 4-element identity

$$((u, x, y), (v, x, y), v) = 1. \tag{6.8}$$

If θ is defined as in the proof of Lemma 5.1, (6.8) becomes $(u\theta, v\theta, v) = 1$. Replacing v by vw and using (3.8), we get $1 = (u\theta, (vw)\theta, vw) = (u\theta, st, vw)$ where $s = (v\theta)(w\theta)$, $t = (v\theta, v, w)(w\theta, w, v)$. If H is the subloop of G generated by u, v, w, x, y, Lemma 6.1, with u, v, w, x, y suitably permuted, shows that $(v\theta, v, w)$ is in $Z_2(H)$. Similarly, $(w\theta, w, v)$ is in $Z_2(H)$ and hence t is in $Z_2(H)$. By Lemma 5.3 (a), $Z_2(H) \subset D(H)$. Consequently, by (ii) and (iii) of Lemma 5.2, $1 = (u\theta, st, vw) = (u\theta, s, vw)(u\theta, t, vw) = (u\theta, s, vw)$. Since $s = (v\theta)(w\theta)$ and since, by (6.8), $(u\theta, v\theta, v) = (u\theta, w\theta, w) = 1$, three uses of (3.3) give

$$1 = [(u\theta, w\theta, v)\beta^{-1}\alpha][(u\theta, v\theta, w)\beta\alpha^{-1}] \tag{6.9}$$

where $\alpha = R(v, w)$, $\beta = R(v\theta, w\theta)$. Since $((u\theta, w\theta, v), w\theta, v\theta) = ((u\theta, w\theta, v\theta), w\theta, v)^{-1} = (1, w\theta, v)^{-1} = 1$ by Lemmas 3.9, 5.1, the automorphism β^{-1} may be omitted from the left hand bracket in (6.9). Similarly, β may be omitted from the right hand bracket. Hence, after operating on (6.9) by α^{-1}, we get $1 = (u\theta, w\theta, v)[(u\theta, v\theta, w)\alpha]$ and thence

$$(u\theta, w\theta, v)^{-1}(u\theta, v\theta, w)^{-1} = ((u\theta, v\theta, w), v, w). \tag{6.10}$$

Replacement of v by v^{-1} replaces the left hand side of (6.9) by its inverse and leaves the right hand side fixed. Therefore $(u\theta, w\theta, v)(u\theta, v\theta, w) = 1$. This is equivalent to the skew-symmetry of $(u\theta, v\theta, w)$ in u, v, w and hence to (6.5). Moreover, (5.3), (6.5) and (2.4) imply (6.6). Since the right hand side of (6.10) is the identity by (6.5), we use (6.6) to get $(((u\theta, v, w), x, y), v, w) = ((u\theta, v\theta, w), v, w)^{-1} = 1$. This proves (6.7) and completes the proof of Lemma 6.2.

Theorem 6.1. *Every commutative Moufang loop G generated by five elements is centrally nilpotent of class at most* 4.

Proof. Let G be generated by a subset S consisting of the five elements u, v, w, x, y. By Lemma 3.8 and symmetry, we need only prove that Z_2 contains the elements $p = ((u, v, w), u, x)$ and $q = ((u, v, w), x, y)$. However, $p \in Z_2$ by Lemma 6.1. Since q is skew-symmetric in u, v, w and in x, y, then q will be in Z_2 provided (q, x, y), (q, u, x) and (q, v, w) are in Z. However, $(q, x, y) = (((u, v, w), x, y), x, y) = 1$ and $(q, u, x) = (((u, v, w), x, u), y, x)^{-1} = (p, y, x)$ are in Z, so we are left with (q, v, w). On interchanging v with x, w with y in Lemma 6.2, we see that $(q, v, w) = (((u, v, w), x, y), v, w)$ is skew-symmetric in u, x, y and satisfies $((q, v, w), x, y) = 1$. Moreover, (q, v, w) is symmetric in v, w and, obviously, $((q,v,w),v,w) = 1$. Since, further, $((q,v,w),v,x) = ((q,v,x,)\,v,w)^{-1}$ and $(q, v, x) = (((u, v, w), x, y), v, x) = (((u, v, w), x, v,), v, y)^{-1}$ and $((u, v, w), x, v)$ is in Z_2 by Lemma 6.1, we have $((q, v, w), v, x) = 1$. Therefore $((q, v, w), S, S) = 1$, so $(q, v, w) \in Z$ and, consequently, $q \in Z_2$. This completes the proof of Theorem 6.1.

With the help of Theorem 6.1 we may obtain many new identities.

Lemma 6.3. *If the commutative Moufang loop G can be generated by five elements then $(G', G', G') = (G, G', G_2) = 1$. In particular, the associator subloop G' is an abelian group.*

Proof. By Lemma 4.3, $(G', G', G') = (G_1, G_1, G_1) \subset G_4$ and $(G, G', G_2) = (G_0, G_1, G_2) \subset G_4$. By Theorem 6.1, $G_4 = 1$. This completes the proof of Lemma 6.3.

Lemma 6.4. *Every commutative Moufang loop G satisfies the identities*

$$((u, v, w), x, y) = ((u, x, y), v, w)\,((v, x, y), w, u)\,((w, x, y), u, v), \tag{6.11}$$

$$((u, v, w), x, y) = ((x, v, w), u, y)\,((u, x, w), v, y)\,((u, v, x), w, y). \tag{6.12}$$

Proof. We work within the subloop H generated by u, v, w, x, y. By Lemma 6.3, H' is an abelian group and $(H, H_1, H_2) = 1$. In particular, the right hand sides of (6.11), (6.12) are unambiguous. Define the mapping $\theta = \theta(x, y)$ by $u\theta = (u, x, y)$ for all u and set $a = (u, v, w)$, $b_1 = (u\theta, v, w)$, $b_2 = (v\theta, w, u)$, $b_3 = (w\theta, w, u)$. By (6.5), $(u\theta, v\theta, w) = (v\theta, w\theta, u) = (w\theta, u\theta, v) = c$, say. By (3.8), $(u(u\theta), v, w) = ab_1(a, u, u\theta)\,(b_1, u\theta, u)$. However, $(a, u, u\theta) = ((u, v, w), u, (u, x, y)) = 1$ by Lemma 3.10, and $(b_1, u\theta, u) \in (H_2, H', H) = 1$. Hence $(u(u\theta), v, w) = ab_1$; and similarly with u, v, w permuted. A like argument gives $(u\,(u\theta), v\theta, w) = bc$; and similarly with u, v, w permuted. From the identity $(u(u\theta), v, w) = ab_1$, on replacement of v by $v(v\theta)$, we get $(u\,(u\theta), v\,(v\theta), w) = (u, v\,(v\theta, w)\,(u\theta, v\,(v\theta), w) = (ab_2)\,(b_1c) = ab_1b_2c$. From the identity $(u(u\theta), v(v\theta), w) = ab_1b_2c$, on replacement of w by $w(w\theta)$, we get, in a similar way, $(u(u\theta), v(v\theta), w(w\theta)) = (ab_3)\,(b_1c)\,(b_2c)\,c = ab_1b_2b_3c^3 = ab_1b_2b_3$. However, $a(a\theta) = aR(x, y) = (uR(x, y), vR(x, y), wR(x, y)) = (u(u\theta), v(v\theta), w(w\theta))$. Hence $a\theta = b_1b_2b_3$, and this is (6.11). By (6.11), $((y, u, x), v, w) = ((y, v, w), u, x)\,((u, v, w), x, y)\,((x, v, w), y, u)$, whence, after rearrangement,

$$((y, u, x), v, w)\,((y, v, w), u, x)^{-1} = ((u, v, w), x, y)\,((x, v, w), u, y)^{-1}. \tag{6.13}$$

We apply the permutation $(uv)\,(xw)$ to (6.13), take inverses, and get

$$((y,u,x),v,w)\;((y,v,w),u,x)^{-1} = ((u,v,x),w,y)\;((u,x,w),v,y). \qquad (6.14)$$

Since the left hand sides of (6.13), (6.14) are equal, the right hand sides are also equal, whence we get (6.12). This completes the proof of Lemma 6.4.

Lemma 6.5. *In every commutative Moufang loop G, the associator $(((x, y, a), a, b), b, z)$ is symmetric in a, b and skew-symmetric in x, y, z. Equivalently, G satisfies the identity*

$$(((x, y, a), a, b), b, z) = (((z, x, b), b, a), a, y)\,. \qquad (6.15)$$

Proof. We evaluate the associator $E = ((x, y, a), a, bz)$ in two ways. On the one hand, by (3.8),

$$E = ((x, y, a), a, b)\;((x, y, a), a, z) \qquad (6.16)$$
$$(((x, y, a), a, b), b, z)\;(((x, y, a), a, z), z, b)\,.$$

On the other hand, by Lemma 3.9, $E = ((bz, x, a), a, y)$. In evaluating the latter associator we note that we may work within the subloop H generated by the five elements b, z, x, a, y. By Lemma 6.3, H' is an abelian group. Consequently, if s, t, are arbitrary elements of H', $((s, a, y), s, t) = 1$. From this and (3.8), $(st, a, y) = (s, a, y)\,(t, a, y)$ for all s, t in H'. Therefore, since, by (3.8) and Lemma 3.9, $(bz, x, a) = (b,x,a)\,(z,x,a)\,((b,x,a),b,z)\,((z,x,a),z,b) = (b,x,a)\,(z,x,a)\,((z,x,b),b,a)\,((b, x, z), z, a)$, we see that

$$E = ((b, x, a), a, y)\;((z, x, a), a, y)$$
$$(((z, x, b), b, a), a, y)\;(((b, x, z), z, a), a, y)\,.$$

By Lemma 3.9, $((b, x, a), a, y) = ((x, y, a), a, b)$ and $((z, x, a), a, y) = ((x, y, a), a, z)$. Hence, by comparison with (6.16),

$$(((x, y, a), a, b), b, z)\;(((x, y, a), a, z), z, b) \qquad (6.17)$$
$$= (((z, x, b), b, a), a, y)\;(((b, x, z), z, a), a, y)\,.$$

When z is replaced by z^{-1} in (6.17) the first factors on left and right are replaced by their inverses and the other two factors are left unchanged. Therefore, by multiplication, we split (6.17) into two identities, one of which is (6.15). Since the left hand side of (6.15) is skew-symmetric in x, y and the right hand side is skew-symmetric in x, z, we see that both sides are skew-symmetric in x, y, z. In particular, the right hand side is equal to $(((x, y, b), b, a), a, z)$; and thus we see that the left hand side is symmetric in a, b.

Lemma 6.6. *Every commutative Moufang loop G satisfies the identity*

$$((x, y, ab), ab, z) = ((x, y, a), a, z)\;((x, y, b), b, z) \qquad (6.18)$$
$$(((x, y, a), a, b), b, z)\;((x, y, a), b, z)\;((x, y, b), a, z)\,.$$

Proof. We expand the left hand side of (6.18) by (3.8), working entirely within the subloop H generated by x, y, z, a, b. Since $(st, ab, z) = (s, ab, z)\,(t, ab, z)$ for all s, t in H', we get a product of four factors (s, ab, z) where s ranges over (x, y, a), (x, y, b), $((x, y, a), a, b)$, $((x, y, b), b, a)$. Expanding each of the terms (s, ab, z) by (3.8), using Lemmas 3.9, 6.5 and identities such as

$$(((x, y, a)\ a, b), a, z) = ((z, a, b), a, (x, y, a))^{-1}$$

together with (3.14), we finally get (6.18).

7. The functions f_i

For each non-negative integer i we define a function f_i on the commutative Moufang loop G by

$$f_0(x, y, z) = (x, y, z), \tag{7.1}$$

$$f_{i+1}(x, y, z;\, a_1, a_2, \ldots, a_i, u) = (f_i(x, y, u;\, a_1, \ldots, a_i), u, z), \qquad i \geqq 0. \tag{7.2}$$

The following lemma assures us of a certain amount of associativity.

Lemma 7.1. *For each fixed pair p, q of elements of the commutative Moufang loop G, let $K(p, q)$ be the subloop of G generated by the set of all associators (p, q, z), z in G. Then $K(p, q)$ is an abelian group containing $f_i(p, q, z;\, a_1, \ldots, a_i)$ for each non-negative integer i and all $z, a_1, a_2, \ldots, a_i$ in G.*

Proof. That $K(p, q)$ contains $f_i(p, q, z;\, a_1, \ldots, a_i)$ is the content of Lemma 3.5. That $K(p, q)$ is associative follows from Lemma 5.1 and the fact that every associative subset of a commutative Moufang loop is contained in an associative subloop.

Lemma 7.2. *In any commutative Moufang loop G the function $f_i = f_i(x, y, z;\, a_1, \ldots, a_i)$ has the following properties:* (I) *f_i is skew-symmetric in x, y, z.* (II) *For $i \geqq 2$, f_i is symmetric in $a_1, a_2, \ldots, a_i$.* (III) *For $i \geqq 1$, f_i satisfies $(f_i, a_p, G) = 1$* for $p = 1, 2, \ldots, i$. (IV) *f_i lies in the centre of the subloop generated by its arguments, namely x, y, z if $i = 0$ and $x, y, z, a_1, \ldots, a_i$ if $i \geqq 1$.* (V) *G satisfies the identities*

$$f_i(w, x, yz;\, a_1, \ldots, a_i) = f_i(w, x, y;\, a_1, \ldots, a_i)\, f_i(w, x, z;\, a_1, \ldots, a_i) \tag{7.3}$$
$$f_{i+1}(w, x, y;\, a_1, \ldots, a_i, z)\, f_{i+1}(w, x, z;\, a_1, \ldots, a_i, y)\,,$$

$$f_{i+1}(x, y, z;\, a_1, \ldots, a_i, uv) = f_{i+1}(x, y, z;\, a_1, \ldots, a_i, u) \tag{7.4}$$
$$f_{i+1}(x, y, z;\, a_1, \ldots, a_i, v)\, f_{i+2}(x, y, z;\, a_1, \ldots, a_i, u, v)$$
$$((f_i(x, y, u;\, a_1, \ldots, a_i), v, z)\, ((f_i(x, y, v;\, a_1, \ldots, a_i), u, z)\,.$$

Remark. It will appear from the proof that the five factors on the right of (7.4) lie in an associative subloop.

Proof. We begin with (V). For $i = 0$, (7.3) is a variant of (3.8). If $i = j + 1$ where j is non-negative, then, by (7.2), $f_i(w, x, yz;\, a_1, \ldots, a_i)$

$-(p, a_i, yz)$ where $p = f_j(x, y, a_i; a_1, \ldots, a_j)$. Hence application of (3.8) gives (7.3). Next we prove (7.4). Keeping the integer i and the elements $x, y, z, a_1, \ldots, a_i$ fixed, we define a function $F(u)$ by $F(u) = f_{i+1}(x, y, z; a_1, \ldots, a_i, u)$. Then $F(u) = ((p, q, u), u, z)$ where p, q are defined as follows: If $i = 0$, then $p = x, q = y$. If $i = j + 1$ for $j \geqq 0$, then $p = f_j(x, y, a_i; a_1, \ldots, a_j), q = a_i$. We observe that p, q, z are independent of u. Evaluating $F(uv)$ by Lemma 6.6, we get

$$F(uv) = F(u)\,F(v)\,(((p, q, u), u, v), v, z)\,((p, q, u), v, z)\,((p, q, v), u, z). \tag{7.5}$$

And (7.5) may be translated into (7.4). This proves (V). In particular, the six associators displayed in (7.5) all lie in the abelian group H' where H is the subloop generated by p, q, u, v, z. Noting that $F(u^{-1}) = F(u)$, we use equation (7.5) to obtain

$$\begin{aligned} F(uv)\,F(u^{-1}v)\,F(u)\,F(v) &= ((p, q, u), u, v), v, z)^{-1} \\ &= f_{i+2}(x, y, z; a_1, \ldots, a_i, u, v)^{-1}. \end{aligned} \tag{7.6}$$

In view of (7.6), if (I) holds for $i + 1$, then (I) also holds for $i + 2$. Since, moreover, $f_0(x, y, z) = (x, y, z)$ and $f_1(x, y, z; a_1) = ((x, y, a_1), a_1, z)$, so that (I) holds for $i = 0, 1$, we deduce that (I) holds for every non-negative integer i. By Lemma 6.5, $f_2 = (((x, y, a_1), a_1, a_2), a_2, z)$ is symmetric in a_1, a_2. By (7.6) and Lemma 6.5, $f_{i+2}(x, y, z; a_1, \ldots, a_i, u, v)$ is symmetric in u, v for $i \geqq 0$. Consequently, by (7.2), for each $i \geqq 2$, $f_i(x, y, z; a_1, \ldots, a_i)$ is symmetric in each adjacent pair in the sequence $a_1, a_2, \ldots, a_i$; and this is enough to prove (II). To prove (III) we suppose $i = j + 1, j \geqq 0$, and note that $f_i = ((s, t, a_i), a_i, z)$ where $s = x$, $t = y$ or $s = f_{j-1}(x, y, a_j; a_1, \ldots, a_{j-1})$, $t = a_j$ according as $j = 0$ or $j \geqq 1$. Then, by Lemmas 3.9, 3.10, for any w in G,

$$(f_i, a_i, w) = (((s, t, a_i), a_i, z), a_i, w) = ((w, a_i, z), a_i, (s, t, a_i))^{-1} = 1.$$

Thus (III) holds for $p = i$. Therefore, by (II), we see that (III) holds for $p = 1, 2, \ldots, i$. To prove (IV), we must show that $(f_i, k, k') = 1$ for all choices of k, k' from the arguments of f_i. In view of (III), we need only consider the case that k, k' are chosen from x, y, z. In view of (I), we are reduced to the case that $k = x, k' = y$. Since, by Theorem 2.1, G satisfies the identity $((x, y, z), x, y) = 1$, then, by Lemma 3.6, G satisfies the identity $(f_i(x, y, z; a_1, \ldots, a_i), x, y) = 1$ for each non-negative integer i. Thus $(f_i, x, y) = 1$, completing the proof of (IV) and the proof of Lemma 7.2.

8. Two lemmas

The importance of the next two lemmas stems from their connection with (III) of Lemma 7.2.

Lemma 8.1. *Let k be a non-negative integer and let c, s be elements of the commutative Moufang loop G such that $(G, c, s) \subset Z_k$. Then*

$$(G_i, c, s) \subset Z_{k-i} \tag{8.1}$$

for every non-negative integer i.

Proof. (8.1) holds for $i = 0$ by hypothesis. If (8.1) holds for some i, then, by (6.11) and Lemma 4.3,

$$(G_{i+1}, c, s) = ((G_i, G, G), c, s) \subset ((G_i, c, s), G, G)\ ((G, c, s), G, G_i) \subset$$
$$\subset (Z_{k-i}, G, G)\ (Z_k, G, G_i) \subset Z_{k-i-1}.$$

This completes the proof of Lemma 8.1.

Lemma 8.2. *Let k be a non-negative integer and let c, s, s' be elements of the commutative Moufang loop G such that $(G, c, s) \subset Z_k$, $(G, c, s') \subset Z_k$. Then*

$$((G_i, G_j, c), s, s') \subset Z_{k-i-j-1} \tag{8.2}$$

for all non-negative integers i, j.

Proof. By (6.12) and Lemmas 4.3, 8.1,

$$((G_i, G_j, c), s, s') \subset ((s, G_j, c), G_i, s')\ ((G_i, s, c), G_j, s')\ ((G_i, G_j, s), c, s') \subset$$
$$\subset (Z_{k-j}, G_i, G)\ (Z_{k-i}, G_j, G)\ (G_{i+j+1}, c, s') \subset Z_{k-i-j-1}.$$

9. The sets $\mathfrak{L}(n)$

Our proof of Slaby's theorem (see § 1) for the general case of n generators will proceed by induction over the elements of a certain finite simply ordered set $\mathfrak{L}(n)$ which we now define and explain.

Let $\mathfrak{L}$ be the (infinite) set consisting of all ordered sets $(p_1, p_2, \ldots, p_k)$ of non-negative integers $p_1, \ldots, p_k$, where k is to take on all positive integral values. Equality in $\mathfrak{L}$ is to be componentwise: $(p_1, \ldots, p_k) = (q_1, \ldots, q_t)$ if and only if $k = t$ and $p_i = q_i$ for $i = 1, 2, \ldots, k$. We introduce an ordering on $\mathfrak{L}$ as follows:

$$(p_1, \ldots, p_k) > (q_1, \ldots, q_t) \tag{9.1}$$

if and only if

either (I) There exists an integer w, with $1 \leqq w \leqq \mathrm{Min}\,(k, t)$, such that

$$p_i = q_i \text{ if } 1 \leqq i \leqq w-1 \text{ but } p_w > q_w,$$

or (II) $p_i = q_i$ for $i = 1, 2, \ldots, \mathrm{Min}\,(k, t)$ but $k > t$.

We may verify that $\mathfrak{L}$ is now simply ordered. In fact, if λ, μ, ν denote elements of $\mathfrak{L}$, we have: (i) Exactly one of the relation $\lambda > \mu$, $\lambda = \mu$, $\mu > \lambda$ holds. (ii) If $\lambda > \mu$ and $\mu > \nu$, then $\lambda > \nu$.

The set $\mathfrak{L}$ can be used to define a set of functions f_λ, $\lambda \in \mathfrak{L}$, on the commutative Moufang loop G, such that $f_{(i)}$ is the function f_i of § 7. If $\lambda = (p_1, \ldots, p_k, q)$, $\mu = (p_1, \ldots, p_k)$, then f_λ has the form $f_q(f_\mu, x, y; z_1, \ldots, z_q)$ where the $q+2$ arguments $x, y, z_1, \ldots, z_q$ are elements of G to be chosen independently of the arguments of f_μ. Thus, if $v(\lambda)$ denotes the number of arguments of f_λ, we have $v(\lambda) = v(\mu) + q + 2$. Since $\imath(i) = v((i)) = i + 3$, a simple induction gives

$$v(p_1, p_2, \ldots, p_k) = p_1 + p_2 + \cdots + p_k + 2k + 1 \,. \tag{9.2}$$

In proving the n-generator theorem we shall be concerned, in essence, with those functions f_λ which have n or less arguments. For this reason, we define $\mathfrak{L}(n)$ to be the subset of $\mathfrak{L}$ consisting of all λ in $\mathfrak{L}$ such that $v(\lambda) \leqq n$. We order $\mathfrak{L}(n)$ according to the ordering induced by the ordering (9.1) of $\mathfrak{L}$. It is easily verified that $\mathfrak{L}(n)$ is finite for each positive integer n. The size of $\mathfrak{L}(n)$ increases rapidly with n, as the first few cases show:

$\mathfrak{L}(3)$: (0). $\quad \mathfrak{L}(4)$: $(1) > (0)$. $\quad \mathfrak{L}(5)$: $(2) > (1) > (0, 0) > (0)$.
$\mathfrak{L}(6)$: $(3) > (2) > (1, 0) > (1) > (0, 1) > (0, 0) > (0)$.
$\mathfrak{L}(7)$: $(4) > (3) > (2, 0) > (2) > (1, 1) > (1, 0) > (1) > (0, 2) > (0, 1) >$
$> (0, 0, 0) > (0, 0) > (0)$.

The length of a detailed (non-inductive) proof of the n-generator theorem seems roughly proportional to the number of elements of $\mathfrak{L}(n)$. Hence the necessity for a proof by induction. A clear idea of the nature of the latter may be obtained by considering the proof for the case $n = 20$. The greatest and lowest elements of $\mathfrak{L}(20)$ are (17) and (0) respectively and, for example, the complete chain leading from (11) to (10) in $\mathfrak{L}(20)$ is as follows:

$$\begin{aligned}
&(11) > (10, 5) > (10, 4) > (10, 3, 0) > (10, 3) > (10, 2, 1) > (10, 2, 0) > \\
&\quad > (10, 2) > (10, 1, 2) > (10, 1, 1) > (10, 1, 0, 0) > (10, 1, 0) > \\
&\quad > (10, 1) > (10, 0, 3) > (10, 0, 2) > (10, 0, 1, 0) > (10, 0, 1) > \\
&\quad > (10, 0, 0, 1) > (10, 0, 0, 0) > (10, 0, 0) > (10, 0) > (10) \,.
\end{aligned}$$

We state a certain proposition $\mathfrak{R}(\lambda)$ for each λ in $\mathfrak{L}(20)$; $\mathfrak{R}(17)$ is easily seen to be true and our object is to establish $\mathfrak{R}(0)$. Suppose we assume inductively that $\mathfrak{R}(\lambda)$ is true for each $\lambda > (10, 2, 1)$. We wish to establish $\mathfrak{R}(10, 2, 1)$; we do this by first establishing two other propositions $\mathfrak{P}(\lambda), \mathfrak{Q}(\lambda)$ for $\lambda = (10, 2, 1)$. For example, to establish $\mathfrak{P}(10, 2, 1)$ we must first descend further and establish $\mathfrak{P}(\lambda)$ by working up along the inverted chain

$$(10) < (10, 0) < (10, 1) < (10, 2) < (10, 2, 0) < (10, 2, 1).$$

To sum up, we have the main "downward" induction for $\mathfrak{R}(\lambda)$ and, at each stage, two subsidiary "upward" inductions for $\mathfrak{P}(\lambda), \mathfrak{Q}(\lambda)$.

10. The main theorem

Let n be a positive integer and let G be a commutative Moufang loop which can be generated by n distinct elements. Each such set of n

generators gives rise to $n!$ ordered sets of n generators and, for present purposes, we shall consider only ordered sets of n generators. Let S be the generating set consisting of the n distinct elements

$$s_1, s_2, \ldots, s_n, \tag{10.1}$$

ordered from left to right. We need to observe that each of the following operations leads from an ordered set of n generators to an ordered set of n generators:

(I) *Perform any permutation of the ordered set* (10.1) *of generators.*

(II) *For some* i, j, *with* $1 \leqq i \leqq n$, $1 \leqq j \leqq n$, $i \neq j$, *replace the* j*th generator* s_j *by* $s_i s_j$ *and leave the remaining generators* s_k $(k \neq j)$ *unchanged.*

If S is a given ordered set of n generators of G, we define, inductively, for each λ in $\mathfrak{L}(n)$, an associator $A(\lambda) = A_S(\lambda)$ as follows:

(i) *If* $(p) \in \mathfrak{L}(n)$, $A(p) = f_p(s_1, s_2, s_3; s_4, \ldots, s_{p+3})$.

(ii) *If* $\lambda = (p_1, p_2, \ldots, p_k, q)$ *and* $\mu = (p_1, \ldots, p_k)$ *are in* $\mathfrak{L}(n)$ *and if* $v(\mu) = t$, *then*

$$A(\lambda) = f_q(A(\mu), s_{t+1}, s_{t+2}; s_{t+3}, \ldots, s_{t+q+2}).$$

In particular, $A(\lambda)$ is formed from the $v(\lambda)$ elements $s_1, s_2, \ldots, s_{v(\lambda)}$; these we call the arguments of $A(\lambda)$. The $v(\lambda)$ arguments of $A(\lambda)$ are partitioned into two sets, the set of *skew* arguments of $A(\lambda)$ and the set of *symmetric* arguments of $A(\lambda)$. This is done inductively as follows:

(iii) *Let* $(p) \in \mathfrak{L}(n)$. *The skew arguments of* $A(p)$ *are* s_1, s_2, s_3. *If* $p = 0$, *then* $A(p)$ *has no symmetric arguments. If* $p > 0$, *the symmetric arguments of* $A(p)$ *are* $s_4, \ldots, s_{p+3}$.

(iv) *Let* $\lambda = (p_1, \ldots, p_k, q)$ *and* $\mu = (p_1, \ldots, p_k)$ *be in* $\mathfrak{L}(n)$ *and let* $t = v(\mu)$. *The skew arguments of* $A(\lambda)$ *are* s_{t+1}, s_{t+2} *together with the skew arguments of* $A(\mu)$. *If* $q = 0$, *the symmetric arguments of* $A(\lambda)$ *are the symmetric arguments of* $A(\mu)$. *If* $q > 0$, *the symmetric arguments of* $A(\lambda)$ *are* $s_{t+3}, \ldots, s_{t+q+2}$ *together with the symmetric arguments of* $A(\mu)$.

We wish to establish the following propositions for each λ in $\mathfrak{L}(n)$:

$\mathfrak{P}(\lambda)$. *If* $\lambda = (p_1, \ldots, p_k)$, *then for each ordered set* S *of* n *generators of the commutative Moufang loop* G, $A(\lambda)$ *is skew-symmetric mod* $Z_{k+n-v(\lambda)}$ *in its skew arguments.*

$\mathfrak{Q}(\lambda)$. *If* $\lambda = (p_1, \ldots, p_k)$, *then for each ordered set* S *of* n *generators of the commutative Moufang loop* G *and for each symmetric argument* s *of* $A(\lambda)$, $(A(\lambda), s, G) \subset Z_{k+n-1-v(\lambda)}$.

$\mathfrak{R}(\lambda)$. *If* $\lambda = (p_1, \ldots, p_k)$, *then, for each ordered set* S *of* n *generators of the commutative Moufang loop* G, $A(\lambda) \in Z_{k+n-v(\lambda)}$.

Some comment is necessary. First, if $A(\lambda)$ has no symmetric arguments, we make the convention that $\mathfrak{Q}(\lambda)$ is true. Secondly, $\mathfrak{P}(\lambda)$ is to be interpreted as follows. Let θ be any permutation of the skew arguments of $A(\lambda)$, let $sgn(\theta)$ be 1 or -1 according as θ is even or odd, let $S\theta$ be the ordered set obtained from S by applying θ and let $A'(\lambda)$ be the "canonical" associator corresponding to $S\theta$. Then $\mathfrak{P}(\lambda)$ means that

$$A(\lambda) \equiv (A'(\lambda))^{sgn(\theta)} \mod Z_{k+n-v(\lambda)}.$$

In what follows we shall assume $n \geqq 5$.

Lemma 10.1. *$\mathfrak{R}(\lambda)$ implies both $\mathfrak{P}(\lambda)$ and $\mathfrak{Q}(\lambda)$.*

Proof. Obvious.

Lemma 10.2. *$\mathfrak{R}(n-3)$ is true.*

Proof. If $\lambda = (n-3)$ then $v(\lambda) = n$ and $k = 1$, so $k + n - v(\lambda) = 1$. Also $A(n-3) = f_{n-3}(s_1, s_2, s_3; s_4, \ldots, s_n)$ is in $Z(G) = Z_1$ by Lemma 7.2 (IV) and the hypothesis that S generates G.

Lemma 10.3. *$\mathfrak{R}(n-4)$ is true.*

Proof. Here $\lambda = (n-4)$, $v(\lambda) = n-1$, $k = 1$ and $k + n - v(\lambda) = 2$. We are to show that $A(\lambda) = f_{n-4}(s_1, s_2, s_3; s_4, \ldots, s_{n-1})$ is in Z_2. To do so, we must verify that $(A(\lambda), s_i, s_j) \in Z_1$ for $i, j = 1, 2, \ldots, n$. However, by Lemma 7.2, $A(\lambda)$ is skew-symmetric in s_1, s_2, s_3, is symmetric in $s_4, \ldots, s_{n-1}$, satisfies $(A(\lambda), s_2, s_3) = 1$ and $(A(\lambda), s_{n-1}, G) = 1$, and does not have s_n as an argument. Hence we need only prove that $(A(\lambda), s_3, s_n) \in Z_1$. However, by (7.2), $(A(\lambda), s_3, s_n) = f_{n-3}(s_1, s_2, s_n; s_4, \ldots, s_{n-1}, s_3)$. The latter is a "canonical" associator $A'(n-3)$ for the ordered set of generators obtained from S by interchanging s_3, s_n. Hence, by Lemma 10.2, $(A(\lambda), s_3, s_n) \in Z_1$. This completes the proof of Lemma 10.3.

Lemma 10.4. *If $\lambda = (p) \in \mathfrak{L}(n)$ and if $\mathfrak{R}(\mu)$ is true for each μ in $\mathfrak{L}(n)$ such that $\mu > \lambda$, then $\mathfrak{R}(\lambda)$ is true.*

Proof. Here $\lambda = (p)$, $v(\lambda) = p + 3$, $k = 1$ and $k + n - v(\lambda) = n - p - 2$. Hence we are to prove that $A(\lambda) \in Z_{n-p-2}$. In view of Lemmas 10.2, 10.3, we may assume that $p < n - 4$. Thus $p + 5 \leqq n$. We must verify that $(A(\lambda), s_i, s_j) \in Z_{n-p-3}$ for $i, j = 1, 2, \ldots, n$. However, $A(\lambda) = f_p(s_1, s_2, s_3; s_4, \ldots, s_{p+3})$ is skew-symmetric in s_1, s_2, s_3, satisfies $(A(\lambda), s_2, s_3) = 1$, and, for $p > 0$, is symmetric in $s_4, \ldots, s_{p+3}$ and satisfies $(A(\lambda), s_{p+3}, G) = 1$. Furthermore, $A(\lambda)$ is independent of $s_{p+4}, s_{p+5}, \ldots, s_n$. Hence it will be enough to show that both of $(A(\lambda), s_3, s_{p+4})$, $(A(\lambda), s_{p+4}, s_{p+5})$ are in Z_{n-p-3}. Now $(A(\lambda), s_3, s_{p+4}) = f_{p+1}(s_1, s_2, s_{p+4}; s_4, \ldots, s_{p+3}, s_3)$ is a "canonical" associator $A'(\mu)$, where $\mu = (p+1)$, for a suitable permutation of S. We note that $v(\mu) = p + 4 < n$, so that μ is indeed in $\mathfrak{L}(n)$. Moreover, since $\mu > \lambda$, $\mathfrak{R}(\mu)$ is true by hypotheses. Since $\mu = (p+1)$ and $1 + n - [p+4] = n - p - 3$, $\mathfrak{R}(\mu)$ implies that $(A(\lambda), s_3, s_{p+4}) = A'(\mu)$ is in Z_{n-p-3}. Again, $(A(\lambda), s_{p+4}, s_{p+5}) = f_0(A(\lambda), s_{p+4}, s_{p+5}) = A(\mu')$ where $\mu' = (p, 0)$. Here $v(\mu') = p + 5 \leqq n$ and $\mu' > \lambda$ and $2 + n - v(\mu') = n - p - 3$, so, by our hypothesis, $A(\mu') \in Z_{n-p-3}$. This completes the proof of Lemma 10.4.

Lemma 10.5. *If $\lambda \in \mathfrak{L}(n)$ and if $\mathfrak{R}(\mu)$ is true for each μ in $\mathfrak{L}(n)$ such that $\mu > \lambda$, then $\mathfrak{P}(\lambda)$ is true.*

Proof. In view of Lemmas 10.4, 10.1, we may assume that $\lambda = (p_1, \ldots, p_k)$ where $k > 1$. In view of Lemmas 10.2, 10.3 we may assume that $(n-4) > \lambda$

Our method of establishing $\mathfrak{P}(\lambda)$ involves an induction over the ascending chain

$$(p_1) < (p_1, 0) < \cdots < (p_1, p_2) < (p_1, p_2, 0) < \cdots < (p_1, \cdots, p_{k-1}, 0) < \cdots < (p_1, \cdots, p_k).$$

First we note that $\mathfrak{P}(p_1)$ is trivial since f_{p_1} is skew-symmetric in its first three arguments. Next we consider some integer t with $1 \leqq t \leqq k-1$ and assume inductively that $\mathfrak{P}(p_1, \ldots, p_t)$ is true. We begin by proving $\mathfrak{P}(p_1, \ldots, p_t, 0)$.

First we note that $v(p_1, \ldots, p_t, 0) = p_1 + \cdots + p_t + 2(t+1) + 1 \leqq \leqq p_1 + \cdots + p_k + 2k + 1 = v(\lambda) \leqq n$, so that $(p_1, \ldots, p_t, 0)$ is in $\mathfrak{L}(n)$. For convenience we set $\alpha = (p_1, \ldots, p_{t-1})$ with the understanding that, in case $t = 1$, so that α is meaningless, the definitions $A(\alpha) = s_1$, $v(\alpha) = 1$ shall be used and $A(\alpha)$ shall have no skew arguments. Then $v(\alpha) = p_1 + \cdots + p_{t-1} + 2(t-1) + 1$ in all cases. We define $w = v(\alpha), q = p_t$, so that $v(p_1, \ldots, p_t) = w + q + 2 = m$, say. Then

$$\begin{aligned} A(p_1, \ldots, p_t, 0) &= (A(p_1, \ldots, p_t), s_{m+1}, s_{m+2}) \\ &= (f_q(A(\alpha), s_{w+1}, s_{w+2}; s_{w+3}, \ldots, s_m), s_{m+1}, s_{m+2}). \end{aligned}$$

Since the skew arguments of $A(p_1, \ldots, p_t, 0)$ are s_{m+1}, s_{m+2} and those of $A(p_1, \ldots, p_t)$; since s_{w+2} is a skew argument of $A(p_1, \ldots, p_t)$; since $\mathfrak{P}(p_1, \ldots, p_t)$ holds and $t + n - v(p_1, \ldots, p_t)$ is greater by one than $t + 1 + n - v(p_1, \ldots, p_t, 0) = t + n - m - 1$; we may establish $\mathfrak{P}(p_1, \ldots, p_t, 0)$ by proving the congruence

$$\begin{aligned} &(f_q(A(\alpha), s_{w+1}, s_{w+2}; s_{w+3}, \ldots, s_m), s_{m+1}, s_{m+2}) \\ &\equiv (f_q(A(\alpha), s_{w+1}, s_{m+1}; s_{w+3}, \ldots, s_m), s_{w+2}, s_{m+2})^{-1} \quad \bmod Z_{t+n-m-1}. \end{aligned} \tag{10.2}$$

To prove (10.2) we proceed as follows: If $\mu = (p_1, \ldots, p_{t-1}, 1 + p_t)$, then, certainly, μ is in $\mathfrak{L}$ and $\mu > \lambda$. Moreover, in our preceding notation $v(\mu) = v(p_1, \ldots, p_t) + 1 = m + 1 < n$, so μ is in $\mathfrak{L}(n)$. Consequently, $\mathfrak{R}(\mu)$ is true by hypothesis. Moreover, $t + n - v(\mu) = t + n - m - 1$, so $A(\mu)$ is in $Z_{t+n-m-1}$ for each ordered set S of n generators. Finally, $A(\mu) = f_{q+1}(A(\alpha), s_{w+1}, s_{w+2}; s_{w+3}, \ldots, s_{m+1})$. We alter S by replacing s_{w+2} by s_{m+2} and s_{m+1} by the product $s_{w+2}s_{m+1}$; this can be done by performing two admissible operations on S. Therefore, by $\mathfrak{R}(\mu)$ and (7.4), $Z_{t+n-m-1}$ contains the left hand side of

$$\begin{aligned} &f_{q+1}(A(\alpha), s_{w+1}, s_{m+2}; s_{w+3}, \ldots, s_m, s_{w+2}s_{m+1}) \\ &= f_{q+1}(A(\alpha), s_{w+1}, s_{m+2}; s_{w+3}, \ldots, s_m, s_{w+2}) \\ &\quad f_{q+1}(A(\alpha), s_{w+1}, s_{m+2}; s_{w+3}, \ldots, s_m, s_{m+1}) \\ &\quad f_{q+2}(A(\alpha), s_{w+1}, s_{m+2}; s_{w+3}, \ldots, s_m, s_{w+2}, s_{m+1}) \\ &\quad (f_q(A(\alpha), s_{w+1}, s_{w+2}; s_{w+3}, \ldots, s_m), s_{m+1}, s_{m+2}) \\ &\quad (f_q(A(\alpha), s_{w+1}, s_{m+1}; s_{w+3}, \ldots, s_m), s_{w+2}, s_{m+2}) \end{aligned} \tag{10.3}$$

Moreover, $Z_{t+n-m-1}$ contains the first two terms on the right of (10.3) by $\mathfrak{R}(\mu)$ and contains the third by $\mathfrak{R}(\mu)$, (7.2) and the fact that $Z_{t+n-m-1}$ is a normal subloop. Therefore (10.3) yields a congruence equivalent to (10.2).

Now that $P(p_1, \ldots, p_t, 0)$ is established, we must prove $P(p_1, \ldots, p_t, i)$ for $0 \leqq i \leqq p_{t+1}$. We need only consider the case that $i > 0$. Then, in our previous notation, by (7.2),

$$A(p_1, \ldots, p_t, i) = f_i(A(p_1, \ldots, p_t), s_m, s_{m+1}; s_{m+2}, \ldots, s_{m+1+i})$$
$$= (\ldots((A(p_1, \ldots, p_t), s_m, s_{m+2}), \ldots).$$

Clearly $A(p_1, \ldots, p_t, i)$ is skew-symmetric in s_m, s_{m+1}. Moreover, by $R(p_1, \ldots, p_t, 0)$, $((A(p_1, \ldots, p_t), s_m, s_{m+2})$ is (in particular) skew-symmetric mod $Z_{t+n-m-1}$ in s_m and the skew arguments of $A(p_1, \ldots, p_t)$. Consequently, if $x = t+n-m-1-i$, $A(p_1, \ldots, p_t, i)$ is skew-symmetric mod Z_x in s_m, s_{m+1}; and in s_m together with the skew arguments of $A(p_1, \ldots, p_t)$; and hence in the skew arguments of $A(p_1, \ldots, p_t, i)$. Since $t+1+n-v(p_1, \ldots, p_t, i) = t+1+n-m-i-2 = t+n-m-1-i = x$, we have proved $\mathfrak{P}(p_1, \ldots, p_t, i)$ for $0 \leqq i \leqq p_{t+1}$. Therefore $\mathfrak{P}(p_1, \ldots, p_t)$, for $1 \leqq t \leqq k-1$, implies $\mathfrak{P}(p_1, \ldots, p_{t+1})$. Now we can assert $\mathfrak{P}(\lambda)$. And this completes the proof of Lemma 10.5.

Lemma 10.6. *If $\lambda \in \mathfrak{L}(n)$ and if $\mathfrak{R}(\mu)$ is true for each μ in $\mathfrak{L}(n)$ such that $\mu > \lambda$, then $\mathfrak{Q}(\lambda)$ is true. If, in addition, $v(\lambda) = n$, then $\mathfrak{R}(\lambda)$ is true.*

Proof. We assume that $\lambda = (p_1, \ldots, p_k)$. We note that $\mathfrak{Q}(p_1)$ is true, either by default, because $p_1 = 0$, or by (III) of Lemma 7.2. Hence we need only treat the case that $k > 1$. We consider some t, with $1 \leqq t \leqq k-1$, and assume inductively that $\mathfrak{Q}(p_1, \ldots, p_t)$ is true. We wish to establish $\mathfrak{Q}(p_1, \ldots, p_{t+1})$. If s is a symmetric argument of $A(p_1, \ldots, p_{t+1})$ but not of $A(p_1, \ldots, p_t)$, then $(A(p_1, \ldots, p_{t+1}), s, G) = 1$ by (III) of Lemma 7.2. Hence we need only prove $\mathfrak{Q}(p_1, \ldots, p_{t+1})$ under the assumption that s is a symmetric argument of $A(p_1, \ldots, p_t)$.

Case I. $v(p_1, \ldots, p_{t+1}) \leqq n-1$.

We consider $\mu = (p_1, \ldots, p_t, 1+p_{t+1})$ and note that $v(\mu) = 1 + v(p_1, \ldots, p_{t+1}) \leqq n$. Hence μ is in $\mathfrak{L}(n)$. Moreover, since $\lambda = (p_1, \ldots, p_t, p_{t+1}, \ldots, p_k)$, $\mu > \lambda$. Therefore $\mathfrak{R}(\mu)$ is true by hypothesis. Now we set $c = A(p_1, \ldots, p_t)$, $w = v(p_1, \ldots, p_t)$, $q = p_{t+1}$, $m = w+q+2$, so that

$$A(p_1, \ldots, p_{t+1}) = f_q(c, s_{w+1}, s_{w+2}; s_{w+3}, \ldots, s_m);$$
$$A(p_1, \ldots, p_t, 1+p_{t+1}) = f_{q+1}(c, s_{w+1}, s_{w+2}; s_{w+3}, \ldots, s_m, s_{m+1}).$$

Where s is a symmetric argument of c, we replace s_{m+1} by the product $s_{m+1}s$, use $\mathfrak{R}(\mu)$ with (7.4) and deduce that Z_{t+n-m} contains the left hand side of

$$\begin{aligned} & f_{q+1}(c, s_{w+1}, s_{w+2}; s_{w+3}, \ldots, s_m, s_{m+1}s) \\ = & f_{q+1}(c, s_{w+1}, s_{w+2}; s_{w+3}, \ldots, s_m, s_{m+1}) \\ & f_{q+1}(c, s_{w+1}, s_{w+2}; s_{w+3}, \ldots, s_m, s) \\ & f_{q+2}(c, s_{w+1}, s_{w+2}; s_{w+3}, \ldots, s_m, s_{m+1}, s) \\ & ((f_q(c, s_{w+1}, s_{m+1}; s_{w+3}, \ldots, s_m), s, s_{w+2}) \\ & ((f_q(c, s_{w+1}, s; s_{w+3}, \ldots, s_m), s_{m+1}, s_{w+2}) . \end{aligned} \tag{10.4}$$

The first factor on the right of (10.4) is in Z_{t+n-m} by $\mathfrak{R}(\mu)$. Each of the second, third and fifth factors can be written in the form

$$(\ldots((c, s, x), y_1, z_1), \ldots,), y_r, z_r) \tag{10.5}$$

for suitable $x, y_1, z_1, \ldots, y_r, z_r$ in G, where $r = q + 1$ in two instances and $r = q + 2$ in the remaining instance. Since, by our inductive assumption, (c, s, G) is in $Z_{t+n-1-w}$, then the associator (10.5) is in Z_{t+n-m}. Therefore the second last factor on the right of (10.4) is in Z_{t+n-m}; whence, by interchanging s_{m+1}, s_{w+2}, we see that

$$(A(p_1, \ldots, p_{t+1}), s, s_{m+1}) \in Z_{t+n-m}. \tag{10.6}$$

We complete the argument for Case I as follows. Let H be the set of all h in G such that $(A(p_1, \ldots, p_{t+1}), s, h) \in Z_{t+n-m}$. Clearly H is a subloop of G. By (10.6), H contains s_{m+1}. Moreover, since s_{m+1} is not one of the arguments of $A(p_1, \ldots, p_{t+1})$, (10.6) will still hold with s_{m+1} replaced by the product $s_i s_{m+1}$ for any choice of $i = 1, 2, \ldots, n$. Thus H contains $s_i s_{m+1}$ along with s_{m+1}, so H contains s_i. That is, $H = G$. This completes the proof of $\mathfrak{Q}(p_1, \ldots, p_{t+1})$ in Case I.

Case II. $v(p_1, \ldots, p_{t+1}) = n$.

We begin by showing that $t + 1 = k$. Suppose on the contrary that $t + 1 \leqq k - 1$. Then $n = v(p_1, \ldots, p_{t+1}) = p_1 + \cdots + p_{t+1} + 2(t+1) + 1 \leqq \leqq p_1 + \cdots + p_k + 2(k-1) + 1 < v(\lambda) \leqq n$, a contradiction. Therefore $t + 1 = k$ and we are treating the case $v(\lambda) = n$. Consequently, if $v(\lambda) < n$, Case II cannot occur, the inductive step is complete, and we can assert $\mathfrak{Q}(\lambda)$.

We continue with the assumption that $t + 1 = k, v(\lambda) = n$. By Lemma 10.1, $\mathfrak{R}(\lambda)$ implies $\mathfrak{Q}(\lambda)$, so we merely prove $\mathfrak{R}(\lambda)$. As in Case I, we set $w = v(p_1, \ldots, p_t)$, $q = p_{t+1} = p_k$, $c = A(p_1, \ldots, p_t)$. Here, however, $w + q + 2 = n$ and we are to prove that $A(\lambda) = f_q(c, s_{w+1}, s_{w+2}; s_{w+3}, \ldots, s_n)$ is in Z_k. Moreover, by our inductive assumption, since $t + n - 1 - w = k + n - 2 - w = k + q$, we have $(c, s, G) \subset Z_{k+q}$ for every symmetric argument s of c. Our proof will consist in showing that $(A(\lambda), s_i, s_j) \in Z_{k-1}$ for all $i, j = 1, 2, \ldots, n$. By the properties of f_q, $A(\lambda)$ lies in the centre of the subloop generated by $c, s_{w+1}, s_{w+2}, \ldots, s_n$. In particular, $(A(\lambda), s_{w+1}, s_{w+2}) = 1$. Since s_{w+1}, s_{w+2} are two of the skew arguments of $A(\lambda)$ and since, in view of Lemma 10.5, $\mathfrak{P}(\lambda)$ is true, we have $(A(\lambda), s_i, s_j) \in Z_{k-1}$ for every two skew-arguments s_i, s_j of $A(\lambda)$. Again, if $q > 0$, $A(\lambda)$ is symmetric in $s_{w+3}, \ldots, s_n$ and satisfies $(A(\lambda), s_n, G) = 1$. It remains only to show that, if s, s' denote symmetric arguments of $A(\lambda)$, then $(A(\lambda), s_{w+2}, s)$ and $(A(\lambda), s, s')$ are in Z_{k-1}. Now, by (7.2),

$$(A(\lambda), s_{w+2}, s) = f_{q+1}(c, s_{w+1}, s; s_{w+3}, \ldots, s_n, s_{w+2}),$$

whence, since $(c, s, G) \subset Z_{k+q}$, we easily see that $(A(\lambda), s_{w+2}, s) \in Z_{k-1}$. Finally,

$$\begin{aligned} A(\lambda) &= f_q(s_{w+1}, s_{w+2}, c; s_{w+3}, \ldots, s_n) \\ &= (f_{q-1}, s_n, c) \in (G_q, G, c) \end{aligned}$$

and therefore, by Lemma 8.2, since $(c, s, G) \subset Z_{k+q}$, $(c, s', G) \subset G_{k+q}$, we have $(A(\lambda), s, s') \in Z_{k-1}$. This disposes of Case II and completes the proof of Lemma 10.6.

Lemma 10.7. *If $\lambda \in \mathfrak{L}(n)$ and if $\mathfrak{R}(\mu)$ is true for each μ in $\mathfrak{L}(n)$ such that $\mu > \lambda$, then $\mathfrak{R}(\lambda)$ is true.*

Proof. In view of the second sentence of Lemma 10.6, we may assume $v(\lambda) = m < n$. By Lemmas 10.5, 10.6, we may assume $\mathfrak{P}(\lambda)$, $\mathfrak{Q}(\lambda)$. Without loss of generality we may assume that $A(\lambda) = f_q(c, s_{w+1}, s_{w+2}; s_{w+3}, \ldots, s_m)$ for integers q, w, where $w > 0$. Here $\lambda = (p_1, \ldots, p_{k-1}, q)$ for some non-negative integer k and we must prove that $(A(\lambda), s_i, s_j) \in Z_{k+n-m-1}$ for all $i, j = 1, 2, \ldots, n$. If one or both of s_i, s_j is a symmetric argument of $A(\lambda)$, this follows by $\mathfrak{Q}(\lambda)$. If s_i, s_j are skew arguments of $A(\lambda)$, it follows by $\mathfrak{P}(\lambda)$ and $(A(\lambda), s_{w+1}, s_{w+2}) = 1$. If one of s_i, s_j is a skew argument of $A(\lambda)$ and the other is not an argument, it follows by $\mathfrak{P}(\lambda)$ and the fact that $(A(\lambda), s_{w+2}, s_{m+1}) = f_{q+1}(c, s_{w+1}, s_{m+1}; s_{w+3}, \ldots, s_m, s_{w+2}) = A'(\mu)$, where $\mu = (p_1, \ldots, p_{q-1}, q+1) > \lambda$, $v(\mu) = v(\lambda) + 1$. If neither of s_i, s_j is an argument of $A(\lambda)$, and if $i \neq j$, it follows by symmetry and the fact, that $m \leqq n-2$ and $(A(\lambda), s_{m+1}, s_{m+2}) = A'(\mu)$ where $\mu = (p_1, \ldots, p_{k-1}, q, 0)$, $k + 1 - v(\mu) = k - v(\lambda) - 1$. This completes the proof of Lemma 10.7

Now we are ready for the proof of Slaby's theorem:

Theorem 10.1. *Let n be a positive integer, $n \geqq 3$. Then every commutative Moufang loop G which can be generated by n elements is centrally nilpotent of class at most $n-1$.*

Proof. Without loss of generality we may assume that G is generated by an ordered set S of n distinct elements. Since the greatest element of $\mathfrak{L}(n)$ is $(n-3)$, we see from Lemmas 10.2, 10.7 that $\mathfrak{R}(0)$ holds. That is, $A(0) = (s_1, s_2, s_3)$ is contained in Z_{n-2}. As a consequence, $(S, S, S) \subset Z_{n-2}$. Therefore, by Lemma 3.7, $G = Z_{n-1}$. This completes the proof of Theorem 10.1.

11. Local properties

Theorem 10.1 implies that *every commutative Moufang loop is locally centrally nilpotent and* hence *is a local Z-loop.*

Theorem 11.1. *Every chief composition system of a commutative Moufang loop is a central system. In particular, every nontrivial simple commutative Moufang loop is a cyclic group of prime order.*

Proof. By the above remark and Theorem 4.1 of Chapter VI.

Theorem 11.2. *If G is a finitely generated commutative Moufang loop, then the associator subloop G' is finite.*

Proof. By Theorem 10.1, G is centrally nilpotent, say of class c. The lemma is trivial for $c = 0$ or 1, so we consider the case that $c = k+1$, $k \geqq 1$, and assume the lemma for commutative Moufang loops of class at most k. The loop $H = G/G_k$ has class k. Hence, by our inductive assumption, $H' = G'/G_k$ is finite. We now must show that G_k is finite. If $k > 1$, then G_{k-1}/G_k is a subloop of H' and therefore is finite. If $k = 1$, then $G_{k-1}/G_k = G/G_k$ is finitely generated. In either case, there exists a finite non-empty subset T of G such that T and G_k generate G_{k-1}. Moreover, G is generated by a finite subset S. Let F be the subloop of G_k generated by the finite set (T, S, S). Since G_k is part of the centre, $Z(G)$, of G, F is normal in G. Moreover, F is a finitely generated abelian group of exponent 3, and hence is finite. Since $(T, S, S) \equiv 1 \bmod F$ and since S generates G, we readily deduce that $(T, G, G) \equiv 1 \bmod F$. Since $(G_k, G, G) = G_{k+1} = 1$ and since T, G_k generate G_{k-1}, then $G_k = (G_{k-1}, G, G) \subset F \subset G_k$. Therefore $G_k = F$ is finite and the proof is complete.

Theorem 11.3. *Every commutative Moufang loop G without elements of infinite order is locally finite.*

Proof. If H is a finitely generated subloop of G, then H/H' is a finitely generated abelian group without elements of infinite order and hence is finite. By Theorem 11.2, H' is finite also. Therefore H is finite.

If G is a commutative Moufang loop, let $\mathfrak{M} = \mathfrak{M}(G)$ and $\mathfrak{J} = \mathfrak{J}(G)$ denote the multiplication group and inner mapping group, respectively, of G. Let $\{\mathfrak{J}_i\}$, $\{\mathfrak{M}_i\}$, $\{Z_i(\mathfrak{J})\}$, $\{Z_i(\mathfrak{M})\}$ denote the respective lower and upper central series of these groups. By Lemma 1.2 of IV, each element of $\mathfrak{M}$ has the form $\theta R(x)$ where θ is in $\mathfrak{J}$ and x is in G. The following commutator relations are well known from group theory:

$$\mathfrak{J}_0 = \mathfrak{J}; \quad \mathfrak{J}_{i+1} = (\mathfrak{J}_i, \mathfrak{J}), \qquad i = 0, 1, 2, \ldots ; \tag{11.1}$$

$$(\mathfrak{J}_i, \mathfrak{J}_j) \subset \mathfrak{J}_{i+j+1}, \qquad i, j = 0, 1, 2, \ldots . \tag{11.2}$$

It will prove convenient to define an additional series $\{\mathfrak{J}(i)\}$ as follows: for each non-negative integer i, $\mathfrak{J}(i)$ is the subset of $\mathfrak{J}$ consisting of every θ in $\mathfrak{J}$ which induces an identity automorphism on each of the quotient loops G_k/G_{k+i+1}, where k ranges over the non-negative integers. Since G_k/G_{k+1} lies in the centre of G/G_{k+1} for each k, we see that

$$\mathfrak{J}(0) = \mathfrak{J} = \mathfrak{J}_0. \tag{11.3}$$

And, in general, it is clear that $\mathfrak{J}(i)$ is a subgroup of $\mathfrak{J}$.

Lemma 11.1. *For all non-negative integers i, j and for every commutative Moufang loop G with inner mapping group $\mathfrak{J}$:*

$$(\mathfrak{J}(i), \mathfrak{J}(j)) \subset \mathfrak{J}(i+j+1), \tag{11.4}$$

$$\theta \in \mathfrak{J}(i) \to \theta^3 \in \mathfrak{J}(2i+1). \tag{11.5}$$

Hence $\mathfrak{J}_i \subset \mathfrak{J}(i)$ and, for every θ in $\mathfrak{J}$, $\theta^{f(i)} \in \mathfrak{J}(g(i))$ where

$$f(i) = 3^i, \quad g(i) = 2^i - 1. \tag{11.6}$$

Proof. For arbitrary non-negative integers i, j, k, let $\theta \in \mathfrak{J}(i)$, $\varphi \in \mathfrak{J}(j)$, $x \in G_k$. By definition, $x\theta \equiv x \bmod G_{k+i+1}$. Thus and similarly, $x\theta = xy$, $x\varphi = xz$ where $y \in G_{k+i+1}$, $z \in G_{k+j+1}$. Write $H = G_{k+i+j+2}$. Then $y\varphi \equiv y \bmod H$. Moreover, $(x, y, z) \in G_p$ where $p = k + (k+i+1) + (k+j+1) + 1 > k+i+j+2$; consequently, $(x, y, z) \equiv 1 \bmod H$. Therefore $x\theta\varphi = (xy)\varphi = (x\varphi)(y\varphi) \equiv (xz)y \equiv (xy)z \equiv (x\theta)z \bmod H$. Again, $z\theta^{-1} \equiv z \bmod H$, so $x\theta\varphi\theta^{-1} \equiv (x\theta\theta^{-1})(z\theta^{-1}) \equiv xz \equiv x\varphi \bmod H$ and $x\theta\varphi\theta^{-1}\varphi^{-1} \equiv x \bmod H$. We deduce that $\theta\varphi\theta^{-1}\varphi^{-1}$ is in $\mathfrak{J}(i+j+1)$; and this is enough to prove (11.4). With x, θ, y as before, define $K = G_{k+2i+2}$. Then $y\theta \equiv y \bmod K$, so $x\theta^2 = (xy)\theta = (x\theta)(y\theta) \equiv (xy)y \equiv xy^2 \bmod K$ and $x\theta^3 \equiv (x\theta)y^2 \equiv xy^3 \bmod K$. However, since $k+i+1 \geq 1$, $y \in G'$ and hence $y^3 = 1$. Therefore $x\theta^3 \equiv x \bmod K$, proving (11.5). If $\mathfrak{J}_i \subset \mathfrak{J}(i)$ for some i, then, by (11.1), (11.3), (11.4), $\mathfrak{J}_{i+1} = (\mathfrak{J}_i, \mathfrak{J}) \subset (\mathfrak{J}(i), \mathfrak{J}(0)) \subset \mathfrak{J}(i+1)$. If $\theta^{f(i)} \in \mathfrak{J}(g(i))$ for some i, then, by (11.6), (11.5), (11.6), $\theta^{f(i+1)} = (\theta^{f(i)})^3 \in \mathfrak{J}(2g(i)+1) = \mathfrak{J}(g(i+1))$. Since $\mathfrak{J}_0 \subset \mathfrak{J}(0)$ and $\theta^{f(0)} \in \mathfrak{J}(g(0))$, the proof of Lemma 11.1 is complete.

Lemma 11.2. *Let G be a commutative Moufang loop and let i, p be integers with $0 \leq i \leq 2p$. Then, for each θ in $\mathfrak{J}(i)$ and x in G_p,*

$$[\theta R(x)]^3 = \theta' R(x') R(x^3) \tag{11.7}$$

where θ' is in $\mathfrak{J}(2i+1)$ and x' is in G_{p+2i+2}.

Proof. Since θ is an automorphism of G, the left-hand side of (11.7) is equal to $\theta^3 R(x\theta^2) R(x\theta) R(x) = \theta^3\varphi\psi R(c)$ where $\varphi = R(x\theta^2, x\theta)$, $\psi = R((x\theta^2)(x\theta), x)$, $c = [(x\theta^2)(x\theta)]x$. By (11.5), $\theta^3 \in \mathfrak{J}(2i+1)$. Since $\theta \in \mathfrak{J}(i)$ and $x \in G_p$, then $x\theta = xy$ where $y \in G_{p+i+1}$. Hence $y\theta \equiv y \bmod G_{p+2i+2}$, so $x\theta^2 \equiv (x\theta)y \equiv xy^2 \bmod G_{p+2i+2}$. In particular, $x\theta^2 = (x\theta)a$ where a is in G_{p+i+1}. Thus, for each non-negative integer k and every w in G_k, $(w, x\theta^2, x\theta) = (w, (x\theta)a, x\theta) = (w, a, x\theta) \in (G_k, G_{p+i+1}, G_p) \subset G_q$ where $q = k + (p+i+1) + p + 1 = k + 2p + i + 2 \geq k + 2i + 2$. Therefore $w\varphi \equiv w \bmod G_{k+2i+2}$, whence we see that φ is in $\mathfrak{J}(2i+1)$. Moreover, $(x\theta^2)(x\theta) \equiv (xy^2)(xy) \equiv x^2y^3 \equiv x^2 \bmod G_{p+2i+2}$, so $(x\theta^2)(x\theta) = x^2b$ for some b in G_{p+2i+2}. Then, with w as before, $(w, (x\theta^2)(x\theta), x) = (w, x^2b, x) = (w, b, x) \in G_{k+2i+2}$. Hence ψ is in $\mathfrak{J}(2i+1)$. Therefore $[\theta R(x)]^3 = \theta' R(c)$ where $\theta' = \theta^3\varphi\psi$ is in $\mathfrak{J}(2i+1)$. Finally, $c = [(x\theta^2)(x\theta)]x \equiv x^2x \equiv x^3 \bmod G_{p+2i+2}$, so $c = x'x^3$ for some x' in G_{p+2i+2}. Since x^3 is in the centre, $Z(G)$, of G, then, for every z in

G, $z(x'x^3) = (zx')x^3$. Hence $R(c) = R(x')R(x^3)$. This completes the proof of Lemma 11.2.

For the next lemma we need the functions f, g defined by (11.6).

Lemma 11.3. *Let G be a commutative Moufang loop. Then, for each θ in $\mathfrak{J}$, each x in G and each non-negative integer k,*

$$[\theta R(x)]^{f(k)} = \theta_k R(x_k)\, R(x^{f(k)}) \tag{11.8}$$

where θ_k is in $\mathfrak{J}(g(k))$ and x_k is in $G_{2g(k)}$.

Proof. If we define $\theta_0 = \theta$, $x_0 = 1$, then (11.8) holds trivially for $k = 0$. Again, by Lemma 11.2 with $i = p = 0$, (11.8) holds for $k = 1$. Now we assume inductively that (11.8) holds for some $k \geqq 1$. We note that $R(x^{f(k)})$ lies in the centre of $\mathfrak{M} = \mathfrak{M}(G)$ and that $R(x^{f(k)})^3 = R(x^{f(k+1)})$. Moreover, x_k is in G' and hence $x_k{}^3 = 1$. Therefore, on cubing both sides of (11.8), applying Lemma 11.2 to $[\theta_k R(x_k)]^3$ with $i = g(k)$, $p = 2g(k)$, and observing that $2i + 1 = g(k+1)$, $p + 2i + 2 = 2g(k+1)$, we derive (11.8) with k replaced by $k + 1$. This completes the proof of Lemma 11.3.

Lemma 11.4. *If G is a commutative Moufang loop with multiplication group $\mathfrak{M} = \mathfrak{M}(G)$, then $\mathfrak{M}(G/Z)$ is isomorphic to $\mathfrak{M}/Z_2(\mathfrak{M})$.*

Proof. We know that $\mathfrak{M}(G/Z)$ is isomorphic to $\mathfrak{M}/\mathfrak{N}$ where $\mathfrak{N}$ is the set of all α in $\mathfrak{M}$ such that $x\alpha \equiv x \bmod Z$ for all x in G. Clearly $\mathfrak{P} = \mathfrak{N} \cap \mathfrak{J}(G)$ is the set of all θ in $\mathfrak{J} = \mathfrak{J}(G)$ such that $x\theta \equiv x \bmod Z$ for all x in G; that is, $\mathfrak{P}$ is the intersection of $\mathfrak{J}$ with the group of centre automorphisms of G. If θ is in $\mathfrak{P}$ and x is in G, then $x\theta = xz$ for some z in Z and hence $R(x)\theta = \theta R(xz) = \theta R(x)R(z)$. Since $Z(\mathfrak{M})$ is the set of all $R(z)$ with z in Z, we see that $R(x)\theta \equiv \theta R(x) \bmod Z(\mathfrak{M})$ for every x in G. This proves that $\mathfrak{P} \subset Z_2(\mathfrak{M})$. If α is in $\mathfrak{N}$, then $1\alpha \equiv 1 \bmod Z$, so $1\alpha = z \in Z$. If $\theta = \alpha R(z)^{-1}$, then θ is in $\mathfrak{P}$ and hence $\alpha = \theta R(z)$ is in $\mathfrak{P}Z(\mathfrak{M})$. Therefore $\mathfrak{N} \subset \mathfrak{P}Z_2(\mathfrak{M}) \subset Z_2(\mathfrak{M})$. Now assume, conversely, that α is in $Z_2(\mathfrak{M})$. Then $\alpha = \theta R(k)$ for some θ in $\mathfrak{J}$, k in G. For any x in G, $\theta R(x\theta)R(k) = R(x)\theta R(k) = R(x)\alpha \equiv \alpha R(x) \equiv \theta R(k)R(x) \bmod Z(\mathfrak{M})$ and hence $R(x\theta) = R(k)R(x)R(k)^{-1}R(z)$ for some z in Z. Thus $x\theta = 1R(x\theta) = (kx)k^{-1}z = xz$. In particular, $x\theta \equiv x \bmod Z$ for each x in G, so θ is in $\mathfrak{P}$. Moreover, $R(x\theta) = R(xz) = R(x)R(z)$ and therefore $R(k)R(x)R(k)^{-1} = R(x)$. Hence $R(x)R(k) = R(k)R(x)$ for all x in G; and this implies that $R(k)$ is in $Z(\mathfrak{M})$. Hence k is in Z and therefore $x\alpha = (x\theta)k \equiv x \bmod Z$ for every x in G. This proves that $Z_2(\mathfrak{M}) \subset \mathfrak{N}$. Hence $\mathfrak{N} = Z_2(\mathfrak{M})$, completing the proof of Lemma 11.4.

Lemma 11.5. *Let G be a commutative Moufang loop which can be generated by n elements, $n \geqq 2$. Let e be the least positive integer such that $g(e) = 2^e - 1 \geqq n - 1$. Then:* (i) *The inner mapping group $\mathfrak{J}(G)$ of G is a 3-group of exponent dividing $f(e) = 3^e$ and is nilpotent of class at most $n - 2$.* (ii) *The mapping $\alpha \to \alpha^{f(e)}$ is a centralizing endomorphism of the multiplication group $\mathfrak{M}(G)$ of G; moreover, $\alpha^{f(e)} = R((1\alpha)^{f(e)})$ for each α in $\mathfrak{M}(G)$.* (iii) *$\mathfrak{M}(G)$ is nilpotent of class at most $2n - 3$.*

Proof. If α is in $\mathfrak{M}(G)$, set $x = 1\alpha$. Then $\alpha = \theta R(x)$ for some θ in $\mathfrak{J}(G)$; moreover, α is in $\mathfrak{J}(G)$ if and only if $x = 1$. By Theorem 10.1, G is centrally nilpotent of class at most $n-1$. Hence, by definition, $\mathfrak{J}(n-2)$ induces the identity mapping on $G_0/G_{n-1} = G$. Therefore, in particular, by Lemma 11.1, $\mathfrak{J}_{n-2} = 1$, showing that $\mathfrak{J}$ is nilpotent of class at most $n-2$. By Lemma 11.3, $\alpha^{f(e)} = \theta_e R(x_e)\, R(x^{f(e)})$ where θ_e is in $\mathfrak{J}(g(e)) \subset \mathfrak{J}(n-1) = 1$ and x_e is in $G_{2g(e)} \subset G_{n+1} = 1$. Hence $\alpha^{f(e)} = R(x^{f(e)})$. In particular, $\mathfrak{J}$ has exponent dividing $f(e)$; this completes the proof of (i). If β is in $\mathfrak{M}(G)$ and $y = 1\beta$, then $\beta = \varphi R(y)$ for φ in $\mathfrak{J}$. As before, $\beta^{f(e)} = R(y^{f(e)})$. Moreover, $\alpha\beta = \theta\varphi\psi R(z)$ where $\psi = R(x\varphi, y)$ and $z = (x\varphi)y$. Hence $(\alpha\beta)^{f(e)} = R(z^{f(e)})$. In addition, since 3 divides $f(e)$, $x^{f(e)}$ is in Z. Therefore $(x\varphi)^{f(e)} = (x^{f(e)})\varphi = x^{f(e)}$ and $z^{f(e)} = (x\varphi)^{f(e)} y^{f(e)} = x^{f(e)} y^{f(e)}$. Finally, since $R(x^{f(e)})$ is in the centre of $\mathfrak{M}(G)$, $R(z^{f(e)}) = R(x^{f(e)})\, R(y^{f(e)}) = \alpha^{f(e)} \beta^{f(e)}$. This completes the proof of (ii). For (iii) we use Lemma 11.4. By a straightforward induction, $\mathfrak{M}(G/Z_i)$ is isomorphic to $\mathfrak{M}/Z_{2i}(\mathfrak{M})$. If G has class $c \geqq 1$ and if $i = c-1$, then G/Z_i is a non-trivial abelian group. Also $\mathfrak{M}/Z_{2i}(\mathfrak{M})$ is isomorphic to G/Z_i and therefore $\mathfrak{M}$ is nilpotent of class $2i + 1 = 2c - 1$. Since $c \leqq n-1$, then $2c - 1 \leqq 2n - 3$. If G has class 0 then so does $\mathfrak{M}(G)$. In either case, we have (iii). This completes the proof of Lemma 11.5.

Theorem 11.4. *Let G be a commutative Moufang loop with associator subloop G', centre $Z = Z(G)$, multiplication group $\mathfrak{M} = \mathfrak{M}(G)$, inner mapping group $\mathfrak{J} = \mathfrak{J}(G)$. Then:* (i) *$G'$ and G/Z are locally finite loops of exponent 3 and are finite if G is finitely generated.* (ii) *$\mathfrak{J}$, $\mathfrak{M}/Z(\mathfrak{M})$ and $\mathfrak{M}' = (\mathfrak{M}, \mathfrak{M})$ are locally finite 3-groups and are finite if G is finitely generated.*

Proof. (i). Since the mapping $x \to x^3$ is an endomorphism of G into Z, the loops G' and G/Z have exponent 3. The rest of the proof follows from Theorem 11.2, 11.3.

(ii) Let $\mathfrak{K}$ be a subgroup of $\mathfrak{M}$ which is generated by a finite non-empty subset T. Each element of T can be expressed as a product of finitely many right multiplications $R(x)$ of G. Hence there exists a finite non-empty subset S of G such that $\mathfrak{K}$ is contained in the subgroup of $\mathfrak{M}$ generated by the right multiplications $R(s)$ of G, where s ranges over S. Let $n-1$ be the number of elements of S. For each x in G, let $H(x)$ be the subloop of G generated by $x \cup S$. Then $\mathfrak{K}$ maps $H(x)$ into itself. The restriction of $\mathfrak{K}$ to $H(x)$ is a homomorphism of $\mathfrak{K}$ upon a subgroup $\mathfrak{L}$ of the multiplication group of $H(x)$ which maps $\mathfrak{K}_i$ upon $\mathfrak{L}_i$ for each non-negative integer i. Since $H(x)$ is generated by n elements, we may apply Lemma 11.5 to $H(x)$. By Lemma 3.5 (iii), $\mathfrak{L}_c = 1$ where $c = 2n - 3$. Hence $\mathfrak{K}_c$ induces the identity mapping on $H(x)$. In particular, $\mathfrak{K}_c$ maps x upon x; and since this is true for each x in G, $\mathfrak{K}_c = 1$. By the same device, using Lemma 3.5 (ii), $x(\alpha\beta)^{f(e)} = x\alpha^{f(e)}\beta^{f(e)}$ and $x\alpha^{f(e)} = xR((1\alpha)^{f(e)})$ for all x in G, α, β in $\mathfrak{K}$. Hence $\alpha^{f(e)} = R((1\alpha)^{f(e)})$

for all α in $\mathfrak{K}$, and the mapping $\alpha \to \alpha^{f(e)}$ is an endomorphism of $\mathfrak{K}$ into $\mathfrak{K} \cap Z(\mathfrak{M})$. Now we know that $\mathfrak{K}/(\mathfrak{K} \cap Z(\mathfrak{M}))$ is a finitely generated nilpotent group of exponent dividing $f(e)$; from this we deduce readily (compare the proofs of Theorems 1.2, 1.3) that $\mathfrak{K}/(\mathfrak{K} \cap Z(\mathfrak{M}))$ is finite.

If $\mathfrak{K} \subset \mathfrak{J}$, then $\mathfrak{K}$ itself has exponent dividing $f(e)$ and hence is finite. Thus $\mathfrak{J}$ is locally finite.

Any finitely generated subgroup of $\mathfrak{M}/Z(\mathfrak{M})$ has form $\mathfrak{K}Z(\mathfrak{M})/Z(\mathfrak{M})$ where $\mathfrak{K}$ is a finitely generated subgroup of $\mathfrak{M}$. Then $\mathfrak{K}Z(\mathfrak{M})/Z(\mathfrak{M})$ is isomorphic to the finite group $\mathfrak{K}/(\mathfrak{K} \cap Z(\mathfrak{M}))$. Hence $\mathfrak{M}/Z(\mathfrak{M})$ is locally finite.

If $\mathfrak{P}$ is a finitely generated subgroup of $\mathfrak{M}' = (\mathfrak{M}, \mathfrak{M})$, there exists at least one finitely generated subgroup $\mathfrak{K}$ of $\mathfrak{M}$ such that $\mathfrak{K}' = (\mathfrak{K}, \mathfrak{K})$ contains $\mathfrak{P}$. Since $\mathfrak{K}$ is nilpotent, so is $\mathfrak{P}$. Moreover, $\alpha^{f(e)} = 1$ for all α in $\mathfrak{K}'$, so $\mathfrak{P}$ is a 3-group. Therefore $\mathfrak{P}$ is finite. Hence $\mathfrak{M}'$ is locally finite.

Finally, let G be generated by a finite set S. Since $\mathfrak{J}$ is a group of automorphisms of G, there is at most one element θ of $\mathfrak{J}$ for which the elements $s^{-1}(s\theta)$, s in S, have preassigned values. Since $s^{-1}(s\theta)$ is in G' for each s and since S and G' are finite, then $\mathfrak{J}$ is finite. Every element of $\mathfrak{M}$ has form $\theta R(x)$ where θ is in $\mathfrak{J}$ and x is in G. If z is in Z, then $\theta R(xz)\, Z(\mathfrak{M}) = \theta R(x)\, Z(\mathfrak{M})$. Since $\mathfrak{J}$ and G/Z are finite, we see that $Z(\mathfrak{M})$ has finite index in $\mathfrak{M}$. That is, $\mathfrak{M}/Z(\mathfrak{M})$ is finite. Then $\mathfrak{M}'$ is finitely generated and hence finite. This completes the proof of Theorem 11.4.

Lemma 11.6. *If G is a commutative Moufang loop with inner mapping group $\mathfrak{J} = \mathfrak{J}(G)$, then $\mathfrak{J}(G/Z)$ is isomorphic to $\mathfrak{J}/Z(\mathfrak{J})$.*

Proof. $\mathfrak{J}(G/Z)$ is isomorphic to $\mathfrak{J}/\mathfrak{P}$ where, as noted in the proof of Lemma 11.4, $\mathfrak{P}$ is the intersection of $\mathfrak{J}$ with the group of centre automorphisms of G. Since centre automorphisms commute with inner mappings, $\mathfrak{P} \subset Z(\mathfrak{J})$. Conversely, let θ be in $Z(\mathfrak{J})$. Then, for all x, y, z in G, $(x\theta)(x\theta, y, z) = x\theta R(y,z) = xR(y,z)\theta = [x(x,y,z)]\theta = (x\theta)[(x,y,z)\theta]$. Hence

$$(x\theta, y, z) = (x, y, z)\theta \tag{11.9}$$

for all x, y, z in G. By (11.9) and skew-symmetry, $(x, y, z)\theta = (x\theta, y\theta, z\theta) = (x, y, z)\theta^3$; hence $(x, y, z) = (x, y, z)\theta^2$. Therefore θ^2 induces the identity mapping on G'. However, by Theorem 11.4, θ has odd order (a power of 3); therefore θ itself induces the identity mapping on G'. Consequently (11.9) implies the identity

$$(x\theta, y, z) = (x, y, z)\,. \tag{11.10}$$

Now we fix x and define the element a by $x\theta = xa$. From (11.10), in view of (3.8), $(x,y,z) = (xa,y,z) = (x,y,z)\ (a,y,z)\ ((x,y,z),x,a)\ ((a,y,z),a,x)$. Thus

$$(a, y, z) = (a, x, (x, y, z))\ (a, x, (a, y, z))^{-1} \tag{11.11}$$

for all y, z in G. From (11.11) with $y = x, z = w$, we deduce that $(a, x, w) = 1$ for every w in G. In particular, both factors on the right hand side of (11.11) are equal to 1. Therefore $a \in Z$. That is, $x\theta \equiv x$ mod Z for every x in G. Hence $Z(\mathfrak{J}) \subset \mathfrak{P}$, so $\mathfrak{P} = Z(\mathfrak{J})$. This completes the proof of Lemma 11.6.

Theorem 11.5. *Let n be a positive integer and let G be a commutative Moufang loop with inner mapping group $\mathfrak{J}$, multiplication group $\mathfrak{M}$. Then the following statements are equivalent:* (i) *G is centrally nilpotent of class n.* (ii) *$\mathfrak{J}$ is nilpotent of class $n-1$.* (iii) *$\mathfrak{M}$ is nilpotent of class $2n-1$.*

Proof. For the equivalence of (i) and (ii), we use Lemma 11.6; for that of (i) and (iii), we use Lemma 11.4.

Let us now consider the infinitely generated commutative Moufang loop H constructed in § 1. It was noted in § 1 that H is not nilpotent. Hence $\mathfrak{J} = \mathfrak{J}(H)$ and $\mathfrak{M} = \mathfrak{M}(H)$ are not nilpotent either. However, $H_\omega = 1$. If θ is in $\mathfrak{J}_\omega$, then, by Lemma 11.1, θ is in $\mathfrak{J}_i \subset \mathfrak{J}(i)$ for every non-negative integer i. Hence, for each x in H, $x^{-1}(x\theta)$ is in H_{i+1} for every i and therefore $x^{-1}(x\theta) \in H_\omega = 1$. Thus $\mathfrak{J}_\omega = 1$. Again, $Z(H) = 1$, and $Z(\mathfrak{J})$ consists of centre automorphisms, so $Z(\mathfrak{J}) = 1$.

Bibliography

A. Books

Baer, Reinhold: Linear Algebra and Projective Geometry. New York, N. Y.: Academic Press, Inc. 1952, MR **14**, 675.

Birkhoff, Garrett: Lattice Theory. Amer. Math. Soc. Coll. Publ. **25** (1948), revised edition. Amer. Math. Soc., New York, N. Y. 1948, MR **10**, 673.

Blaschke, W., u. G. Bol: Geometrie der Gewebe. Berlin 1938.

Bourbaki, N.: Eléments de mathématique. XIV. etc. Actualités Sci. Ind. no. 1179. Paris: Hermann et Cie. 1952. MR **14**, 237.

Burnside, W.: Theory of Groups. 2nd ed. Cambridge/England: Cambridge Univ. Press 1911. Reprinted by Dover Publications Inc., New York, N. Y. 1955.

Dubreil, Paul: Algèbre. Tome I. Équivalences, Opérations, Groupes, Anneaux, Corps. Paris: Gautier-Villars 1946. MR **8**, 192.

Dubreil-Jacotin, M.-L., L. Lesieur et R. Croisot: Leçons sur la théorie des structures algébriques ordonnés et des treillis géométriques. Paris: Gautier-Villars 1953. MR **15**, 279.

Hille, Einar: Functional Analysis and Semigroups. Amer. Math. Soc. Coll. Publ. **31** (1948). Amer. Math. Soc., New York, N. Y. 1948. MR **9**, 594.

Jacobson, Nathan: Theory of Rings. Amer. Math. Soc. Math. Surveys **1** (1943). Amer. Math. Soc., New York, N. Y. 1943. MR **5**, 31.

Kleene, S. C.: Introduction to Metamathematics. New York, N. Y.: D. Van Nostrand Co. Inc. 1952. MR **14**, 525.

Kuroš, A. G.: The Theory of Groups. 2 vols. Translated and edited by K. A. Hirsch. Chelsea Publ. Co., New York, N. Y. 1955, 1956. MR **15**, 501; **17**, 124.

Pickert, Günter: Projektive Ebenen. Berlin-Göttingen-Heidelberg: Springer-Verlag 1955. MR **17**, 399.

Séminaire A: Chatelet et P. Dubreil de la Faculté des Sciences de Paris, 1953/54. Partie Complémentaire: Demi-groupes. etc. Paris 1956. MR **18**, 282.

SHÔDA, KENJIRÔ: General Algebra. Tokyo 1947. (Japanese.) MR 12, 313.

SPECHT, WILHELM: Gruppentheorie. Berlin-Göttingen-Heidelberg: Springer-Verlag 1956. MR 18, 189.

ZASSENHAUS, HANS: The Theory of Groups. Trans. by SAUL KRAVETZ, New York: Chelsea Publ. Co. 1949. MR 11, 77.

B. Selected Papers

B1 BAER, REINHOLD: The decomposition of enumerable, primary groups into direct summands. Quart. J. Math. Oxford Ser. **6**, 217—221 (1935).

B2 — Direct decomposition into infinitely many summands. Trans. Amer. Math. Soc. **64**, 519—551 (1948). MR **10**, 425.

B3 BURNSIDE, W.: On an unsettled question in the theory of discontinuous groups. Quart. J. Math. Oxford Ser. **33**, 230—238 (1902).

B4 — On groups in which every two conjugate elements are permutable. Proc. London Math. Soc. **35**, 28—37 (1902—1903).

B5 HALL, MARSHALL: Projective planes. Trans. Amer. Math. Soc. **54**, 229—277 (1943). MR **5**, 72.

B6 HALL, PHILIP: A contribution to the theory of groups of prime power order. Proc. London Math. Soc. (2) **36**, 29—95 (1934).

B7 KRULL, W.: Matrizen, Moduln und verallgemeinerte Abelsche Gruppen im Bereich der ganzen algebraischen Zahlen. Sitzgsber. Heidelberger Akad. Wiss., Math.-naturwiss. Kl., **1932**, 13—38.

B8 KUROŠ, A.: Isomorphisms of direct decompositions. II. Bull. Acad. Sci. URSS, **10**, 47—72 (1946). (Russian.) MR **8**, 309.

B9 LEVI, F. W.: Groups in which the commutator satisfies certain algebraic relations. J. Indian Math. Soc. (N. S.) **6**, 87—97 (1942). MR **4**, 133.

B10 LEVI, F., u. B. L. VAN DER WAERDEN: Über eine besondere Klasse von Gruppen. Abh. Math. Sem. Hansische Univ. **9**, 154—158 (1933).

B11 NEUMANN, B. H.: Some remarks on infinite groups. J. London Math. Soc. **12**, 120—127 (1937).

B12 — An essay on free products of groups with amalgamations. Philos. Trans. Roy. Soc. London Ser. A **246**, 503—554 (1954). MR **16**, 10.

B13 — On ordered groups. Amer. J. Math. **71**, 1—18 (1949). MR **10**, 428.

B14 BAER, REINHOLD: Factorization of *n*-soluble and *n*-nilpotent groups. Proc. Amer. Math. Soc. **4**, 15—26 (1953). MR **14**, 722.

B15 HALL, PHILIP: The splitting properties of relatively free groups. Proc. London Math. Soc. (3) **4**, 343—356 (1954). MR **16**, 217.

C. The main Bibliography

I. Algebras (General)

1 BIRKHOFF, G.: On the structure of abstract algebras. Proc. Cambridge Philos. Soc. **31**, 433—454 (1935).

2 BUCK, R. CREIGHTON: A factoring theorem for homomorphisms. Proc. Amer. Math. Soc. **2**, 135—137 (1951). MR **12**, 669.

3 DUBREIL-JACOTIN, MARIE-LOUISE, et R. CROISOT: Équivalences régulières dans un ensemble ordonné. Bull. Soc. Math. France **80**, 11—35 (1952). MR **14**, 529.

4 DUFFIN, R. J., and ROBERT S. PATE: An abstract theory of the Jordan-Hölder composition series. Duke Math. J. **10**, 743—750 (1943). MR **5**, 170.

5 EVANS, TREVOR: The word problem for abstract algebras. J. London Math. Soc. **26**, 64—71 (1951). MR **12**, 475.

6 EVANS, TREVOR: Embedding theorems for multiplicative systems and projective geometries. Proc. Amer. Math. Soc. 3, 614—620 (1952). MR 14, 347.
7 — Embeddability and the word problem. J. London Math. Soc. 28, 76—80 (1953). MR 14, 839.
8 FELL, J. M. G., and ALFRED TARSKI: On algebras whose factor algebras are Boolean. Pacific J. Math. 2, 297—318 (1952). MR 14, 130.
9 FUCHS, L.: On subdirect unions. I. Acta Math. Acad. Sci. Hungar. 3, 103—120 (1952). (Russian summary.) MR 14, 612.
10 GOLDIE, A. W.: On direct decompositions. I, II. Proc. Cambridge Philos. Soc. 48, 1—22 and 23—34 (1952). MR 14, 9 and 10.
11 — The scope of the Jordan-Hölder theorem in abstract algebra. Proc. London Math. Soc. (3) 2, 349—368 (1952). MR 14, 129.
12 GREEN, J. A.: A duality in abstract algebra. J. London Math. Soc. 27, 67—73 (1952). MR 14, 133.
13 HIGMAN, GRAHAM: Ordering by divisibility in abstract algebras. Proc. London Math. Soc. (3) 2, 326—336 (1952). MR 14, 238.
14 JAKUBÍK, JÁN: Congruence relations on abstract algebras. Czechoslovak Math. J. 4 (79), 314—317 (1954). (Russian. English summary.) MR 16, 787.
15 JÓNSSON, B., and A. TARSKI: Direct decomposition of finite algebraic systems. Univ. Notre Dame, 1947. MR 8, 560.
16 KUROŠ, A. G.: Radicals of rings and algebras. Mat. Sbornik N. S. 33 (75), 13—26 (1953) (Russian). MR 15, 194.
17 LORENZEN, PAUL: Über die Korrespondenzen einer Struktur. Math. Z. 60, 61—65 (1954). MR 16, 787.
18 LÖWIG, HENRY F. J.: On the properties of freely generated algebras. J. reine angew. Math. 190, 65—74 (1952). MR 14, 443.
19 — Gesetzrelationen über frei erzeugten Algebren. J. reine angew. Math. 193, 129—142 (1954). MR 16, 786.
20 LYAPIN, E. (LIAPIN): Systems with an infinite operation. Doklady Acad. Nauk SSSR (NS) 50, 49—51 (1945). (Russian). MR 14, 951.
21 — Free systems with an infinite univalent operation. C. R. (Doklady) Acad. Sci. URSS (N. S.) 51, 493—496 (1946). MR 8, 501.
22 LYNDON, R. C.: Identities in finite algebras. Proc. Amer. Math. Soc. 5, 8—9 (1954). MR 15, 676.
23 MAL'CEV, A. I.: On a class of algebraic systems. Uspehi Matem. Nauk (N. S.) 8, no. 1 (53), 165—171 (1953) (Russian). MR 14, 839.
24 — On the general theory of algebraic systems. Mat. Sb. N. S. 35 (77), 3—20 (1954) (Russian). MR 16, 440.
25 PEREMANS, W.: Some theorems on free algebras and on direct products of free algebras. Simon Stevin 29, 51—59 (1952). MR 14, 347.
26 POST, EMIL L.: Polyadic groups. Trans. Amer. Math. Soc. 48, 208—350 (1940). MR 2, 128.
27 SIKORSKI, R.: Products of abstract algebras. Fund. Math. 39 (1952), 211—228 (1953). MR 14, 839.
28 TCHOUNKINE, S. A.: On the theory of non-associative n-groups satisfying postulate K. C. R. (Doklady) Acad. Sci. URSS (N. S.) 48, 7—10 (1945). MR 7, 375.
29 THURSTON, H. A.: A note on continued products. J. London Math. Soc. 27, 239—241 (1952). MR 14, 238.
30 — The structure of an operation. J. London Math. Soc. 27, 271—279 (1952). MR 14, 239.
31 TVERMOES, HELGE: Über eine verallgemeinerung des Gruppenbegriffs. Math. Scand. 1, 18—30 (1953). MR 15, 98.

II. Groupoids

32 Borůvka, O.: Gruppoidentheorie. I. Publ. Fasc. Sci. Univ. Masaryk **1939**, no. 275, 17 pp. (1939). (Czech. German summary.) MR **8**, 14.
33 — Über Ketten von Faktoroiden. Math. Ann. **118**, 41—64 (1941). MR **3**, 200.
34 — Introduction to the Theory of Groups. Královski Česká Společnost Nauk, Praha, 1944. 80 pp. (Czech.) MR **7**, 510.
35 — Uvod do theorie grup. (Introduction to the theory of groups) 2nd ed. Přírodoveděcké Vyadavatelství, Prague, 1952. 154 pp. MR **15**, 7.
36 Brandt, H.: Über die Axiome des Gruppoids. Vierteljschr. naturforsch. Ges. Zürich **85**, Beiblatt (Festschrift Rudolph Fueter) 95—104 (1940). MR **2**, 218.
37 Climescu, Al. C.: Études sur la theorie des systèmes multiplicatifs uniformes. I. L'indice de non-associativité. Bull. École Polytech. Jassy **2**, 347—371 (1947). MR **10**, 100.
38 Conkling, Randall, and David Ellis: On metric groupoids and their completions. Portugaliae Math. **12**, 99—103 (1953). MR **15**, 684.
39 Croisot, Robert: Une interpretation des relations d'équivalence dans un ensemble. C. R. Acad. Sci. (Paris) **226**, 616—617 (1948). MR **9**, 406.
40 Dubreil, Paul: Remarques sur les théorèmes d'isomorphisme. C. R. Acad. Sci. (Paris) **215**, 239—241 (1942). MR **5**, 144.
41 Eilenberg, Samuel, and Saunders MacLane: Homology theories for multiplicative systems. Trans. Amer. Math. Soc. **71**, 294—330 (1951). MR **13**, 314.
42 Evans, T.: A note on the associative law. J. London Math. Soc. **25**, 196—201 (1950). MR **12**, 75.
43 Evans, Trevor, and B. H. Neumann: On varieties of groupoids and loops. J. London Math. Soc. **28**, 342—350 (1953). MR **15**, 284.
44 Faragó, Tibor: Contribution to the definition of a group. Publ. Math. Debrecen **3** (1953), 133—137 (1954). MR **15**, 851.
45 Higman, Graham, and B. H. Neumann: Groups as groupoids with one law. Publ. Math. Debrecen **2**, 215—221 (1952). MR **15**, 284.
46 Mintuhisa, Takasaki: Abstraction of symmetric transformations. Tôhoku Math. J. **49**, 145—207 (1943) (Japanese). MR **9**, 8.
47 Novotný, Miroslav: Les systèmes à deux compositions avec une loi distributive. Publ. Fac. Sci. Univ. Masaryk no. **321**, 49—68 (1951). MR **13**, 718.
48 Richardson, A. R.: Groupoids and their automorphisms. Proc. London Math. Soc. (2) **48**, 83—111 (1943). MR **5**, 60.
49 — Congruences in multiplicative systems. Proc. London Math. Soc. (2) **49**, 195—210 (1946). MR **8**, 439.
50 Sholander, Marlow: On the existence of the inverse operation in alternation groupoids. Bull. Amer. Math. Soc. **55**, 746—757 (1949). MR **11**, 159.
51 Smiley, M. F.: Notes on left division systems with left units. Amer. J. Math. **74**, 679—682 (1952). MR **14**, 10.
52 Szász, G.: Die Unabhängigkeit der Associativitätsbedingungen. Acta Sci. Mat. Szeged **15**, 20—28 (1953). MR **15**, 95.
53 — Über die Unabhängigkeit der Associativitätsbedingungen kommutativer multiplikativer Strukturen. Acta Sci. Math. Szeged **15**, 130—142 (1954). MR **15**, 773.

III. Loops and Quasigroups

54 Albert, A. A.: Quasigroups. I. Trans. Amer. Math. Soc. **54**, 507—519 (1943); II. Trans. Amer. Math. Soc. **55**, 401—419 (1944). MR **5**, 229; MR **6**, 42.
55 Artzy, Raphael: A note on the automorphisms of special loops. Riveon Lematematika **8**, 81 (1954) (Hebrew. English summary). MR **16**, 670.

56 Artzy, Raphael: On loops with a special property. Proc. Amer. Math. Soc. **6**, 448—453 (1955). MR **16**, 1083.
57 Baer, Reinhold: The homomorphism theorems for loops. Amer. J. Math. **67**, 450—460 (1945). MR **7**, 7.
58 — Splitting endomorphisms. Trans. Amer. Math. Soc. **61**, 508—516 (1947). MR **8**, 563.
59 — Endomorphism rings of operator loops. Trans. Amer. Math. Soc. **61**, 517—529 (1947). MR **8**, 564.
60 — Direct decompositions. Trans. Amer. Math. Soc. **62**, 62—98 (1947). MR **9**, 134.
61 — The rôle of the center in the theory of direct decompositions. Bull. Amer. Math. Soc. **54**, 167—174 (1948). MR **9**, 410.
62 Bateman, P. T.: A remark on infinite groups. Amer. Math. Monthly **57**, 623—624 (1950). MR **12**, 670.
63 Bates, Grace E.: Free loops and nets and their generalizations. Amer. J. Math. **69**, 499—550 (1947). MR **9**, 8.
64 — Decompositions of a loop into characteristic free summands. Bull. Amer. Math. Soc. **54**, 566—574 (1948). MR **10**, 12.
65 Bates, Grace E., and Fred Kiokemeister: A note on homomorphic mappings of quasigroups into multiplicative systems. Bull. Amer. Math. Soc. **54**, 1180—1185 (1948). MR **10**, 353.
66 Bol, G.: Gewebe und Gruppen. Math. Ann. **114**, 414—431 (1937).
67 Bruck, R. H.: Some results in the theory of quasigroups. Trans. Amer. Math. Soc. **55**, 19—52 (1944). MR **5**, 229.
68 — Some results in the theory of linear non-associative algebras. Trans. Amer. Math. Soc. **56**, 141—199 (1944). MR **6**, 116.
69 — Simple quasigroups. Bull. Amer. Soc. **50**, 769—781 (1944). MR **6**, 147.
70 — Contributions to the theory of loops. Trans. Amer. Math. Soc. **60**, 245—354 (1946). MR **8**, 134.
71 — An extension theory for a certain class of loops. Bull. Amer. Math. Soc. **57**, 11—26 (1951). MR **12**, 585.
72 — On a theorem of R. Moufang. Proc. Amer. Math. Soc. **2**, 144—145 (1951). MR **13**, 8.
73 — Loops with transitive automorphism groups. Pacific J. Math. **1**, 481—483 (1951). MR **13**, 621.
74 — Pseudo-automorphisms and Moufang loops. Proc. Amer. Math. Soc. **3**, 66—72 (1952). MR **13**, 905.
75 — Analogues of the ring of rational integers. Proc. Amer. Math. Soc. **6**, 50—58 (1955). MR **16**, 1083.
76 — and Lowell J. Paige: Loops whose inner mappings are automorphisms. Ann. of Math. (2) **63**, 308—323 (1956). MR **17**, 943.
77 Choudhury, A. C.: Quasigroups and nonassociative systems. I. Bull. Calcutta Math. Soc. **40**, 183—194 (1948); II. Bull. Calcutta Math. Soc. **41**, 211—219 (1949). MR **10**, 591; **11**, 417.
78 Cockcroft, W. H.: Interpretation of vector cohomology groups. Amer. J. Math. **76**, 599—619 (1954). MR **16**, 110.
79 Eilenberg, Samuel, and Saunders MacLane: Algebraic cohomology groups and loops. Duke Math. J. **14**, 435—463 (1947). MR **9**, 132.
80 Ellis, David: Remarks on isotopies. Publ. Math. Debrecen **2**, 175—177 (1952). MR **14**, 945.
81 Evans, T.: Homomorphisms of non-associative systems. J. London Math. Soc. **24**, 254—260 (1949). MR **11**, 327.

82 EVANS, T.: On multiplicative systems defined by generators and relations. I. Normal form theorems. Proc. Cambridge Philos. Soc. **47**, 637—649 (1951). MR **13**, 312.
83 — On multiplicative systems defined by generators and relations. II. Monogenic loops. Proc. Cambridge Philos. Soc. **49**, 579—589 (1953). MR **15**, 283.
84 GARRISON, G. N.: Quasigroups. Ann. of Math. (2) **41**, 474—487 (1940). MR **2**, 7.
85 — Note on invariant complexes of a quasigroup. Ann. of Math. (2) **47**, 50—55 (1946). MR **7**, 375.
86 GRIFFIN, HARRIET: The abelian quasigroup. Amer. J. Math. **62**, 725—737 (1940). MR **2**, 127.
87 HAUSMANN, B. A., and OYSTEIN ORE: Theory of quasigroups. Amer. J. Math. **59**, 983—1004 (1937).
88 ISEKI, KYOSHI: Structure of special ordered loops. Portugaliae Math. **10**, 81—83 (1951). MR **13**, 313.
89 KIOKEMEISTER, FRED: A theory of normality for quasigroups. Amer. J. Math. **70**, 99—106 (1948). MR **9**, 330.
90 MAL'CEV, A. I.: Analytic loops. Mat. Sb. N. S. **36** (78), 569—576 (1955) (Russian). MR **16**, 997.
91 MOUFANG, RUTH: Zur Struktur von Alternativkörpern. Math. Ann. **110**, 416—430 (1935).
92 MURDOCH, D. C.: Quasigroups which satisfy certain generalized associative laws. Amer. J. Math. **61**, 509—522 (1939).
93 — Note on normality in quasigroups. Bull. Amer. Math. Soc. **47**. 134—138 (1941). MR **2**, 218.
94 — Structure of abelian quasigroups. Trans. Amer. Math. Soc. **49**, 392—409 (1941). MR **2**, 218.
95 NORTON, D. A.: Hamiltonian loops. Proc. Amer. Math. Soc. **3**, 56—65 (1952). MR **13**, 720.
96 PAIGE, LOWELL J.: Neofields. Duke Math. J. **16**, 39—60 (1949). MR **10**, 430.
97 — A theorem on commutative power associative loop algebras. Proc. Amer. Math. Soc. **6**, 279—280 (1955). MR **16**, 897.
98 PEREMANS, W.: Abstract algebraic systems. Math. Centrum Amsterdam. Rapport ZW **1949—003**, 12 pp. (1949). (Dutch). MR **11**, 156.
99 POPOVA, HÉLÈNE: Logarithmétiques des quasi-groupes finis. C. R. Acad. Sci. (Paris) **234**, 1936—1937 (1952). MR **13**, 906.
100 — Sur les quasigroupes dont les logarithmétiques sont groupes. C. R. Acad. Sci. (Paris) **234**, 2582—2583 (1952). MR **14**, 131.
101 — Sur les vecteurs dérivés des quasi-groupes unis. C. R. Acad. Sci. (Paris) **235**, 1360—1362 (1952). MR **14**, 444.
102 — Logarithmétiques réductibles de quasigroupes. C. R.. Acad. Sci. (Paris) **235**, 1589—1591 (1952). MR **14**, 615.
103 — L'isotopies des logarithmétiques des quasi-groupes finis. C. R. Acad. Sci. (Paris) **236**, 769—771 (1953). MR **14**, 841.
104 — Sur la logarithmétique d'une boucle. C. R. Acad. Sci. (Paris) **236**, 1220 to 1222 (1953). MR **14**, 842.
105 — Logarithmetics of finite quasigroups. I. Proc. Edinburgh Math. Soc. (2) 9, 74—81 (1954). MR **16**, 564.
106 SADE, ALBERT: Quasigroupes. Published by the author, Marseille, 1950. 16 pp. MR **13**, 301.
107 — Contribution à la théorie des quasi-groupes: diviseurs singulières. C. R. Acad. Sci. (Paris) **237**, 372—374 (1953); quasi-groupes obéissant à la "loi des keys" ou automorphes par certains groupes de permutation de leur support. C. R. Acad. Sci. (Paris) **237**, 420—422 (1953). MR **15**, 98.

108 SCHÖNHARDT, E.: Über lateinische Quadrate und Unionen. J. reine angew. Math. **163**, 183—230 (1930).
109 SMILEY, M. F.: An application of lattice theory to quasi-groups. Bull. Amer. Math. Soc. **50**, 782—786 (1944). MR **6**, 147.
110 SUŠKEVIČ, A. K. (SUSCHKEWITSCH): On a generalization of the associative law. Trans. Amer. Math. Soc. **31**, 204—214 (1931).
111 THURSTON, H. A.: Certain congruences on quasigroups. Proc. Amer. Math. Soc. **3**, 10—12 (1952). MR **13**, 621.
112 — Non-commuting quasigroup congruences. Proc. Amer. Math. Soc. **3**, 363—366 (1952). MR **14**, 241.
113 TOYODA, KOSCHICHI: On axioms of linear functions. Proc. Imp. Acad. Tokyo **17**, 221—227 (1941). MR **7**, 241.
114 TREVISAN, GIORGIO: A proposita delle relazioni di congruenza sui quasigruppi. Rend. Sem. Mat. Univ. Padova **19**, 367—370 (1950). MR **12**, 313.
115 — Costruzione di quasigruppi con relazione di congruenza non permutabili Rend. Sem. Mat. Univ. Padova **22**, 11—22 (1953). MR **15**, 861.
116 ZELINSKY, DANIEL: On ordered loops. Amer. J. Math. **70**, 681—697 (1948). MR **10**, 233.

IV. Multigroups

117 CAMPAIGNE, HOWARD: Partition hypergroups. Amer. J. Math. **62**, 599—612 (1940). MR **2**, 7.
118 — A lower limit on the number of hypergroups of a given order. J. Washington Acad. Sci. **44**, 5—7 (1954). MR **15**, 598.
119 CHÂTELET, ALBERT: Algèbre des relations de congruence. Ann. Sci. École Norm. Sup. (3) **64** (1947), 339—368 (1948). MR **10**, 181.
120 CROISOT, ROBERT: Hypergroupes partiels. C. R. Acad. Sci. (Paris) **228**, 1090—1092 (1949). MR **10**, 508.
121 — Algèbres de relations et hypergroupes partiels. C. R. Acad. Sci. (Paris) **228**, 1181—1182 (1949). MR **10**, 508.
122 DIČMAN, A. P. (DIETZMANN): On the multigroups of complete conjugate sets of elements of a group. C. R. (Doklady) Acad. Sci. URSS (N. S.) **49**, 315—317 (1946). MR **7**, 511.
123 DRESHER, MELVIN, and OYSTEIN ORE: Theory of multigroups. Amer. J. Math. **60**, 705—733 (1938).
124 EATON, J. E.: Associative multiplicative systems. Amer. J. Math. **62**, 222—232 (1940). MR **1**, 105.
125 — Theory of cogroups. Duke Math. J. **6**, 101—107 (1940). MR **1**, 164.
126 EATON, J. E., and OYSTEIN ORE: Remarks on multigroups. Amer. J. Math. **62**, 67—71 (1940). MR **1**, 105.
127 GRIFFITHS, L. W.: On hypergroups, multigroups and product systems. Amer. J. Math. **60**, 345—354 (1938).
128 KRASNER, MARC: Sur la primitivité des corps P-adiques. Mathematica **13**, 72—191 (1937).
129 — La loi de JORDAN-HÖLDER dans les hypergroupes et les suites génératrices des corps de nombres P-adiques. I. Duke Math. J. **6**, 120—140 (1940); II. Duke Math. J. **7**, 121—135 (1940). MR **1**, 260; **2**, 123.
130 — La charactérisation des hypergroupes de classes et le problème de SCHREIER dans ces hypergroupes. C. R. Acad. Sci. (Paris) **212**, 948—950 (1941); **218**, 483—484 et 542—544 (1944). MR **3**, 37; **6**, 202.
131 — Une généralisation de la notion de corps-corpoïde. Un corpoïde remarquable de la théorie des corps valués. C. R. Acad. Sci. (Paris) **219**, 345—347 (1944). MR **7**, 363.

132 Krasner, Marc: Hypergroupes extramoduliformes et moduliformes. C. R. Acad. Sci.(Paris) **219**, 473—476 (1944). MR **7**, 363.

133 — Théorie de ramifications dans les extensions finies des corps valués. C. R. Acad. Sci. (Paris) **219**, 539—541 (1944); **220**, 28—30 et 761—763 (1945); **221**, 737—739 (1945). MR **7**, 364.

134 — Le produit complet et la théorie de ramification, etc. C. R. Acad. Sci. (Paris) **229**, 1103—1105 et 1287—1289 (1949); **230**, 162—164 (1950). MR **11**, 579.

135 Krasner, Marc, et Jean Kuntzman: Remarques sur les hypergroupes. C. R. Acad. Sci. (Paris) **224**, 525—527 (1947). MR **8**, 368.

136 Kuntzman, Jean: Opérations multiformes. Hypergroupes. C. R. Acad. Sci. (Paris) **204**, 1787—1788 (1937).

137 — Homomorphie entre systèmes multiformes. C. R. Acad. Sci. (Paris) **205**, 208—210 (1937).

138 — Systèmes multiformes et systèmes hypercomplexes. C. R. Acad. Sci. (Paris) **208**, 493—495 (1939).

139 — Contribution à l'étude des systèmes multiformes. Ann. Fac. Sci. Univ. Toulouse (4) **3**, 155—194 (1939). MR **8**, 439.

140 — Représentations sur un système multiforme. Ann. Univ. Grenoble Sec. Sci. Math. Phys. (N. S.) **21** (1945), 95—99 (1946). MR **8**, 134.

141 — Opérations multiformes qui s'obtiennent à partir d'opérations uniformes. C. R. Acad. Sci. (Paris) **224**, 177—179 (1947). MR **8**, 312.

142 Marty, F.: Sur une généralisation de la notion de groupe. Huitième Congrès des mathématiciens scand. Stockholm **1934**, 45—49.

143 — Rôle de la notion de hypergroupe dans l'étude de groupes non abéliens. C. R. Acad. Sci. (Paris) **201**, 636—638 (1935).

144 — Sur les groupes et les hypergroupes attachés à une fraction rationelle. Ann. Sci. École Norm. Sup. (3) **53**, 82—123 (1936).

145 Prenowitz, Walter: Projective geometries as multigroups. Amer. J. Math. **65**, 235—256 (1943). MR **4**, 251.

146 — Descriptive geometries as multigroups. Trans. Amer. Math. Soc. **59**, 333—380 (1946). MR **7**, 375.

147 — Partially ordered fields and geometries. Amer. Math. Monthly **53**, 439—449 (1946). MR **8**, 249.

148 — Spherical geometries and multigroups. Canad. J. Math. **2**, 100—119 (1950). MR **11**, 327.

149 San Juan, R.: A generalization of the concept of a group. Revista Mat. Hisp.-Amer. (4) **3**, 354—356 (1943) (Spanish). MR **7**, 375.

150 Utumi, Yuzo: On hypergroups of group right cosets. Osaka Math. J. **1**, 73—80 (1949). MR **11**, 158.

151 Vikhrov, A.: Theory of extensions of ultragroups. Učenye Zapiski Moskov. Gos. Univ. **100**, Matematika, Tom I, 3—19 (1946) (Russian. English summary). MR **11**, 713.

152 Wall, H. S.: Hypergroups. Amer. J. Math. **59**, 77—98 (1937).

V. Semigroups

153 Alimov, N. G.: On ordered semigroups. Izvestiya Akad. Nauk. SSSR Ser. Mat. **14**, 569—576 (1950) (Russian). MR **12**, 481.

154 Andersen, Olaf: Ein Bericht über die Struktur abstrakter Halbgruppen. Thesis, Hamburg, 1952, unpublished.

155 Andreoli, Giulio: Sulla teoria delle sostituzioni generalizzate e dei loro gruppi generalizzati. Rend. Acad. Sci. Fis. Mat. Napoli (4) **10**, 115—127 (1940). MR **8**, 439.

156 Asano, Keizo, and Kentaro Murata: Arithmetic ideal in semigroups. J. Inst. Polytech. Osaka City Univ. Ser. A Math. **4**, 9—33 (1953). MR **15**, 502.

157 Aubert, Karl Egil: On the ideal theory of commutative semigroups. Math. Scand. **1**, 39—54 (1953). MR **15**, 7.

158 — Généralisation de la théorie des *r*-idéaux de Prüfer-Lorenzen. C. R. Acad. Sci. (Paris) **238**, 2214—2216 (1954). MR **15**, 848.

159 Baer, Reinhold: Free sums of groups and their generalizations. An analysis of the associative law. I, Amer. J. Math. **71**, 706—742 (1950); II, III, Amer. J. Math. **72**, 625—646 and 647—670 (1950). MR **11**, 78; **12**, 478.

160 Baer, R., u. F. Levi: Vollständige irreduzible Systeme von Gruppenaxiomen. Sitzgsber. Heidelberger Akad Wiss., Math.-Naturwiss. Kl. Beiträge zur Algebra no. **18** (1932). 12 pp.

161 Ballieu, Robert: Sur les groupes de parties d'un demi-groupe. Ann. Soc. Sci. Bruxelles. Sér. I **64**, 139—147 (1950). MR **12**, 586.

162 Chamberlin, Eliot, and James Wolfe: Multiplicative homomorphisms of matrices. Proc. Amer. Math. Soc. **4**, 37—42 (1953). MR **14**, 611.

163 Chehata, C. G.: On an ordered semigroup. J. London Math. Soc. **28**, 353—356 (1953). MR **14**, 944.

164 Clifford, A. H.: A system arising from a weakened set of group postulates. Ann. of Math. **34**, 865—871 (1933).

165 — Partially ordered abelian groups. Ann. of Math. (2) **41**, 465—473 (1940). MR **2**, 4.

166 — Semigroups admitting relative inverses. Ann. of Math. (2) **42**, 1037—1049 (1941). MR **3**, 199.

167 — Matrix representation of completely simple semigroups. Amer. J. Math. **64**, 327—342 (1942). MR **4**, 4.

168 — Semigroups containing minimal ideals. Amer. J. Math. **70**, 521—526 (1948). MR **10**, 12.

169 — Semigroups without nilpotent ideals. Amer. J. Math. **71**, 834—844 (1949). MR **11**, 327.

170 — Extensions of semigroups. Trans. Amer. Math. Soc. **68**, 165—173 (1950). MR **11**, 499.

171 — A class of *d*-simple semigroups. Amer. J. Math. **75**, 547—556 (1953). MR **15**, 98.

172 — Bands of semigroups. Proc. Amer. Math. Soc. **5**, 499—504 (1954). MR **15**, 930.

173 — Naturally totally ordered commutative semigroups. Amer. J. Math. **76**, 631—646 (1954). MR **15**, 930.

174 Clifford, A. H., and D. D. Miller: Semigroups having zeroid elements. Amer. J. Math. **70**, 117—125 (1948). MR **9**, 330.

175 Climescu, Al. C.: Sur les quasicycles. Bull. École Polytech. Jassy **1**, 5—14 (1946). MR **8**, 134.

176 Cotlar, M., and E. Zarontonello: Semiordered groups and Riesz-Birkhoff *L*-ideals. Fac. Ci. Mat. Univ. Nac. Litoral Publ. Inst. Mat. **8**, 105—192 (1948) (Spanish). MR **10**, 99.

177 Croisot, Robert: Holomorphies d'un semi-groupe. C. R. Acad. Sci. (Paris) **227**, 1134—1136 (1948). MR **10**, 353.

178 — Autre généralisation de l'holomorphie dans un semi-groupe. C. R. Acad. Sci. (Paris) **227**, 1195—1197 (1948). MR **10**, 430.

179 — Propriétés des complexes forts et symétriques des demi-groupes. Bull. Soc. Math. France **80**, 217—223 (1952). MR **14**, 842.

180 — Demi-groupes et axiomatiques des groupes. C. R. Acad. Sci. (Paris) **237**, 778—780 (1953). MR **15**, 195.

181 Croisot, Robert: Demi-groupes inversifs et demi-groupes réunions de demi-groupes simples. Ann. Sci. École Norm. Sup. (3) **70**, 361—379 (1953). MR **15**, 680.

182 — Automorphismes intérieurs d'un semi-groupe. Bull. Soc. Math. France **82**, 161—194 (1954). MR **16**, 215.

183 — Demi-groupes simples inversifs à gauche. C. R. Acad. Sci. (Paris) **239**, 845—847 (1954). MR **16**, 215.

184 Doss, Raouf: Sur l'immersion d'un semi-groupe dans un groupe. Bull. Sci. Math. (2) **72**, 139—150 (1948). MR **10**, 591.

185 Dubreil, Paul: Contribution à la théorie des demi-groupes. Mém. Acad. Sci. Inst. France (2) **63**, no. 3, 52 pp. (1941). MR **8**, 15.

186 — Sur les problèmes d'immersion et la théorie des modules. C. R. Acad. Sci. (Paris) **216**, 625—627 (1943). MR **5**, 144.

187 — Contribution à la théorie des demi-groupes. II. Univ. Roma. Ist Naz. Alta Mat. Rend. Mat. e Appl. (5) **10**, 183—200 (1951). MR **14**, 12.

188 — Contribution à la théorie des demi-groupes. III. Bull. Math. France **81**, 289—306 (1953). MR **15**, 680.

189 Dubreil, P., et M.-L. Dubreil-Jacotin: Équivalences et opérations. Ann. Univ. Lyon. Sect. A (3) **3**, 7—23 (1940). MR **8**, 254.

190 Dubreil-Jacotin, Marie-Louise: Sur l'immersion d'un semi-groupe dans un groupe. C. R. Acad. Sci. (Paris) **225**, 787—788 (1947). MR **9**, 174.

191 — Quelques propriétés arithmétiques dans un demi-groupe demi-réticulé entier. C. R. Acad. Sci. (Paris) **232**, 1174—1176 (1951). MR **12**, 670.

192 — et Robert Croisot: Sur les congruences dans les ensembles où sont définés plusieurs opérations. C. R. Acad. Sci. (Paris) **233**, 1162—1164 (1951). MR 13, 620.

193 Evans, Trevor: An imbedding theorem for semigroups with cancellation. Amer. J. Math. **76**, 399—413 (1954). MR **15**, 681.

194 Fuchs, L.: On semigroups admitting relative inverses and having minimal ideals. Publ. Math. Debrecen **1**, 227—231 (1950). MR **12**, 473.

195 Forsythe, George E.: SWAC computes 126 distinct semigroups of order **4**. Proc. Amer. Math. Soc. **6**, 443—447 (1955). MR **16**, 1085.

196 Gardaschnikoff, M.: Über einen Typus endlicher Gruppen ohne das Associativgesetz. Comm. Inst. Sci. Math. Mec. Univ. Kharkoff (4) **17**, 29—33 (1940) (Russian. German summary). MR **3**, 37.

197 Green, J. A.: On the structure of semigroups. Ann. of Math. (2) **54**, 163—172 (1951). MR **13**, 100.

198 — and D. Rees: On semigroups in which $x^r = x$. Proc. Cambridge Philos. Soc. **48**, 35—40 (1952). MR 13, 720.

199 Hall, Marshall jr.: The word problem for semigroups with two generators. J. Symbolic Logic **14**, 115—118 (1949). MR **11**, 1.

200 Hashimoto, Hiroshi: On the kernel of semigroups. J. Math. Soc. Japan **7**, 59—66 (1955). MR **16**, 670.

201 Huntington, E. V.: Simplified definition of a group. Bull. Amer. Math. Soc. **8**, 296—300 (1901—1902).

202 Iséki, Kiyosi: Sur les demi-groupes. C. R. Acad. Sci. (Paris) **236**, 1524—1525 (1953). MR **14**, 842.

203 Ivan, Jan: On the direct product of semigroups. Mat. Fyz. Casopis. Slovensk. Akad. Vied. **3**, 57—66 (1953) (Slovak. Russian summary). MR **16**, 9.

204 Jaffard, Paul: Théorie axiomatique des groupes définis par des systèmes de générateurs. Bull. Sci. Math. (2) **75**, 114—128 (1951). MR **13**, 430.

205 Kawada, Yukiyosi, u. Kôiti Kondô: Idealtheorie in nichtkommutativen Halbgruppen. Jap. J. Math. **16**, 37—45 (1939). MR **1**, 164.

206 Klein-Barmen, Fritz: Über gewisse Halbverbände und kommutative Semigruppen. I, II. Math. Z. **48**, 275—288 (1942) und 715—734 (1943). MR **5**, 31.
207 Klein-Barmen, F.: Ein Beitrag zur Theorie der linearen Holoide. Math. Z. **51**, 355—366 (1948). MR **10**, 353.
208 — Zur Theorie der Operative und Associative. Math. Ann. **126**, 23—30 (1953). MR **15**, 95.
209 Kontorovič, P.: On the theory of semigroups in a group. Doklady Akad. Nauk SSSR (N. S.) **93**, 229—231 (1953) (Russian) MR **15**, 681.
210 Krasner, Marc.: Théorie non-abélienne des corps de classe, etc. C. R. Acad. Sci. (Paris) **225**, 785—787 et 973—975 (1947). MR **9**, 223.
211 Lambek, J.: The immersibility of a semigroup into a group. Canad. J. Math. **3**, 34—43 (1951). MR **12**, 481.
212 Lesieur, L.: Théorèmes de décomposition dans certains demi-groupes réticulés satisfaisant à la condition de chaîne descendante affaiblie. C. R. Acad. Sci. (Paris) **234**, 2250—2252 (1952). MR **13**, 906.
213 Levi, F. W.: On semigroups. Bull. Calcutta Math. Soc. **36**, 141—146 (1944). MR **6**, 202.
214 — On semigroups II. Calcutta Math. Soc. **38**, 123—124 (1946). MR **8**, 368.
215 Liber, A. E.: On symmetric generalized groups. Mat. Sbornik N. S. **33** (**75**), 531—544 (1953) (Russian). MR **15**, 502.
216 — On the theory of generalized groups. Doklady Akad. Nauk SSSR (N. S.) **97**, 25—28 (1954) (Russian). MR **16**, 9.
217 Lyapin, E.: The kernels of homomorphisms of associative systems. Rec. Math. N. S. **20** (62), 497—515 (1947) (Russian. English summary). MR **9**, 134.
218 — Normal complexes of associative systems. Izvestiya Akad. Nauk SSSR Ser. Mat. **14**, 179—192 (1950) (Russian). MR **11**, 575.
219 — Semisimple commutative associative systems. Izvestiya Akad. Nauk SSSR Ser. Mat. **14**, 367—380 (1950) (Russian). MR **12**, 154.
220 — Associative systems of all partial transformations. Doklady Akad. Nauk SSSR (N. S.) **88**, 13—15; errata **92**, 692 (1953) (Russian). MR **15**, 395.
221 — Semigroups in all of whose representations the operators have fixed points. I. Mat. Sbornik (N. S.) **34** (**76**), 289—306 (1954) (Russian). MR **15**, 850.
222 MacKenzie, Robert E.: Commutative semigroups. Duke Math. J. **21**, 471—477 (1954). MR **16**, 8.
223 McLean, David: Idempotent semigroups. Amer. Math. Monthly **61**, 110—113 (1954). MR **15**, 681.
224 Mal'cev, A.: Über die Einbettung von assoziativen Systemen in Gruppen. I, Rec. Math. (N. S.) **6** (**48**), 331—336 (1939); II, Rec. Math. (N. S.) **8** (**50**), 251—264 (1940) (Russian. German summary). MR **2**, 7; **2**, 128.
225 — Symmetric groupoids. Mat. Sbornik (N. S.) **31** (73), 136—151 (1952) (Russian). MR **14**, 349.
226 Mann, Henry B.: On certain systems which are almost groups. Bull. Amer. Math. Soc. **50**, 879—881 (1944). MR **6**, 147.
227 Markov, A.: The impossibility of certain algorithms in the theory of associative systems. I, Doklady Akad. Nauk SSSR (N. S.) **55**, 583—586 (1947); II, Doklady Akad. Nauk SSSR (N. S.) **58**, 353—356 (1947); III, Doklady Akad. Nauk SSSR (N. S.) **77**, 19—20 (1951) (Russian). MR **8**, 558; **9**, 321; **12**, 661.
228 — The impossibility of algorithms for the recognition of certain properties of associative systems. Doklady Akad. Nauk SSSR (N. S.) **77**, 953—956 (1951) (Russian). MR **13**, 4.
229 Morse, Marston, and Gustav A. Hedlund: Unending chess, symbolic dynamics and a problem in semigroups. Duke Math. J. **11**, 1—7 (1944). MR **5**, 202.

230 Munn, W. D., and R. Penrose: A note on inverse semigroups. Proc. Cambridge Philos. Soc. **51**, 396—399 (1955). MR **17**, 10.
231 Murata, Kentaro: On the quotient semi-group of a non-commutative semi-group. Osaka Math. J. **2**, 1—5 (1950). MR **12**, 155.
232 Nakada, Osamu: Partially ordered Abelian semigroups. I. On the extension of the strong partial order defined on Abelian semigroups. J. Fac. Sci. Hokkaido Univ. Ser. I **11**, 181—189 (1951). MR **13**, 817.
233 — II. On the strongness of the linear order defined on abelian semigroups. J. Fac. Sci. Hokkaido Univ. **12**, 73—86 (1952). MR **14**, 945.
234 Numakura, Katsumi: A note on the structure of commutative semigroups. Proc. Jap. Acad. **30**, 262—265 (1954). MR **16**, 214.
235 Parker, E. T.: On multiplicative semigroups of residue classes. Proc. Amer. Math. Soc. **5**, 612—616 (1954). MR **16**, 9.
236 Petropavlovskaya, R. V.: Structural isomorphism of free associative systems. Mat. Sbornik (N. S.) **28** (**70**), 589—602 (1951) (Russian). MR 13, 100.
237 — On the determination of a group by the structure of its subsystems. Mat. Sbornik (N. S.) **29** (**71**), 63—78 (1951) (Russian). MR **13**, 203.
238 — On the decomposition into a direct sum of the structure of subsystems of an associative system. Doklady Akad. Nauk SSSR (N. S.) **81**, 999—1002 (1951) (Russian). MR **13**, 525.
239 Pierce, R. S.: Homomorphisms of semigroups. Ann of Math. (2) **59**, 287—281 (1954). MR **15**, 930.
240 Post, Emil: Recursive unsolvability of a problem of Thue. J. Symbolic Logic **12**, 1—11 (1947). MR **8**, 558.
241 Preston, G. B.: Inverse semi-groups. J. London Math. Soc. **29**, 396—403 (1954). MR **16**, 215.
242 — Inverse semigroups with minimal right ideals. J. London Math. Soc. **29**, 404—411 (1954). MR **16**, 215.
243 — Representations of inverse semigroups. J. London Math. Soc. **29**, 411—419 (1954). MR **16**, 216.
244 Pták, Vlastimil: Immersibility of semigroups. Acta Fac. Nat. Univ. Carol., Prague no. **192** (1949) 16 pp. (1949). MR **12**, 155.
245 — (Extract from the above.) Cehoslovak. Mat. Z. **2** (**77**), 247—271 (1952) (Russian. English summary).
246 Rédei, L., u. O. Steinfield: Über Ringe mit gemeinsamer multiplikativer Halbgruppe. Comment. Math. Helv. **26**, 146—151 (1952). MR **14**, 10.
247 Rees, D.: On semi-groups. Proc. Cambridge Philos. Soc. **36**, 387—400 (1940). MR **2**, 127.
248 — Note on semi-groups. Proc. Cambridge Philos. Soc. **37**, 434—435 (1941). MR 3, 199.
249 — On the ideal structure of a semigroup satisfying a cancellation law. Quart. J. Math. Oxford Ser. **19**, 101—108 (1948). MR **9**, 567.
250 — On the group of a set of partial transformations. J. London Math. Soc. **22** (1947), 281—284 (1948). MR **9**, 568.
251 Rich, R. P.: Completely simple ideals of a semigroup. Amer. J. Math. **71**, 883—885 (1949). MR **11**, 327.
252 Rybakoff, L.: Sur une classe de semi-groupes commutatifs. Rec. Math. (Moscou) **5** (**47**), 521—536 (1939) (Russian. French summary). MR **1**, 164.
253 Schwarz, Štefan: Zur Theorie der Halbgruppen. Sbornik Prác Priredovedeckej Fak. Slov. Univ. v Bratislave no **6**, 64 pp. (1943) (Slovakian. German summary). MR **10**, 12.
254 — On the structure of simple semigroups without zero. Czechoslovak Math. J. **1** (**76**), 41—53 (1951). MR **14**, 12.

255 SCHWARZ, ŠTEFAN: On semigroups having a kernel. Czechoslovak Math. J. **1** (76), (1951), 229—264 (1952). MR **14**, 444.

256 — On maximal ideals in the theory of semigroups. I, II. Čehoslovak Mat. Ž. **3** (78), 139—153 and 365—383 (1953) (Russian. English summary). MR **15**, 850.

257 — Contributions to the theory of torsion semigroups. Čehoslovak. Mat. Ž. **3** (78), 7—21 (1953) (Russian. English summary). MR **15**, 850.

258 — The theory of characters of finite commutative semigroups. Czechoslovak Math. J. **4** (79), 219—247 (1954) (Russian. English summary). MR **16**, 1085.

259 — Characters of commutative semigroups as class functions. Czechoslovak Math. J. **4** (79), 291—295 (1954) (Russian. English summary). MR **16**, 1086.

260 — On a Galois connection in the theory of characters of commutative semigroups. Czechoslovak Math. J. **4** (79), 296—313 (1954) (Russian. English summary). MR **16**, 1086.

261 SIVERCEVA, N.: On the simplicity of the associative system of singular square matrices. Mat. Sbornik (N. S.) **24** (66), 101—106 (1949) (Russian). MR **10**, 508.

262 SKOLEM, TH.: Theory of divisibility in some commutative semigroups. Norsk Mad. Tidsskr. 33, 82—88 (1951). MR **13**, 430.

263 — Some remarks on semigroups. Norske Vid. Selsk. Forh., Trondheim **24**, 42—47 (1951). MR **13**, 906.

264 — Theorems of divisibility in some semigroups. Norske Vid. Selsk. Forh., Trondheim **24**, 48—53 (1951); **25** (1952), 72—77 (1953). MR **13**, 906; **14**, 842.

265 STOLL, R. R.: Representations of finite simple semigroups. Duke Math. J. **11**, 251—265 (1944). MR **5**, 229.

266 — Homomorphisms of a semigroup onto a group. Amer. J. Math. **73**, 475—481 (1951). MR **12**, 799.

267 SUŠKEVIČ, A. K. (SUSHKEVICH, SUSCHKEWITSCH, SUSCHKEWITZ): Über die endlichen Gruppen ohne das Gesetz der eindeutigen Umkehrbarkeit. Math. Ann. **99**, 30—50 (1928).

268 — Über die Matrizendarstellung der verallgemeinerten Gruppen. Commun. Soc. Kharkow **6**, 27—38 (1933).

269 — Über eine Verallgemeinerung der Semigruppen. Comm. Soc. Kharkow **12**, 89—97 (1935).

270 — Investigations on infinite substitutions. Memorial volume dedicated to D. A. GRAVE, Moscow, 1940, 245—253 (Russian). MR **2**, 217.

271 — Groupes géneralisés des matrices singulières. Comm. Inst. Sci. Mat. Méc. Univ. Kharkoff (4) **16**, 3—11 (1940) (Russian. French summary). MR 3, 99.

272 — Groupes géneralisés de quelques types des matrices infinies. Comm. Inst. Sci. Mat. Méc. (4) **16**, 115—120 (1940) (Ukrainian. French summary). MR 3, 99.

273 — Über einen Typus der verallgemeinerten Semigruppen. Comm. Inst. Sci. Mat. Méc. Univ. Kharkoff (4) **17**, 19—28 (1940) (Russian. German summary). MR 3, 37.

274 — Untersuchungen über unendliche Substitutionen. Comm. Inst. Sci. Math. Méc. Univ. Kharkoff (4) **18**, 27—37 (1940) (Russian. German summary). MR 3, 99.

275 — On the construction of some types of groups of infinite matrices. Zapiski Naučno-Issled. Inst. Mat. Mec. Har'kov. Mat. Obšč. (4) **19**, 27—33 (1948) (Russian). MR **12**, 241.

276 TAMARI, DOV.: Groupoïdes reliés et demi-groupes ordonnés. C. R. Acad. Sci. (Paris) **228**, 1184—1186 (1949). MR **10**, 508.

277 — Groupoïdes ordonnés. L'ordre lexicographique pondéré. C. R. Acad. Sci. (Paris) **228**, 1909—1911 (1949). MR **11**, 9.

278 — Ordres pondérés. Charactérisation de l'ordre naturel comme l'ordre du semi-groupe multiplicatif des nombres naturels. C. R. Acad. Sci. (Paris) **229**, 98—100 (1949). MR 11, 80.

279 TAMARI, DOV: Les images homomorphes des groupoïdes de BRANDT et l'immersion des semi-groupes. C. R. Acad. Sci. (Paris) **229**, 1291—1293 (1949). MR **11**, 327.
280 — Monoides préordonnés et chaînes de Malcev. Thèse. Université de Paris, 1951. MR **14**, 532.
281 TAMURA, TAKAYUKI: Characterization of groupoids and semilattices by ideals in a semigroup. J. Sci. Gakugei Fac. Tokushima Univ. **1**, 37—44 (1950). MR **13**, 430.
282 — On the system of semigroup operations defined in a set. J. Gakugei Coll. Tokushima Univ. **2**, 1—12 (1952). MR **14**, 616.
283 — Some remarks on semigroups and all types of semigroups of order 2, 3. J. Gakugei Tokushima Univ. **3**, 1—11 (1953). MR **15**, 7.
284 — On finite one-idempotent semigroups. I. J. Gakugei, Tokushima Univ. (Nat. Sci.) **4**, 11—20 (1954). MR **15**, 850.
285 — On a monoid whose submonoids form a chain. J. Gakugei Tokushima Univ. Math. **5**, 8—16 (1954). MR **16**, 1085.
286 — Notes on finite semigroups and determination of finite semigroups of order 4. J. Gakugei. Tokushima Univ. Math. **5**, 17—27 (1954). MR **16**, 1085.
287 TAMURA, TAKAYUKI, and NAOKI KIMURA: On decompositions of a commutative semigroup. Kōdai Math. Sem. Rep. **1954**, 109—112 (1954). MR **16**, 670.
288 TEISSIER, MARIANNE: Sur les équivalences regulières dans les demi-groupes. C. R. Acad. Sci. (Paris) **232**, 1987—1989 (1951). MR **12**, 799.
289 — Sur la théorie des idéaux dans les demi-groupes. C. R. Acad. Sci. (Paris) **234**, 386—388 (1952). MR **13**, 620.
290 — Sur l'algèbre d'un demi-groupe fini simple. C. R. Acad. Sci. (Paris) **234**, 2413—2414 et 2511—2513 (1952). MR **14**, 10.
291 — Sur quelques propriétés des idéaux dans les demi-groupes. C. R. Acad. Sci. (Paris) **235**, 767—769 (1952). MR **14**, 445.
292 — Sur les demi-groupes admettant l'existence du quotient d'un côté. C. R. Acad. Sci. (Paris) **236**, 1120—1122 (1953). MR **14**, 721.
293 — Sur les demi-groupes ne contenant pas d'élément idempotent. C. R. Acad. Sci. (Paris) **237**, 1375—1377 (1953). MR **15**, 598.
294 THIERRIN, GABRIEL: Sur une condition nécessaire et suffisante pour qu'un semi-groupe soit un groupe. C. R. Acad. Sci. (Paris) **232**, 376—378 (1951). MR **12**, 389.
295 — Sur les éléments inversifs et les éléments unitaires d'un demi-groupe inversif. C. R. Acad. Sci. (Paris) **234**, 33—34 (1952). MR **13**, 621.
296 — Sur une classe de demi-groupes inversifs. C. R. Acad. Sci. (Paris) **234**, 177—179 (1952). MR **13**, 621.
297 — Sur une classe de transformations dans les demi-groupes inversifs. C. R. Acad. Sci. (Paris) 1015—1017 (1952). MR **13**, 621.
298 — Sur les demi-groupes inversés. C. R. Acad. Sci. (Paris) **234**, 1336—1338 (1952). MR **14**, 12.
299 — Sur les homogroupes. C. R. Acad. Sci. (Paris) **234**, 1519—1521 (1952). MR **13**, 902.
300 — Sur les homodomaines et les homocorps. C. R. Acad. Sci. (Paris) **234**, 1595—1597 (1952). MR **13**, 902.
301 — Sur quelques classes de demi-groupes. C. R. Acad. Sci. (Paris) **236**, 33—35 (1953). MR **14**, 616.
302 — Sur quelques équivalences dans les demi-groupes. C. R. Acad. Sci. (Paris) **236**, 565—567 (1953). MR **14**, 616.
303 — Quelques propriétés des équivalences réversibles généralisées dans un demi-groupe D. C. R. Acad. Sci. (Paris) **236**, 1399—1401 (1953). MR **14**, 842.

304 THIERRIN, GABRIEL: Sur une équivalence en relation avec l'équivalence réversible généralisée. C. R. Acad. Sci. (Paris) **236**, 1723—1725 (1953). MR **14**, 944
305 — Quelques propriétés des sous-groupoïdes consistants d'un demi-groupe abélien D. C. R. Acad. Sci. (Paris) **236**, 1837—1839 (1953). MR **14**, 944.
306 — Sur la caractérisation des équivalences régulières dans les demi-groupes. Acad. Roy. Belgique. Bull. Cl. Sci. (5) **39**, 942—947 (1953). MR **15**, 680.
307 — Sur quelques classes de demi-groupes possédant certaines propriétés des semi-groupes. C. R. Acad. Sci. (Paris) **238**, 1765—1767 (1954). MR **15**, 849.
308 — Sur la caractérisation des groupes par leurs équivalences régulières. C. R. Acad. Sci. (Paris) **238**, 1954—1956 (1954). MR **15**, 850.
309 — Sur la caractérisation des groupes par leurs équivalences simplifiables. C. R. Acad. Sci. (Paris) **238**, 2046—2048 (1954). MR **15**, 850.
310 — Demi-groupes inversés et rectangulaires. Acad. Roy. Belg. Bull. Cl. Sci. (5) **41**, 83—92 (1955). MR **17**, 10.
311 TURING, A. M.: The word problem in semi-groups with cancellation. Ann. of Math. (2) **52**, 491—505 (1950). MR **12**, 239.
312 THURSTON, H. A.: Equivalences and mappings. Proc. London Math. Soc. (3) **2**, 175—182 (1952). MR **14**, 241.
313 VAGNER, V. V.: On the theory of partial transformations. Doklady Akad. Nauk SSSR (N. S.) **84**, 653—656 (1952) (Russian). MR **14**, 10.
314 — Generalized groups. Doklady Akad. Nauk SSSR (N. S.) **84**, 1119—1122 (1952) (Russian). MR **14**, 12.
315 — The theory of generalized heaps and generalized groups. Mat. Sbornik N. S. **32** (74), 545—632 (1953) (Russian). MR **15**, 501.
316 VANDIVER, H. S.: On the imbedding of one semi-group in another, with application to semi-rings. Amer. J. Math. **62**, 72—78 (1940). MR **1**, 105.
317 — The elements of a theory of abstract discrete semigroups. Vierteljschr. naturforsch. Ges. Zürich **85**, Beiblatt (Festschrift RUDOLPH FUETER) 71—86 (1940). MR **2**, 310.
318 — On a *p*-adic representation of rings and abelian groups. Ann. of Math. (2) **48**, 22—28 (1947). MR **8**, 311.
319 VOROB'EV, N. N.: Normal subsystems of finite symmetric associative systems. Doklady Akad. Nauk SSSR (N. S.) **58**, 1877—1879 (1947) (Russian). MR **9**, 330.
320 — On ideals of associative systems. Doklady Akad. Nauk SSSR (N. S.) **83**, 641—644 (1952) (Russian). MR **14**, 10.
321 — Associative systems of which every subsystem has a unity. Doklady Akad. Nauk SSSR (N. S.) **88**, 393—396 (1953) (Russian). MR **14**, 718.
322 WEAVER, MILO: Cosets in a semi-group. Math. Mag. **25**, 125—136 (1952). MR **14**, 12.
323 WOIDISLAWSKI, M.: Ein konkreter Fall einiger Typen der verallgemeinerten Gruppen. Comm. Inst. Sci. Math. Méc. Univ. Kharkoff (4) **17**, 127—144 (1940) (Russian. German summary). MR **3**, 37.
324 ZAVOLA, S. T.: Free operator groups. Doklady Akad. Nauk SSSR (N. S.) **85**, 949—951 (1952) (Russian). MR **14**, 351.

VI. Other Associative Systems

325 BAER, REINHOLD: Zur Einordnung der Theorie der Mischgruppen in die Gruppentheorie. S.-B. Heidelberg. Akad. Wiss., Math.-naturwiss. Kl. **1928**, **4**, 13 pp. (1928).
326 — Über die Zerlegungen einer Mischgruppe nach einer Untermischgruppe. S.-B. Heidelberg. Akad. Wiss., Math.-naturwiss. Kl. **1928**, **5**, 13 pp. (1928).
327 — Beiträge zur Galoisschen Theorie. S.-B. Heidelberg. Akad. Wiss., Math.-naturwiss. Kl. **1928**, **14**, 1—22 (1928).

328 BAER, REINHOLD: Zur Einführung des Scharbegriffs. J. reine angew. Math. **160**, 199—207 (1929).
329 CERTAINE, J.: The ternary operation of a group. Bull. Amer. Math. Soc. **49**, 869—877 (1943). MR **5**, 227.
330 LOEWY, A.: Über abstrakt definierte Transmutationssysteme oder Mischgruppen. J. reine angew. Math. **157**, 239—254 (1927).
331 — Neue elementare Begründung und Erweiterung der Galloisschen Theorie. S.-B. Heidelberg. Akad. Wiss., Math.-naturwiss. Kl. **1927**, **1**, **27** pp. (1927).
332 PRÜFER, H.: Theorie der Abelschen Gruppen. Math. Z. **20**, 166—187 (1924).
333 DÖRNTE, W.: Untersuchungen über einen verallgemeinerten Gruppenbegriff. Math. Z. **29**, 1—19 (1928).

D. Supplement

Recent Papers on Group Axioms

334 FURSTENBERG, HARRY: The inverse operation in groups. Proc. Amer. Math. Soc. **6**, 991—997 (1955). MR **17**, 1053.
335 STOLT, BENGT: Über Axiomsysteme, die eine abstrakte Gruppe bestimmen. Thesis, University of Uppsala, 1933. Uppsala: Almqvist & Wiksells 1953. MR **15**, 99.
336 — Über irreduzible Axiomsysteme, die endliche abstrakte Gruppen bestimmen. Ark. Mat. **3**, 113—115 (1955). MR **17**, 939.
337 — Zur Axiomatik endlicher Gruppen. I, II. Ark. Mat. **3**, 171—180, 229—238 (1955—1956). MR **17**, 939.
338 — Über gewisse Axiomssysteme, die abstrakte Gruppen bestimmen. Ark. Mat. **3**, 187—191 (1955). MR **17**, 939.
339 — Abschwächung einer klassischen Gruppen-definition. Math. Scand. **3**, 303—305. MR **17**, 1052.

Algebras (General)

340 FOSTER, ALFRED L.: The identities of — and unique factorization within — classes of universal algebras. Math. Z. **62**, 171—188 (1955). MR **17**, 452.
341 JAKUBÍK, JÁN: On the existence algebras. Časopis Pěst. Mat. **81**, 43—54 (1956). MR **18**, 275.
342 KOLIBIAR, MILAN: On permutable relations. Mat.-Fyz. Časopis. Slovensk. Akad. Vied. **5**, 137—139 (1955) (Slovak. Russian summary). MR **17**, 1184.
343 NEUMANN, B. H.: An imbedding theorem for algebraic systems. Proc. London Math. Soc. (3) **4**, 138—153 (1954). MR **17**, 448.
344 ROBINSON, A.: Note on an imbedding theorem for algebraic systems. J. London Math. Soc. **30**, 249—252 (1955). MR **17**, 449.
345 SKOLEM, TH.: The abundance of arithmetical functions satisfying some simple functional equations. Norske Vid. Selsk. Forh. Trondheim **29**, 47—53 (1956). MR **18**, 275.

Groupoids

346 HOSSZÜ, M.: Some functional equations related with the associative law. Publ. Math. Debrecen 3 (1954), 205—214 (1955). MR **17**, 236.
347 SCHWARZ, ŠTEFAN: On various generalizations of the notion of a group. Časopis Pest Mat. Fys. **74**, (1949), 95—113 (1950) (Czech. English summary). MR **12**, 389.

Loops and Quasigroups

348 BAER, REINHOLD: Nets and groups. I, II. Trans. Amer. Math. Soc. **46**, 110 to 141 (1939); **47**, 435—439 (1940).
349 COWELL, W. R.: Concerning a class of permutable congruence relations on a loop. Proc. Amer. Math. Soc. **7**, 583—588 (1956). MR **18**, 14.

350 Evans, Trevor: Some remarks on a paper by R. H. Bruck. Proc. Amer. Math. Soc. **7**, 211—220 (1956). MR **18**, 10.
351 Hall, Marshall jr., and J. D. Swift: Determination of Steiner triple systems of order 15. Math. Tables Aids Comput. **9**, 146—152 (1955). MR **18**, 192.
352 Hughes, D. R.: Additive and multiplicative loops of planar ternary rings. Proc. Amer. Math. **6**, 973—980 (1955). MR **17**, 451.
353 Hughes, D. R.: Planar division neo-rings. Trans. Amer. Math. Soc. **80**, 502—527 (1955). MR **17**, 451.
354 Norton, D. A., and Sherman K. Stein: An integer associated with latin squares. Proc. Amer. Math. Soc. **7**, 331—334 (1956). MR **17**, 1043.
355 Paige, Lowell J.: A class of simple Moufang loops. Proc. Amer. Math. Soc. **7**, 471—482 (1956). MR **18**, 110.
356 Šik, František: Sur les décompositions créatrices sur les quasigroupes. Publ. Fac. Sci. Univ. Masaryk **1951**, 169—186 (1951) (Russian summary). MR **15**, 7.
357 — Über abgeschlossene Kongruenzen auf Quasigruppen. Publ. Fac. Sci. Univ. Masaryk **1954**, 103—112 (1954). MR **18**, 193.
358 Stein, Sherman K.: On the foundations of quasigroups. Trans. Amer. Math. Soc. **85**, 228—256 (1957).

Multigroups

359 Dičman, A. P.: On multigroups whose elements are subsets of a group. Moskov. Gos. Ped. Inst. Uč. Zap. **71**, 71—79 (1953) (Russian). MR **17**, 826.

Semigroups

360 Boccini, Domenico: *P*-gruppoide dei quotienti di un gruppoide con operati. Rend. Sem. Mat. Univ. Padova **25**, 176—195 (1956). MR **18**, 283.
361 Climescu, Al. C.: Sur l'équation fonctionelle de l'associativité. Bull. École Polytech. Jassy **1**, 211—224 (1946). MR **8**, 440.
362 Cohn, P. M.: Embeddings in semigroups with one-sided division. J. London Math. Soc. **31**, 169—181 (1956). MR **18**, 14.
363 — Embeddings in semigroups with sesquilateral division semigroups. J. London Math. Soc. **31**, 181—191 (1956). MR **18**, 14.
364 Dickson, L. E.: On semigroups and the general isomorphism between infinite groups. Trans. Amer. Math. Soc. **6**, 205—208 (1905).
365 Gluskin, L. M.: Homomorphisms of unilaterally simple semigroups on groups. Doklady Akad. Nauk SSSR (N. S.) **102**, 673—676 (1955) (Russian). MR **17**, 237.
366 — Simple semigroups with zero. Doklady Akad. Nauk SSSR (N. S.) **103**, 5—8 (1955) (Russian). MR **17**, 237.
367 Griffiths, H. B.: Infinite products of semigroups and local connectivity. Proc. London Math. Soc. (3) **6**, 455—480 (1956). MR **18**, 192.
368 Halezov, E. A.: Isomorphisms of matrix semigroups. Ivanov. Gos. Ped. Inst. Uč. Zap. Fiz.-Mat. Nauki **5**, 42—56 (1954) (Russian). MR **17**, 825.
369 Hashimoto, Hiroshi: On the structure of semigroups containing minimal left ideals and minimal right ideals. Proc. Jap. Acad. **31**, 264—266 (1955). MR **17**, 459.
370 Hewitt, Edwin, and Herbert S. Zuckerman: Finite dimensional convolution algebras. Acta math. **93**, 67—119 (1955). MR **17**, 1048.
371 Iséki, Kyoshi: Contributions to the theory of semigroups. I, III, IV. Proc. Jap. Acad. **32**, 174—175, 323—324, 430—435 (1956). MR **17**, 1184; **18**, 282.
372 Ivan, Ján: On the decomposition of a simple semigroup into a direct product. Mat. Fyz. Časopis. Slovensk. Akad. Vied **4**, 181—202 (1954) (Slovak. Russian summary). MR **17**, 346.

373 KÁLMAR, LÁSZLÓ: Another proof of the Markov-Post theorem. Acta. Math. Acad. Sci. Hungar. **3**, 1—27 (1952) (Russian summary). MR **14**, 528.

374 KAUFMAN, A. M.: Associative systems with an ideally solvable series of length two. Leningrad. Gos. Ped. Inst. Uč. Zap. **89**, 67—93 (1953) (Russian). MR **17**, 942.

375 KIMURA, NAOKI: Maximal subgroups of a semigroup. Kōdai Math. Sem. Rep. **1954**, 85—88 (1954). MR **16**, 443.

376 — On some examples of semigroups. Kōdai Math. Sem. Rep. **1954**, 88—92 (1954). MR **16**, 443.

377 KLEIN-BARMEN, FRITZ: Über eine weitere Verallgemeinerung des Verbandsbegriffes. Math. Z. **46**. 472—480 (1940). MR **1**, 327.

378 KONTOROVIČ, P. G.: On the theory of semigroups in a group. Kazan Gos. Univ. Uč. Zap. **114**, 35—43 (1954) (Russian). MR **17**, 942.

379 — and A. D. KACMAN: Some types of elements of a semigroup invariant in a group. Uspehi Mat. Nauk (N. S.) **11**, 145—150 (1956) (Russian). MR **18**, 193.

380 LESIER, L.: Sur les idéaux irréductibles d'un demi-groupe. Rend. Sem. Mat. Univ. Padova **24**, 29—36 (1955). MR **17**, 347.

381 — Sur les demi-groupes réticulés satisfaisant à une condition de chaîne. Bull. Soc. Math. France **83**, 161—193 (1955). MR **17**, 584.

382 LEVITZKI, YAAKOV: On powers with transfinite exponents. I, II. Riveon Lematematika **1**, 8—13 (1946); **2**, 1—7 (1947) (Hebrew). MR **8**, 193; **9**, 173.

383 LYAPIN, E. S.: Simple commutative associative systems. Izvestiya Akad. Nauk SSSR Ser. Mat. **14**, 275—282 (1950) (Russian). MR **12**, 5.

384 — Canonical form of elements of an associative system given by defining relations. Leningrad. Gos. Ped. Inst. Uč. Zap. **89**, 45—54 (1953) (Russian). MR **17**, 942.

385 — Increasing elements of associative systems. Leningrad. Gos. Ped. Inst. Uč. Zap. **89**, 55—65 (1955) (Russian). MR **17**, 942

386 — Potential inversion of elements in semigroups. Mat. Sb. N. S. **38** (**80**), 373—383 (1956) (Russian). MR **17**, 825

387 MAL'CEV, A. I.: Nilpotent semigroups. Ivanov. Gos. Ped. Inst. Uč. Zap. Fiz.-Mat. Nauki **4**, 107—111 (1953) (Russian). MR **17**, 825.

388 MILLER, D. D., and A. H. CLIFFORD: Regular D-classes in semigroups. Trans. Amer. Math. Soc. **82**, 270—280 (1956). MR **17**, 1184.

389 POOLE, A. R.: Finite ova. Amer. J. Math. **59**, 23—32 (1937).

390 PRACHAR, KARL: Zur Axiomatik der Gruppen. Akad. Wiss. Wien, S.-B. IIa. **155**, 97—102 (1947).

391 RAUTER, H.: Abstrakte Kompositionssysteme oder Übergruppen. Crelle **159**, 229—237 (1928).

392 RÉDEI, L.: Die Verallgemeinerung der Schreierschen Erweiterungstheorie. Acta Sci. Math. Szeged **14**, 252—273 (1952). MR **14**, 614.

393 SAGASTUME BERRA, ALBERTO E.: Divisibility in groupoids. I, II. Univ. Nac. Eva Peron. Publ. Fac. Ci. Fisicomat. Serie Segunda. Rev. **5**, 67—122 (1954) (Spanish). MR **17**, 458.

394 SCHENKMAN, EUGENE: A certain class of semigroups. Amer. Math. Monthly **63**, 242—243 (1956). MR **17**, 1055.

395 SCHÜTZENBERGER, MARCEL PAUL: Sur deux représentations des demi-groupes finis. C. R. Acad. Sci. (Paris) **243**, 1385—1387 (1956). MR **18**, 282.

396 SCHÜTZENBERGER, MAURICE-PAUL: Sur certain treillis gauche. C. R. Acad. Sci. (Paris) **224**, 776—778 (1947). MR **8**, 432.

397 SONNEBORN, LEE M.: On the arithmetic structure of a class of commutative semigroups. Amer. J. Math. **77**, 783—790 (1955). MR **17**, 459.

398 STOLT, BENGT: Über eine besondere Halbgruppe. Ark Mat. **3**, 275—286 (1956). MR **17**, 942.

399 SUŠKEVIČ, A. K.: Über die Darstellung der eindeutig nicht umkehrbaren Gruppen mittels der verallgemeinerten Substitutionen. Rec. Math. Moscou 33, 371—372 (1926).
400 — Über den Zusammenhang der Rauterschen Übergruppen mit den gewöhnlichen Gruppen. Math. Z. 38, 643—649 (1934).
401 — Über einen merkwürdigen Typus der verallgemeinerten unendlichen Gruppen. Com. Soc. Math. Karkow 9, 39—44 (1934).
402 — Über Semigruppen. Com. Soc. Math. Karkow 8, 25—27 (1934).
403 — Über die Erweiterung der Semigruppen bis zur ganzen Gruppe Com. Soc. Math. Karkow 12, 81—87 (1935).
404 — Sur quelques propriétés des semigroupes generalisés. Com. Soc. Karkow 13, 29—33 (1936).
405 SZÉP, J.: Zur Theorie der Halbgruppen. Publ. Math. Debrecen 4, 344—346 (1956). MR 18, 110.
406 TAMARI, DOV.: Représentations isomorphes par des systèmes de relations. Systèmes associatives. C.R. Acad. Sci. (Paris) 232, 1332—1334 (1951). MR 12, 583.
407 TAMURA, TAKAYUKI: Note on unipotent inversible semigroups. Kōdai Math. Sem. Rep. 1954, 93—95 (1954). MR 16, 443.
408 — One-sided bases and translations of a semigroup. Math. Japon. 3, 137—141 (1955). MR 17, 1184.
409 — Indecomposable completely simple semigroups except groups. Osaka Math. J. 8, 35—42 (1956). MR 18, 282.
410 TAMURA, TAKAYUKI, and NAOKI KIMURA: Existence of a greatest decomposition of a semigroup. Kōdai Math. Sem Rep. 7, 83—84 (1955). MR 18, 192.
411 TAUSSKY, O.: Zur Axiomatik der Gruppen. Ergebn. math. Kolloquium Wien 4, 2—3 (1933).
412 THIERRIN, GABRIEL: Sur quelques propriétés de certaines classes de demigroupes. C. R. Acad. Sci. (Paris) 239, 1335—1337 (1954). MR 16, 443.
413 — Sur la caractérisation des groupes par certaines propriétés de leur relations d'ordre. C. R. Acad. Sci. (Paris) 239, 1453—1455 (1954). MR 16, 443.
414 — Sur une propriété caractéristique des demi-groupes inversés et rectangulaires. C. R. Acad. Sci. (Paris) 241, 1192—1194 (1955). MR 17, 459.
415 — Contribution à la théorie des équivalences dans les demi-groupes. Bull. Soc. Math. France 83, 103—159 (1955). MR 17, 584.
416 — Sur la théorie des demi-groupes. Comment. Math. Helv. 30, 211—223 (1956). MR 17, 711.
417 VAGNER, V. V.: Representation of ordered semigroups. Mat. Sb. N. S. 38 (80), 203—240 (1956) (Russian). MR 17, 942.
418 VAKSELJ, A.: Eine neue Form der Gruppenpostulate und eine Erweiterung des Gruppenbegriffes. Publ. math. Belgrade 3, 195—211 (1934).
419 VÁZQUEZ GARCÍA, R., and E. VALLE FLORES: A relation between the cardinal number of a set and its possibility of being a group. Bol. Soc. Mat. Mexicana 5 (1948), 1—6 (1950) (Spanish). MR 12, 316.
420 VINOGRADOV, A. A.: On the theory of ordered semigroups. Ivanov. Gos. Ped. Inst. Uč. Zap. Fiz.-Mat. Nauki 4, 19—21 (1953) (Russian). MR 17, 711.
421 VOROB'EV, N. N.: On symmetric associative systems. Leningrad Gos. Ped. Inst. Uč. Zap. 89, 161—166 (1953) (Russian). MR 17, 943.
422 WEAVER, MILO W.: On the imbedding of a finite commutative semigroup of idempotents in a uniquely factorable semigroup. Proc. Nat. Acad. Sci. USA 42, 772—775 (1956). MR 18, 283.
423 YAMADA, MIYUKI: On the greatest semilattice decomposition of a semigroup. Kōdai Math. Sem. Rep. 7, 59—62 (1955). MR 17, 584.
424 — A note on middle unitary semigroups. Kōdai Math. Sem. Rep. 7, 49—52 (1955). MR 17, 585.

Index

Ergebnisse der Mathematik und ihrer Grenzgebiete

1. Bachmann: Transfinite Zahlen. DM 42,–; US $ 11.60
2. Miranda: Partial Differential Equations of Elliptic Type. DM 58,–; US $ 16.00
4. Samuel: Méthodes d'Algébre Abstraite en Géométrie Algébrique. DM 29,–; US $ 8.00
5. Dieudonné: La Geometrie des Groupes Classiques. DM 38,–; US $ 10.50
6. Roth: Algebraic Threefolds with Special Regard to Problems of Rationality. DM 22,–; US $ 6.10
7. Ostmann: Additive Zahlentheorie. 1. Teil: Allgemeine Untersuchungen. DM 42,–; US $ 11.60
8. Wittich: Neuere Untersuchungen über eindeutige analytische Funktionen. DM 31,–; US $ 8.60
11. Ostmann: Additive Zahlentheorie. 2. Teil: Spezielle Zahlenmengen. DM 28,–; US $ 7.70
13. Segre: Some Properties of Differentiable Varieties and Transformations. DM 46,–; US $ 12.70
14. Coxeter/Moser: Generators and Relations for Discrete Groups. DM 35,–; US $ 9.70
15. Zeller/Beckmann: Theorie der Limitierungsverfahren. DM 64,–; US $ 17.60
16. Cesari: Asymptotic Behavior and Stability Problems in Ordinary Differential Equations. 3rd edition in preparation
17. Severi: Il teorema di Riemann-Roch per curve-superficie e varietà questioni collegate. DM 26,–; US $ 7.20
18. Jenkins: Univalent Functions and Conformal Mapping. DM 37,–; US $ 10.20
19. Boas/Buck: Polynomial Expansions of Analytic Functions. DM 18,–; US $ 5.00
20. Bruck: A Survey of Binary Systems. DM 46,–; US $ 12.70
23. Bergmann: Integral Operators in the Theory of Linear Partial Differential Equations. DM 40,–; US $ 11.00
25. Sikorski: Boolean Algebras. DM 42,–; US $ 11.60
26. Künzi: Quasikonforme Abbildungen. DM 43,–; US $ 11.90
27. Schatten: Norm Ideals of Completely Continous Operators. DM 26,–; US $ 7.20
28. Noshiro: Cluster Sets. DM 40,–; US $ 11.00
29. Jacobs: Neuere Methoden und Ergebnisse der Ergodentheorie. DM 55,–; US $ 15.20
30. Beckenbach/Bellman: Inequalities. DM 38,–; US $ 10.50
31. Wolfowitz: Coding Theorems of Information Theory. DM 30,–; US $ 8.30
32. Constantinescu/Cornea: Ideale Ränder Riemannscher Flächen. DM 75,–; US $ 20.70
33. Conner/Floyd: Differentiable Periodic Maps. DM 29,–; US $ 8.00
34. Mumford: Geometric Invariant Theory. DM 24,–; US $ 6.60
35. Gabriel/Zisman: Calculus of Fractions and Homotopy Theory. DM 42,–; US $ 11.60
36. Putnam: Commutation Properties of Hilbert Space Operators and Related Topics. DM 31,–; US $ 8.60
37. Neumann: Varities of Groups. DM 51,–; US $ 14.40
38. Boas: Integrability Theorems for Trigonometric Transforms. DM 20,–; US $ 5.50
39. Sz.-Nagy: Spektraldarstellung linearer Transformationen des Hilbertschen Raumes. DM 20,–; US $ 5.50
40. Seligman: Modular Lie Algebras. DM 43,–; US $ 11.90

41. Deuring: Algebren. DM 26,–; US $ 7.20
42. Schütte: Vollständige Systeme modaler und intuitionistischer Logik. DM 26,–; US $ 7.20
43. Smullyan: First-Order Logic. DM 36,–; US $ 9.90
44. Dembowski: Finite Geometries. DM 68,–; US $ 17.00
45. Linnik: Ergodic Properties of Algebraic Fields. DM 44,–; US $ 12.10
46. Krull: Idealtheorie. DM 28,–; US $ 7.70
47. Nachbin: Topology on Spaces of Holomorphic Mappings. DM 18,–; US $ 5.00
48. A. Ionescu Tulcea/C. Ionescu Tulcea: Topics in the Theory of Lifting. DM 36,–; US $ 9.90
49. Hayes/Pauc: Derivation and Martingales. DM 48,–; US $ 13.20
50. Kahane: Séries de Fourier Absolument Convergents. DM 44,–; US $ 12.10
51. Behnke/Thullen: Theorie der Funktionen mehrerer komplexer Veränderlichen. DM 48,–; US $ 13.20
52. Wilf: Finite Sections of Some Classical Inequalities. DM 28,–; US $ 7.70
53. Ramis: Sous-ensembles analytiques d'une variété banachique complexe. DM 36,–; US $ 9.90
54. Busemann: Recent Synthetic Differential Geometry. DM 32,–; US $ 8.80
55. Walter: Differential and Integral Inequalities. DM 74,–; US $ 20.40
56. Monna: Analyse non-archimédienne. DM 38,–; US $ 10.50
57. Alfsen: Compact Convex Sets and Boundary Integrals. DM 46,–; US $ 12.70
58. Greco/Salmon: Topics in m-Adic Topologies. DM 24,–; US $ 6.60
59. López de Medrano: Involutions on Manifolds. DM 36,–; US $ 9.90
60. Sakai: C*-Algebras and W*-Algebras. In preparation